WITHDRAWN

STORAGE
BATTERIES

STORAGE
BATTERIES

A General Treatise on the Physics
and Chemistry of Secondary Batteries
and their Engineering Applications

GEORGE WOOD VINAL, Sc. D.

Formerly Physicist, National Bureau of Standards
Fellow, American Institute of Electrical Engineers
Member, The Electrochemical Society

FOURTH EDITION

New York · JOHN WILEY & SONS, Inc.
London · CHAPMAN & HALL, Ltd.

Library of Congress Catalog Card Number: 54–12826
PRINTED IN THE UNITED STATES OF AMERICA

Preface to Fourth Edition

Fourteen years have elapsed since the third edition of this book was published. These have been eventful years, marked by technologic advances in battery design and construction brought about in large measure by the stimulus of war conditions but equally important for the domestic economy.

Previous editions were published in 1924, 1930, and 1940. A French translation of the second edition was published by Dunod in Paris in 1936.

The first edition of this book laid emphasis on the scientific principles relating to storage batteries without permitting the text to become too highly technical. Physical and chemical properties of the materials employed in making batteries were discussed, and a general description of manufacturing processes was given. The theory of reactions and principles of operating storage batteries were followed by descriptions of the more important industrial applications. Succeeding editions, including the present fourth edition, have followed the same general outline. They give a progressive account of the notable changes in methods of making and using batteries in this rapidly developing industry.

The published literature on storage batteries is more extensive than is generally supposed, but the valuable papers are scattered and many are in foreign languages. An effort has been made to select the more important of these for mention in the text and copious footnotes of the present edition.

Some of the changes that have taken place in the battery industry since the third edition was published deserve mention: improved lead alloys to withstand corrosion; increased use of lead-calcium alloys; increased use of the uncalcined high-metallic oxides; intensive studies of expanders; development of many new types of separators; increased use of plastic containers; production of nickel-cadmium batteries in this country; new types of silver oxide cells; application of batteries to radio-relay stations for long-distance telephony; battery installations for Diesel starting on railroad locomotives; introduction of 12-volt systems for automobile starting, lighting, and ignition; 24-volt and higher voltages for use on airplanes.

To hold the book to approximately its previous size and to provide space for new material, it has been necessary to cut the less important matter rather drastically and to eliminate obsolete matter completely. I have, however, added to the historical discussion in the first chapter and added illustrations which I believe are of general interest. All chapters have been revised and some parts entirely rewritten. About half of the illustrations are new.

Acknowledgments

I wish to thank Princeton University for granting me the privilege of using its library and its special collections in the departmental libraries of chemistry and engineering.

Men of the industry have responded most generously to my request for comment and criticism of the text. Their suggestions have added materially to the value of the book. They and their respective companies have freely provided many of the illustrations. Those who have placed valuable material, illustrations, or new sources of information at my disposal and especially those who have read and discussed parts of the manuscript (4th ed.) deserve my heartiest thanks:

L. E. Lighton, C. C. Wallace, H. J. Strauss, H. P. Murphy, E. A. Hoxie, E. Grothe, S. K. Lessey, E. L. Lord, A. W. Miley, G. E. Petrosky, F. W. Hopkinson of Electric Storage Battery Co.
Eugene Willihnganz and Harvey Stover of Gould-National Batteries.
U. B. Thomas and R. D. deKay of Bell Telephone Laboratories.
Arthur Fleischer of Nickel-Cadmium Battery Corp.
Paul Howard of Yardney Electric Corp.
D. E. Hasbrouck of United States Rubber Co.
H. B. Birt of Delco-Remy Division of General Motors Corp.
E. W. Allen of Thomas A. Edison.
P. K. McCullough of Mercury Manufacturing Co.
H. E. Jensen and J. F. Rittenhouse of C and D Batteries.
R. R. Richards and Grant Wheat of Koehler Manufacturing Co.

Acknowledgments were made in previous editions for illustrations supplied then, and these will not be repeated for those retained in this edition. Acknowledgments for new illustrations are as follows: National Bureau of Standards, 1, 2, 118; C and D Batteries, parts of 9 and 18, 125, 126; Electric Storage Battery Co., 10, 17, part of 18, 19, 20, 23, 24, 25, 32, 74, 77, 96, 99, 100, 101, 123, 127, 130, 132, 135, 144, 146, 150, 151, 152, 154, 159, 160; U. S. Rubber Co., 12; Gould-National Batteries, 22, 26, 27, 28, 30, 33, 97, 102, 103, 124; Edison Storage Battery Division of Thomas A. Edison, 40, 42, 81, 133, 139,

141, 142, 145, 161, 162; Nickel-Cadmium Battery Corp., 43, 83, 84, 85, 153; Yardney Electric Corp., 44, 45; Bell Telephone Laboratories, 121, 122; Union Switch and Signal Co. through the Association of American Railroads, 131; Mercury Manufacturing Co., 138; Delco-Remy Division of General Motors Corp., 155; Koehler Manufacturing Co., 163.

GEORGE WOOD VINAL

Princeton, N. J.
April 3, 1954

Contents

Contents

Contents

1

Introduction

The storage battery is typically an electrochemical apparatus and as such must be discussed from three points of view. The first is chemical, involving the nature and properties of the materials used in its construction and the reactions that occur during charging and discharging. The second is physical, and this includes a study of the electrical input and output, the factors that affect the capacity, and the theory of the transformation of chemical energy into electrical energy, and vice versa. The third viewpoint is the practical one dealing with the engineering applications of storage batteries. There is no sharp line of demarkation among the chemical, physical, and engineering aspects, but a full discussion of all is necessary to an adequate understanding of the nature and performance of storage batteries.

The scientific principles underlying the operation of storage batteries have now been so thoroughly investigated, both in the United States and abroad, that they may be presented with some degree of confidence. They serve as the groundwork for an intelligent study of batteries as they exist today and as a guide for the future development of the art.

HISTORICAL DEVELOPMENT OF THE LEAD-ACID BATTERY

The storage battery of today grew out of the investigations of many early experimenters in the field of electrochemistry. Volta's[1] discovery of the galvanic battery in 1800 initiated this line of research. Two years later, Gautherot[2] discovered the polarization of platinum wires, produced by the passage of an electric current through a cell which he used for studying the decomposition of water. He found

[1] Alessandro Volta, On the electricity excited by the mere contact of conducting substances of different kinds, *Phil. Trans. Roy. Soc.*, *90*, 403 (1800), in French.

[2] N. Gautherot, Mémoires des Sociétés savantes et littéraires de la République française (1801), *J. phys.*, *56*, 429 (1802).

1

that a feeble current was returned when he connected the wires after having disconnected the source of current. Ritter[3] repeated Gautherot's experiment in 1803 and went a step farther. He constructed small piles from plates of several metals, including gold and silver. Between the sheets of metal he placed moistened layers of cloth. He charged these piles with an electric current and obtained a discharge current from them after disconnecting the charging source. He thought that the piles stored electricity much as a capacitor does, because, like a capacitor, the pile had layers of metal of good conductivity alternating with layers of the poorly conducting cloth. Volta showed that this explanation was not correct and attributed the effect to the decomposition of water.

Grove's well-known gas battery[4] was essentially a storage battery. Each cell consisted of a pair of glass tubes having coaxial electrodes of platinum which were sealed through the glass at the upper ends. One tube of each pair was partially filled with oxygen and the other with hydrogen. The electrolyte into which the open ends of the tubes were dipped was a solution of sulfuric acid. Grove found that the platinum in the hydrogen was positive to the platinum in the oxygen. The cells could produce electricity on discharge, or by the passage of a current from an external source through the cells the corresponding gases were produced. There was little use for storage batteries then and no good reason for using a considerable number of his cells to charge a smaller number of like cells. His battery is interesting, however, because of its remarkable reversibility.

Other experimenters entered the field, but it remained for Planté to develop a valuable form of cell as a result of his study of the properties of metals for the accumulation of oxygen.

In 1859 Planté[5] began his study of electrolytic polarization. As a result of his experiments, he devised a battery for the storage of electrical energy, consisting of two sheets of lead separated by strips of rubber and rolled into the form of a spiral. The element thus formed was immersed in a dilute solution, about 10 per cent of sulfuric acid. He studied the charge and discharge of this simple cell and described it as storing the chemical work of the voltaic pile. He found it possible

[3] J. W. Ritter, Über Ladungsfähigkeit der Metalle elektrische Erdpolarität, *Gilbert's Ann., 15,* 106 (1803); Expériences sur un appareil à charger d'électricité par la colonne électrique de Volta, *J. phys., 57,* 345 (1803).

[4] W. R. Grove, On the voltaic series and the combination of gases by platinum, *Phil. Mag., III, 14,* 447 (1839), *21,* 417 (1842).

[5] G. Planté, Nouvelle pile secondaire d'une grande puissance, *Compt. rend., 50,* 640 (1860); *Recherches sur l'électricité,* Gauthier-Villars, Paris, **1883,** Chapter II, page 29.

to increase materially the capacity of the cell by a process that is now known as formation. After periods of charge, he discharged the cell or allowed it to rest for a time, during which local action transformed the covering of peroxide on the positive plate into lead sulfate.

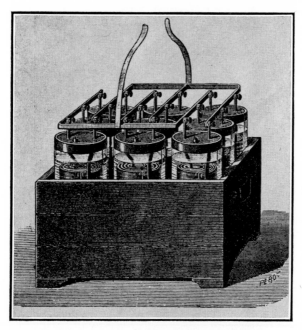

Fig. 1. The first lead-acid storage battery presented in 1860 to the French Academy of Sciences by Gaston Planté.

From time to time he reversed the polarity and repeated the process of charge and discharge to build up the capacity of the cell.

Planté's first battery which he presented to the French Academy of Sciences in 1860 is shown in Fig. 1. At the time Planté was a young man of 26. The battery was remarkable for the large currents which it could deliver and in this respect was markedly superior to the then existing primary batteries. With this emphasis on large currents, it is not surprising that the nine cells were all connected in parallel. Planté describes it as having 10 square meters of active surface area.

As compared with primary batteries, the new storage battery had the disadvantage of requiring much time for the formation of the plates and the expenditure of many primary batteries to charge it. In 1873, however, Planté had a hand-driven Gramme generator with which to charge the battery, and he made an interesting experiment illustrating the transformation of one form of energy into another. The mechanical energy of turning the crank produced electrical energy in the generator, and this in turn was transformed into chemical energy in the battery. When the battery was charged and cranking the generator was stopped, the units were left connected, and the generator ran as a motor. Chemical energy then became electrical energy, which the motor transformed back into mechanical energy. The cycle of transformations was thus completed.

Planté's name has been perpetuated in connection with the storage battery by the so-called Planté plate. This type of plate, as distinguished from the others, consists of a sheet of lead on which the active material is formed electrochemically from the lead of the plate itself.

In 1881 Faure[6] patented a process for pasting the surface of the plates with a compound of lead which could be formed more easily into the active materials of the finished battery. He applied a coating of red lead to the surface of smooth lead plates, rolling them together with a layer of flannel between for a separator. This type of cell possessed marked superiority in capacity and ease of formation over the cell of Planté, but the adherence of the active material to the plates was rather poor. It is stated that a battery of this type capable of exerting 1 horsepower weighed 75 pounds and gave an energy efficiency of 80 per cent.

About the same time, Brush, an American, also discovered the possibility of preparing the active materials from compounds mechanically applied to the plates. He obtained patents covering this principle.

In the latter part of 1881, Volckmar[7] patented the use of lead plates with numerous holes which were filled with a paste made of pulverized lead mixed with sulfuric acid. Swan[8] also obtained a patent on a grid of cellular structure. These supports for the active material were an improvement over the flat plates which Faure used, but the active material still fell out readily. Sellon[9] patented, in 1881, a modification of the grid to make it hold the material better. He designed his grid in such a way as to lock the active material in place. Sellon is said to have made use of the alloy of lead and antimony instead of pure lead for this grid. The Correns[10] grid, devised and patented in 1888, consisted of a double lattice of which the bars were triangular in cross

[6] Faure's Secondary Battery, *Electrician, 6,* 323 (1881); *7,* 122 and 249 (1881); French patents 139,258 (1880) and 141,057 (1881); German 10,926 (1881); British 129 (1881); U. S. 252,002 (1882).

[7] Volckmar, German patent 19,928 (1881).

[8] Swan, British patent 2272 (1881); *Electrician, 8,* 142 (1882).

[9] Sellon, British patent 3987 (1881).

[10] Correns, German patent 51,031 (1888).

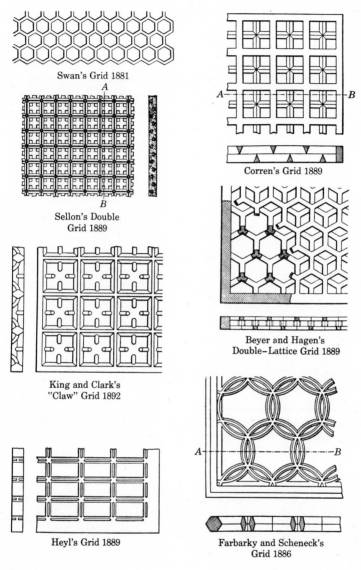

Fig. 2. Early types of grid structures, 1881–1892.

section with the apexes pointing inward so that the active material was held securely in place.

Illustrations of several of the earliest forms of grids are shown in Fig. 2. These grids are no longer used, but some of the principles that they embody are used now.

Since 1881 the development of the storage battery has been rapid because of the decreased time required for formation of the plates and also because of the development of machines for generating electric current. Many of the types of plates that have been devised are of historical interest, but only a few of them are of commercial importance today. Detailed descriptions of many of them have appeared in several of the books[11] previously published on this subject, and it is therefore unnecessary to describe them further.

THE STORAGE-BATTERY INDUSTRY

The storage-battery industry may be said to have had its beginning with the pioneer experiments of Planté in 1859. During the next 20 years the storage battery was little more than a laboratory apparatus, because of the labor and expense involved in preparing and charging the plates. Faure's process of forming the active material from oxides of lead simplified and cheapened the process of manufacture in 1881. About that time, also, dynamo-electric machines became available for charging the batteries. Invention and investigation were greatly stimulated, but this early period was marred by bitter controversies over the relative merits of different types of batteries, over questions of priority of invention, and over theories explaining the chemical reactions that take place when the batteries are charged and discharged. The batteries of this period were mostly of the stationary variety.

The storage battery was tried in a number of different services early in its history. The first installations often were not successful but they led to valuable results later on. Blizard[12] gives the dates and places of the first installations made in this country, including the following: central-station battery at Philipsburg, Pa., in 1885; trolley service at Dover, N. H., in 1893; train-lighting battery on the Pennsylvania Railroad in 1882; isolated lighting plant at Baltimore, Md., in 1883; central-station telephone service at Chicago, Ill., in 1889. The first successful use of batteries for the propulsion of electric vehicles was stated to have been in 1894, but there were many attempts prior to this date.

Beginning with 1900, the industry was characterized by the development of compact, portable types of batteries and a great increase in

---

[11] E. J. Wade, *Secondary Batteries,* Electrician Printing and Publishing Co., London, 1902; L. Jumau, *Les accumulateurs électriques,* Dunod et Pinat, Paris, 1907; A. Treadwell, *The Storage Battery,* The Macmillan Co., New York, 1906; Lamar Lyndon, *Storage Battery Engineering,* McGraw-Hill Book Co., New York, 1911.

[12] Charles Blizard, Development of the use of the storage battery in the United States, *Elec. Rev. N. Y., 38,* 75 (1901).

the number of batteries used. Portable batteries were produced for railway train lighting, propulsion of submarines and electric trucks and tractors, starting and lighting of automobiles, signaling on the railroads, and use in military operations. Stationary batteries were further developed during this period for stand-by and regulating service, telephone exchanges, and isolated lighting plants. The Edison alkaline storage battery was invented during this epoch.

In the next 20-year period ending in 1940, storage batteries found new applications in emergency lighting, air-conditioning of railway cars, starting of Diesel engines, and a variety of services on ships, buses, trucks, and aircraft. This period is significant for the intensive study of materials and construction of batteries for a highly competitive market.

The outbreak of World War II brought new and unprecedented demands for storage batteries of many kinds and for many diverse purposes. Research was stimulated, output increased, and the industry was faced with demands for batteries of lighter weight and smaller volume, and for increased output at low temperatures. Coincident with the new military demands the civilian needs also increased as a result of a growing and intensively active population. Such a situation inevitably caused shortages of various materials which necessitated finding substitutes. Skill in manufacture and the benefits of intensified research have definitely improved storage batteries.

Some of the changes noted in the industry are: increased use of the uncalcined oxides in pasting the plates, new and improved organic expanders to increase output of the batteries at extreme low temperatures, the development of calcium grids for use in telephone batteries which are floated on carefully regulated bus bars, plastic cases in lieu of some hard-rubber or composition cases, new types of separators to replace porous rubber in time of scarcity and to supplement the decreasing availability of Port Orford cedar, and increased use of Fiberglas mats as retainers on the surface of the positive plates.

The well-known nickel-iron or alkaline type of storage battery has continued important. This is the Edison type, which follows along lines standardized effectively for it. Other types of alkaline storage batteries, also well known and hitherto largely confined to use in Europe, began to be manufactured in this country. These included the nickel-cadmium battery and the silver oxide battery.

From a civilian point of view the net result has been to increase enormously the production of storage batteries and to place large numbers of the small sizes in the hands of non-technical people who are quick to appreciate reliable service but who have little knowledge of the theory or construction of the cells. The growth of the industry since 1909 is shown by the statistics in Table 1, compiled by the Census Bureau:

TABLE 1. STATISTICS OF STORAGE-BATTERY MANUFACTURE

	1909	1914	1919	1925	1937	1947
Weight of plates, lb	23,119,331	41,079,047	148,951,766	388,264,038	....	....
Value	$4,243,984	$10,651,150	$56,648,347	$88,870,186	$78,250,221	$293,358,000

Motor vehicles in the United States increased from 7½ millions in 1919 to over 50 millions in 1953. All of these are electrically equipped, requiring a yearly production, for this one purpose, of over 25 millions of batteries per year, if it is assumed that the average life of these batteries is from 18 months to 2 years.

PRIMARY AND SECONDARY CELLS

An electric battery consists of two or more connected cells that convert chemical energy into electrical energy. The cell is the unit part of the battery, but the word "battery" is sometimes used to mean one cell. The essential parts of a cell are two dissimilar electrodes, immersed in an electrolyte in a suitable jar or container. Familiar examples of electrodes are the copper and zinc plates of a simple primary cell, or the lead and lead dioxide plates of a storage cell. The electrolyte is a water solution of certain acids, alkalies, or salts that have been found to be adapted to the purpose.

A number of different kinds of cells are in common use. These may be classified conveniently into two general groups as primary and secondary cells. The most familiar of the primary cells is the "dry cell." Secondary cells are generally spoken of as "storage cells" or "accumulators." The distinction between primary and secondary cells is based on the nature of the chemical reactions that occur in them when they are in use. Primary cells convert chemical energy into electrical energy and in so doing they become exhausted. Dry cells, when no longer serviceable, are discarded, but some of the so-called "wet" cells may be renewed with new electrodes and electrolyte. Storage cells, on the other hand, convert chemical energy into electrical energy by reactions that are essentially reversible, that is, they may be charged by an electric current passing through them in the opposite direction to that of their discharge. During this process, electrical energy is transformed into chemical energy, which may be used again at a later time as electrical energy. Electricity is not stored as electricity by these cells. They store chemical energy and so potentially electricity.

There are other cells, some of which are intermediate between primary and secondary cells. These are generally classed with the primary cells for practical reasons, although they may possess some of the essential characteristics of the secondary cells.

THE GROUPING OF CELLS

For most purposes storage cells are used in groups or batteries, the number of cells and their size depending on the service required. Several arrangements are possible, and it is therefore desirable to

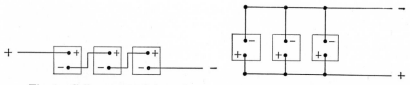

Fig. 3. Cells connected in series.   Fig. 4. Cells connected in parallel.

arrange the cells in such a way as to secure the most economical service. Two factors are involved in arranging the cells: one is the voltage requirement and the other the capacity. When the cells are connected in series, that is, when the positive pole of one cell is connected to the negative pole of the next, and so on to the end of the row, as in Fig. 3, the voltage of the cells is additive. Two cells in series will give twice the voltage of one cell, and five cells will give

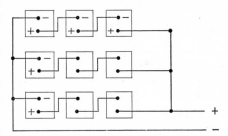

Fig. 5. Parallel of series-connected cells.

five times the voltage of one, assuming that the cells, taken individually, are of the same voltage. The capacity of a row of series-connected cells, however, is no more than the capacity of a single cell.

Cells may also be connected in parallel, by connecting like poles together, as shown in Fig. 4. The voltage of such a group is no more than the voltage of a single cell, but the capacity of the group is equal to the sum of the capacities of the individual cells. Such an arrangement of storage cells is not commonly made, because it is better to use a single cell of the required capacity rather than a group of small ones connected in parallel.

When more than three cells are involved in a series and parallel connection, there is a choice of arrangement, as shown in Figs. 5 and 6. The cells may be arranged in several rows connected in series, and these rows may then be connected in parallel (Fig. 5), or they may be

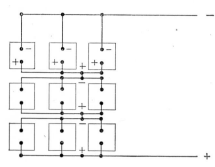

arranged in parallel groups which are then put in series (Fig. 6). The voltage and capacity of either of these groups are the same, but the former is preferred for storage batteries although the latter is the best arrangement for dry cells. The paralleling of series-connected groups of storage batteries is found in cases where exceptional capacity is required, or for charg-

Fig. 6.   Series of parallel-connected cells.

ing when the voltage of the charging source would not be sufficient to charge the cells if they were all connected in series.

In some installations involving a considerable number of cells in a series-parallel arrangement all points of the same nominal potential are connected. This provides a crisscross pattern that assists in maintaining equalization of the cells and a minimum of connector resistance. In effect such an arrangement is a combination of the connections in Figs. 5 and 6.

Ordinarily the internal resistance of a storage cell is small and may be neglected in comparison with the resistance of the external circuit. This is discussed at greater length in Chapter 7, but for the present purpose of grouping cells into the most desirable batteries a brief discussion is given here.

The working voltage of a cell is lower than its electromotive force by an amount equal to the voltage drop within the cell itself. The electrical energy gainfully employed is proportional to working voltage and the current which flows for a specified time.

Let E = the emf of a single cell.
$\quad s$ = the number of cells connected in series.
$\quad p$ = the number of cells connected in parallel.
$\quad R'$ = the external resistance.
$\quad b$ = the internal resistance of a single cell.
$\quad I$ = the current flowing.

Increasing s increases the available emf, the working voltage of the battery, and the current that it delivers, assuming that R' remains constant. The increments of current as each successive cell is added to the circuit are substantially equal unless sb cannot be neglected in comparison with R'. In such a case the increments of current will be progressively less as more cells are added to the circuit. The addition of cells in parallel may then become desirable.

In a series-parallel arrangement the emf is sE and the internal resistance is sb/p. The current furnished by the battery through the external resistance is:

$$I = \frac{sE}{R' + (sb/p)} = \frac{sE}{R'} \times \frac{1}{1 + (sb/pR')}$$

The equation is thus separated into two parts: the first is a simple statement of Ohm's law, and the second part, which modifies the first, shows the effect of the internal resistance of the battery. From this equation it is apparent that sb/pR' should be small.

2

Materials and Methods
of Manufacture

Successful production of storage batteries is the reward of exacting care in the selection of materials and their fabrication into the finished product. Physical and chemical properties of the materials are usually covered by manufacturer's specification, but these provide no complete answer to his need for skill and experience. More than 90 years have elapsed since Planté invented the first successful storage battery, but, now more than before, research is revealing the intricate processes that take place in the battery. The X-ray spectrograph, electron microscope, and other powerful tools for research are being applied with good effect to provide a more scientific basis for hitherto empirical processes. Plastics in a wide variety of fibers, microporous films, and other solids are replacing some of the long-established materials for separators and containers. They have made possible the development of new types of batteries.

Lead-acid storage batteries and the nickel-iron or alkaline batteries are well known and long established in the battery industry. They maintain their importance and will be described in detail. Other types of storage batteries, less familiar to the public perhaps, include the nickel-cadmium and silver oxide storage batteries which will also be described.

MATERIALS FOR LEAD-ACID BATTERIES

Lead

The material used in greatest amount is lead, which is an element having an atomic weight of 207.21. The battery industry is the largest consumer of lead in the United States, about 400,000 tons[1] being used per year. Eighty per cent of this is returnable, however, as secondary metal when the batteries are discarded. Lead is obtained chiefly from the ore, galena, which is the native sulfide of lead, PbS.

[1] R. L. Ziegfeld, Lead survey for the ceramic industry, *Am. Ceram. Soc. Bull.*, *31*, 244 (1952).

Because the storage battery is affected by the presence of small amounts of impurities, it is of interest to note the more common of the metals and sulfides which occur in the ore. These include iron, silver, copper, zinc, and arsenic. Galena is widely distributed, being found in nearly all countries. Other ores of lead of less importance are cerussite, $PbCO_3$, and anglesite, $PbSO_4$.

Lead is a metal, bluish gray in color, with metallic luster. It oxidizes readily in moist air, becoming a dull gray. Lead in its pure state is soft and malleable, but its tensile strength is low. Because of its plasticity it may be extruded into ribbons and other forms by hydraulic presses. The density of cast lead is 11.34, but this may be slightly increased by rolling. The linear coefficient of expansion is 0.0000292 per degree Centigrade. This is greater than the values for copper, iron, tin, and some of the other common metals. The melting point for pure lead is 327.°4 C (621° F). The resistivity of lead is an important factor in the design of storage batteries. The values given in the literature differ somewhat, probably because lead is very sensitive in this respect to cold working, such as bending, hammering, or drawing. A value determined on a cast bar gave 0.0000212 ohm-centimeter at 20° C (68° F), or about 12 times the resistivity of copper.

The chemical properties of lead are of great importance in relation to storage-battery performance. Lead is readily attacked by nitric acid, but not by cold hydrochloric acid, or cold sulfuric acid below 1.700 sp. gr. Lead forms a number of important combinations with oxygen which will be discussed in later paragraphs. Small amounts of many impurities exert a marked influence on the mechanical and electrolytic properties of lead. Arsenic, copper, zinc, and antimony render it harder. Bismuth in small amounts appreciably increases the corrodibility of lead.

The composition of refined pig lead varies with the source of the material and the methods employed in purifying it. Standards for three grades of lead, known as "Corroding," "Chemical," and "Common" lead, which have been established by the American Society for Testing Materials (A.S.T.M.),[2] are given in Table 2, but reference should be made to the complete specification for other grades. The methods of analysis for pig lead are described in the society's specifications.[3]

Very pure lead, prepared by electrolytic methods, is now available. Percentages of 99.99943 have been reported. Such lead is soft, low in tensile strength, and consists of large crystals.

[2] *Standard Specifications*, A.S.T.M., B29–49, 1949.
[3] *A.S.T.M., Methods of Chemical Analysis of Metals*, B35–46.

TABLE 2. THREE GRADES OF PIG LEAD, ACCORDING TO THE STANDARD OF THE AMERICAN SOCIETY FOR TESTING MATERIALS

	1. Corroding Lead, %	2. Chemical Lead, %	5. Common Lead, %
Silver, max. (I, III)	0.0015	0.020 max. 0.002 min.	0.002 ......
Copper, max. (I, III)	0.0015	0.080 max. 0.040 min.	0.0025 ......
Copper and silver together, max.	0.0025	......	......
Arsenic, max.	0.0015	......	......
Antimony and tin together, max.	0.0095	......	......
Antimony, arsenic and tin together, max.	......	0.002	0.015
Zinc, max.	0.0015	0.001	0.002
Bismuth, max.	0.05	0.005	0.15
Iron, max.	0.002	0.0015	0.002
Lead (by difference), min.	99.94	99.90	99.85

Note: In No. 1, bismuth, copper, and tin must not all be present in maximum amounts in the same sample. No. 2 is sometimes known as "undesilverized lead" from southeast Missouri ores. No. 5 is common desilverized A. The A.S.T.M. specification includes four other types: No. 3 Acid lead, No. 4 Copper lead, No. 6 Common desilverized B, and No. 7 Soft undesilverized.

The production of secondary lead from discarded storage batteries has increased to large proportions. To this must be added the metal obtained from the scrap that necessarily accumulates in the course of manufacturing operations. Discarded batteries, minus their rubber or composition case material, and other scrap lead, alloys, and oxides are loaded into blast furnaces together with the reducing agent, coke, and the flux, limestone. Fusion and reduction of the material occur in the smelting zone, and the products separate into three layers: molten metal, matte, and slag. Copper and other impurities are removed, and the remaining antimony content of the recovered alloy is adjusted to the required percentage, but laboratory tests are needed to assure the required degree of purity if the material is to be used for casting grids or other parts of batteries. Specifications of battery manufacturers differ considerably in the amount of impurities permitted. Secondary lead, unalloyed with antimony, is available and subject to nearly the same rigid specifications as the purer grades of primary lead. Methods of reclaiming battery scrap have been published by Hayward[4] and Thews.[5]

[4] C. R. Hayward, How to smelt battery scrap, *Eng. Mining J.*, *145*, 80 (March 1944).

[5] E. R. Thews, Reclaiming storage-battery residues, *Arch. Metallkunde*, *2*, 170, (1948).

Lead-Antimony Alloys

Antimony, which is combined with lead in making the grids, is an element having an atomic weight of 121.76. It is obtained chiefly from the ore, stibnite, which is the native antimony sulfide, Sb_2S_3. This ore consists of prismatic crystals, bluish gray in color with metallic luster. It is often associated with arsenic and bismuth. The color of antimony itself is silver-white with a high metallic luster. It is hard and brittle. The density is 6.684, and the expansion coefficient 0.0000114 per degree Centigrade. Antimony melts at a temperature of 631° C (1168° F). The resistivity of antimony is 39 microhm-centimeters, or about twice that of lead, and the alloys of the two have a higher resistivity than pure lead. Antimony is not readily oxidized by the air, but it combines directly with chlorine, forming $SbCl_5$. Antimony is oxidized by nitric acid to the trivalent state. Arsenic is the principal impurity that is associated with antimony.

In Table 3 are found the results of several investigations made at the Bureau of Standards on the alloys of lead and antimony. The results have been calculated for even percentages of antimony. Undoubtedly, the conditions of casting and the presence of small amounts of impurities may affect the physical properties considerably. The lead used for these investigations was of the grade known as "hardening lead," for which the manufacturer's analysis showed a purity of 99.9947 per cent. The antimony also was the purest that could be obtained.

The amount of antimony in storage-battery grids ranges from 5 to 12 per cent. There are a number of reasons for using antimony: (1) The material flows better in the mold. To increase the fluidity some manufacturers specify a small amount of tin also. (2) The alloy produces sharp castings. Ewen[6] has stated that some alloys of lead and antimony expand when solidification takes place, but Dean[7] denies this. (3) The alloy is less subject to electrochemical formation and can be used as a support for the active material without losing its strength by being "formed" as the battery is used. (4) Antimony increases the stiffness of lead, and also its ductility and tensile strength within limits that are shown in Table 3. (5) The temperature of complete liquefaction of the alloys, within the range of compositions used for the grids, is below the melting point of pure lead. (6) The expansion coefficient of the alloy is less than that of pure lead.

[6] D. Ewen and T. Turner, Shrinkage of antimony-lead alloys during solidification, *J. Inst. Metals*, *4*, 128 (1910).

[7] R. S. Dean, L. Zickrick, and F. C. Nix, Lead-antimony system and hardening lead alloys, *Trans. Am. Inst. Mining Met. Engrs.*, *73*, 505 (1926).

Storage Batteries

TABLE 3. PROPERTIES OF CAST LEAD-ANTIMONY ALLOYS

Per Cent of Antimony	Temp. of Complete Liquefaction		Density	Tensile Strength, lb per sq in.	Elongation, %	Hardness, Brinell Number*	Expansivity Coefficient	Resistivity 20° C, ohm-cm
	° C	° F						
0	327	621	11.34	1780	...	3.0	0.0000292	0.0000212
1	320	608	11.26	....	....	4.2	0.0000288	0.0000220
2	313	596	11.18	....	....	4.8	0.0000284	0.0000227
3	306	583	11.10	4700	15	5.3	0.0000281	0.0000234
4	299	572	11.03	5660	22	5.7	0.0000278	0.0000240
5	292	558	10.95	6360	29	6.2	0.0000275	0.0000246
6	285	545	10.88	6840	24	6.5	0.0000272	0.0000253
7	278	532	10.81	7180	21	6.8	0.0000270	0.0000259
8	271	520	10.74	7420	19	7.0	0.0000267	0.0000265
9	265	509	10.66	7580	17	7.2	0.0000264	0.0000271
10	261	501	10.59	7670	15	7.3	0.0000261	0.0000277
11	256	492	10.52	7620	13	7.4	0.0000258	0.0000283
12	252	485	10.45	7480	12	7.4	0.0000256	0.0000289
13	247	477	10.38	7280	10	.....	0.0000253	0.0000293
14	...	...	10.30	7000	9	.....	0.0000251	0.0000293
15	...	...	10.23	6800	8	.....	0.0000248	0.0000292
16	...	...	.....	6620	6	.....	......	......

* The hardness numbers given in this table are lower than others reported in the technical literature, probably because of the softness of the lead used in the investigation.

Lead and antimony form an alloy of which the eutectic composition is 87 per cent lead and 13 per cent antimony, melting at 247° C (477° F). Lead and antimony are miscible in all proportions in the liquid state. Dean has found evidence of the solubility of antimony in lead to the extent of 2.5 per cent at the melting point of the eutectic. This percentage decreases to about 0.5 per cent at ordinary temperatures. The structure of the alloy used in storage batteries, therefore, consists of the eutectic embedded in a solid solution of lead and antimony, as shown in Fig. 7. The equilibrium diagram is given in Fig. 8. Age-hardening of alloys containing more than 0.5 per cent of antimony has been observed by Dean.

A metallographic examination of the alloy is often valuable in determining the quality. The specimens to be examined are first cut to a smooth surface, polished, and then etched. Various etching reagents that have been recommended include: (1) a solution of 1 part acetic acid, 3 parts hydrogen peroxide, and 3 parts water; (2) nitric acid, 10 per cent; (3) an electrolytic process in a solution of perchloric acid. When the specimen is suitably prepared and examined under

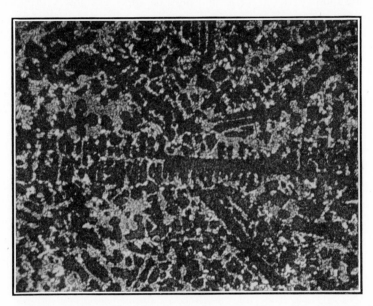

Fig. 7. Alloy of lead and antimony magnified 250 times. The dark portions are the solid solution and the light portions the eutectic. Antimony 8 per cent.

the microscope, its microstructure is revealed as in Fig. 7. Alloys that have been well mixed and cooled quickly after casting exhibit small crystals and are fine grained. Coarser structure indicates slower cooling. Segregated antimony, if present, appears as rectangular crystals. Irregular and unequal distribution of the eutectic is usually the result of poor mixing of the molten metals or uneven cooling. A micrographic study of lead-antimony alloys was made by Vilella and Beregehoff.[8] More recently Simon and Burbank[9] made a study of the dendritic structure of lead crystals as a part of a corrosion study of the grids. Lead-antimony alloys have less tendency to grow and they corrode more uniformly than structures of pure lead.

[8] J. R. Vilella and D. Beregehoff, Polishing and etching lead, tin, and their alloys for microscopic examination, *Ind. Eng. Chem., 19,* 1049 (1927).

[9] A. C. Simon and J. B. Burbank, Subgrain structure in lead and lead-antimony alloys, *Naval Research Lab. Rept. 3941* (1952).

The effects of some impurities in antimonial lead were described by Johnstone.[10] Copper, which ordinarily does not alloy with lead, may be present in small fractional percentages. Traces of it may even improve the alloy, but too much will cause wet-skims in the melting pot and may soften rather than harden the alloy. Tin improves the casting quality of the metal and may be added purposely in amounts of 0.15 to 0.50 per cent. Sulfur even in very small amounts produces wet-skims and brittleness. Iron separates along with a mass of lead as dross and should be avoided.

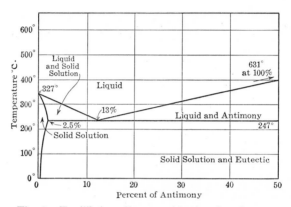

Fig. 8. Equilibrium diagram of lead and antimony.

Other investigations of the effect of impurities in antimonial lead alloys were made by Mashovets and Lyandres,[11] who found lead more resistant to anodic corrosion than the alloy. The resistance of the alloy is increased by the addition of small amounts of silver or arsenic. They found no marked harmful effect of iron, but bismuth in amounts of 0.1 to 1 per cent noticeably increased the rate of corrosion.

Fink and Dornblatt[12] investigated the effect of small additions of silver to lead-antimony alloys. They found anodic corrosion to be lessened by the addition of about 0.1 per cent of silver.

Lead-Calcium Alloys

These alloys have proved to be an important replacement for lead-antimony alloys in telephone batteries. They have distinct advan-

[10] James O. Johnstone, *Facts about Grid Metal*, Metals Refining Co., 1935.

[11] V. P. Mashovets and A. Z. Lyandres. Effect of impurities in Pb-Sb alloys on the functioning of lead storage batteries, *Zhur. Priklad. Khim.*, *21*, 347 (1947).

[12] C. G. Fink and A. J. Dornblatt, Effect of silver on some properties and performance of antimonial lead storage-battery plates, *Trans. Electrochem. Soc.*, *79*, 269 (1941).

tages in telephone central stations, where the service required of them is to float on a carefully regulated bus. Local action is small, and the necessary current to maintain fully charged conditions under normal conditions of floating is said to be of the order of 2.5 milliamperes per 100 ampere-hours of capacity.

Laboratory tests and experimental installations indicate that lead-calcium grids can be used successfully in specially constructed heavy-duty batteries for cycle service where the charging is accomplished by voltage-controlled or tapered charge methods. This is a comparatively new development and still in the experimental stage. Hitherto the general belief was that lead-calcium batteries were not suited to cycle service.

Lead-calcium alloys for battery grids were first described by Haring and Thomas[13] in 1935 and have been the subject of many experiments since then. For grids, the calcium should not exceed 0.10 per cent. Lead-calcium alloys are available commercially, containing about 5 per cent calcium. These must be "diluted" for battery use. This can be done by heating lead under charcoal to 540 to 600° C (1004 to 1112° F) and adding a measured portion of the 5 per cent alloy which melts readily in lead at this temperature. As some of the calcium tends to burn out, it is preferable to make the dilution in two steps, 5 to 1 and 1 to 0.1 per cent or less. The alloy resulting from the first step can then be analyzed for calcium content before determination of the amount required to make the final alloy, which is prepared at a temperature not exceeding 540° C (1004° F). Excessive stirring of the molten metal should be avoided to reduce drossing. When casting thin grids ($\frac{3}{32}$ inch or less) the 0.10 per cent alloy is heated to 480 to 500° C (896 to 932° F), but for thicker grids the casting temperature may be 25° C (45° F) less. After the grids are cast they age-harden in 15 to 18 hours. The maximum tensile strength of the alloy is then about 8100 pounds per square inch.

With 15 years' records of batteries containing calcium grids in float service, Thomas[14] and his coworkers found results best when the percentage of calcium was between 0.065 and 0.090. Anything over 0.1 per cent should be avoided. They found the rate of growth of the positive grids in service to be less than when lead-antimony alloys

---

[13] H. E. Haring and U. B. Thomas, Electrochemical behavior of lead, lead-antimony, and lead-calcium alloys in storage cells, *Trans. Electrochem. Soc.*, *68*, 293 (1935).

[14] U. B. Thomas, F. T. Foster, and H. E. Haring, Corrosion and growth of lead-calcium grids as a function of the calcium content, *Trans. Electrochem. Soc.*, *92*, 313 (1947).

were used, but if the percentage of calcium employed exceeded 0.1 per cent the rate of growth increased proportionately to the excess of calcium above 0.1 per cent.

Calcium alloy grids give rise to slightly different operating conditions than when the lead-antimony grids are used. These will be discussed in later sections of this book. The latest available description[15] of these batteries relates to their application to telephone service.

Other Lead Alloys

Lead-Arsenic Alloys. The property of small amounts of arsenic in hardening lead has been known for a long time. Emicke[16] has studied the properties of such alloys, finding the greatest resistance to chemical corrosion when the arsenic was not in excess of 0.1 per cent.

Lead-Tellurium Alloys. These alloys contain only 0.1 per cent tellurium, but this small amount changes the physical structure. The grain is finer and the alloy more resistant to corrosion.[17]

Lead-Cadmium Alloys. Experimental samples had some of the desired characteristics for use in storage batteries but exhibited an unfortunate tendency to form "trees."[18] In this respect the alloy was improved by the addition of a few per cent of antimony. Cadmium improves the mechanical properties of lead, but these alloys were hardly the equal of the customary lead-antimony alloys.

Because antimony has some deleterious effects in storage batteries various substitutes for it have been proposed. These include the alloys mentioned above and in addition alloys with several of the alkali and alkaline-earth metals. Another method of avoiding the detrimental effects of antimony, in part at least, without sacrificing the mechanical strength of its alloys with lead, is to lead-plate[19] the antimonious lead. On the positive grids this is not a permanent cure, but negative grids so plated behave more like those of lead or calcium alloy.

[15] Anon., New storage battery developed by the Bell Telephone Laboratories, *Elec. Eng.*, *70*, 283 (1951).

[16] O. Emicke, Properties and possible applications of lead-arsenic alloys in comparison with lead-antimony alloys, *Metall*, *4*, 1 and 48 (1950).

[17] Anon., Tellurium lead, *Dutch Boy Quarterly*, *14*, 9 (1936).

[18] G. W. Vinal, D. N. Craig, and C. L. Snyder, Composition of grids for positive plates of storage batteries as a factor influencing the sulfation of the negative plates. *Bur. Standards J. Research*, *10*, 795 (1933).

[19] E. W. Smith, U. S. patents 2,193,782-3 (1942); Anna P. Hauel, U. S. patent 2,282,760 (1942).

Oxides of Lead

Lead combines with oxygen to form a series of oxides. These are important articles of commerce, several of which are utilized as basic materials in the manufacture of storage batteries.

The Monoxide, Commonly Called Litharge or Plumbous Oxide, PbO. This oxide is made by passing air over molten lead in a reverberatory furnace. On cooling, the product solidifies into the familiar yellow modification which, however, is metastable at ordinary temperatures. Its rate of conversion into the stable red modification is slow but may be accelerated somewhat by grinding. X-ray studies have shown the yellow modification to be orthorhombic while the red modification is tetragonal. The transition, therefore, involves important changes in crystal structure. The temperature at which this occurs is variously reported in the literature from about 400 to 600° C. A solidified mass of the yellow modification cooling below the transition temperature is subject to internal stresses and exfoliates, producing a deformation of the crystal lattice and taking on characteristics of the red modification. The composition of the yellow modification agrees with the stoichiometric ratio of the elements and may even carry a slight excess of lead, and then the material may be greenish. Red litharge on the other hand may be deficient in oxygen to satisfy stoichiometric requirements. This and its distorted lattice, if present, are of practical importance in storage batteries because of increased chemical reactivity. Much has been learned about lead oxides by the use of X-rays,[20] and much more remains to be learned as the properties of lead monoxide vary both chemically and physically. These are reflected in battery performance.

Red Lead, Commonly Called Minium, Pb_3O_4. This higher oxide of lead is not to be confused with the red modification of litharge mentioned above. True red lead, corresponding to the formula, is made by a further oxidation of litharge at a temperature of 400 to 500° C. Oxidation is seldom complete, however, and the coarser particles retain a core of the lower oxide. The so-called battery red lead ordinarily contains about 25 per cent of litharge. Red lead does not absorb additional oxygen but may be formed electrolytically with change of composition into the dioxide, PbO_2. Red lead finds its use in the battery industry largely for blending with other oxides. It may be used to adjust the time required for formation of the plates, to

[20] M. LeBlanc and E. Eberius, Untersuchungen über Bleioxyde und deren Systeme mit Sauerstoffe, *Z. physik. Chem.*, *160*, 69 (1932). G. L. Clark and R. Rowan, Studies of lead oxides, *J. Am. Chem. Soc.*, *63*, 1302 and 1305 (1941).

adjust the density of the paste in making the plates, or to provide plates which will quickly reach their maximum capacity in service.

Lead Dioxide, PbO_2. This oxide is also called plumbic oxide and is dark brown or almost black. It is formed electrolytically in the battery by anodic oxidation of a lower oxide. Although it is the active material of the finished positive plate, it is not used in preparing the paste for this plate. It is not available in the quantities required of other oxides, and any paste made of it would rapidly disintegrate when dried. As the electrolytic product there is no higher oxide of lead.[21]

Uncalcined, High-Metallic Oxides. These are variously called litharge, lead powder, or hydroset oxide. The material may be gray or black. Developments in storage-battery manufacture have largely revolved around the use of such oxides. Although Volckmar is said to have filled his plates with powdered lead in 1881, the use of substantially similar material began actively about 1925. One variety of these oxides is now made by abrading small lead spheres or other convenient shapes in a stream of heated air the temperature and quantity of which are carefully regulated. The product is a highly reactive lead powder of fine particles which is partially oxidized to the red or tetragonal form of litharge. The actual metallic content is in the range of 25 to 50 per cent. Because of this finely divided lead the product is extremely sensitive to moisture, which would cause further oxidation and the evolution of heat. The control of this moisture is important in pasting and finishing positive plates. These leady oxides may be used directly in making negative plates, but a blend with about 20 per cent of red lead (minium) is customary in making positives. The advantages of using these oxides, which have brought them into such widespread use, include increased strength of plates, resistance to shedding, and increased life in service.

Barton oxides of high purity and fine particle size are used by some manufacturers interchangeably with the uncalcined, high-metallic oxides.

The So-Called Lead Suboxide, Pb_2O. In the early experiments it was believed that the oxide which was gray or black was truly a suboxide. Identification with X-rays has now shown this to be a mistake. Sidgwick[22] states that there is no suboxide.

Fume Oxides. These are the lightest of the oxides and have the

[21] D. A. MacInnes and E. B. Townsend, An electrovolumetric method for lead, J. Ind. Eng. Chem., 14, 420 (1922).

[22] N. V. Sidgwick, The Chemical Elements and Their Compounds, Vol. 1, p. 624, Clarendon Press, Oxford, 1950.

finest particle size. They are litharge of the orthorhombic modification. Plates made from fume oxides have high capacity.

Properties of Lead Oxides

Impurities. A high state of purity of the oxides used in the manufacture of storage batteries is required. The limiting percentages of impurities for both litharge and the red lead are about the same. A good grade of the oxides for storage-battery purposes would have about the amounts of impurities tabulated.

	Per Cent Not to Exceed		*Per Cent Not to Exceed*
Antimony	0.002	Nickel	0.0001
Arsenic	0.00005	Silver	0.003
Bismuth	0.05	Thallium	0.001
Cadmium	0.003	Zinc	0.002
Copper	0.003	Manganese	0.00003
Iron	0.02		

The methods for determining the impurities are available.[23] Moisture is determined by drying a sample of the material at 105° C.

Relation between Fine and Coarse Particles. A proper balance of the oxide particles of various sizes is a matter of considerable importance in determining the life and capacity of the finished plates. Generally speaking the oxides used for batteries are coarser than those used for paints, but they are so fine, nevertheless, that the problem of classifying them according to size of particles is a difficult one. The Thompson classifier[24] is one of the well-known devices for determining the percentages of the fractions of varying degrees of fineness. The oxide to be analyzed is suspended in oil which flows through a succession of standardized cones, depositing particles within certain limits of size in each. Air classifiers have been used also. The obvious advantages of these are somewhat offset, however, by the difficulties with dust and electrostatic effects. Sedimentation[25] methods are commonly used.

Apparent Density. This is usually measured by the Scott volumeter,

[23] J. A. Schaeffer, B. S. White, and J. H. Calbeck, *Chemical Analysis of Lead and Its Compounds*, 3rd ed., Eagle Picher Lead Co., Chicago, 1926. E. J. Dunn and A. J. Mitteldorf, *Spectroscopy of Lead Oxides for the Storage-Battery Industry*, National Lead Co. Research Laboratories, Brooklyn, N. Y., 1947.

[24] G. W. Thompson, Description of cone classifier, *Am. Soc. Testing Materials, Proc.*, **10**, 601 (1910).

[25] J. H. Calbeck and H. R. Harner, Particle size and distribution by a sedimentation method, *Ind. Eng. Chem.*, **19**, 59 (1927).

and the experiment consists in determining the weight in grams of 1 cubic inch of the oxide as it is sifted into a small box at the bottom of the apparatus. This anomalous unit of grams per cubic inch seems well established in the industry. The values range from 10 to 30 grams per cubic inch as the apparent density of the red lead, and from 12 to 45 grams per cubic inch for litharge. Storage-battery manufacturers endeavor to obtain material that does not differ by more than 5 grams from their standard. This measurement gives an indication of the uniformity of the product but does not give any indication of the true density or the relative percentage of fine and coarse particles. The limit of accuracy of this measurement is about 1 gram per cubic inch.

Acid Absorption. This is an arbitrary test following in a general way the factory procedure for making the paste. Comparable results can be obtained, therefore, only by a strict adherence to some specified directions as to weight of sample, amount and concentration of acid, temperature, and time of contact of the oxide and acid. The following method is recommended:

Weigh 50 grams of the sample into a 500-ml glass-stoppered flask. Add 100 ml (accurately measured) of sulfuric acid, specific gravity 1.100, and shake continuously for 10 minutes; then allow to stand 5 minutes, and decant the clear liquid to a dry filter. Titrate 25 ml of the filtrate with normal KOH. Once the original sulfuric acid solution has been standardized against the KOH, the difference in strength of the acid before and after shaking with the oxide indicates the degree of absorption. The acid absorption number is taken as the number of milligrams of H_2SO_4 absorbed by 1 gram of the oxide at a specified temperature such as 25° C.

In response to factory demands for a quicker test, the periods of shaking and settling are sometimes shortened to 4.5 and 0.5 minutes respectively, but these are hardly sufficient to give consistent results.

True Density. The true density may be calculated from weighings of small portions of the oxides in kerosene in calibrated flasks of 50-ml capacity. It is, of course, necessary that the portion of oxide be weighed before the kerosene is added.

Expanders for Negative Plates

These are substances such as lampblack, barium sulfate (commonly called blanc fixe), and wood flour and organic extracts of wood. Expanders are added in small amounts to the paste for making negative plates, to prevent contraction and solidification of the spongy lead and the consequent loss of capacity and life of the finished battery. When this occurs, the difficulty increases with each succeeding cycle.

Early investigations of the action of expanders were made and reported by Scarpa,[26] and more recently our knowledge of the subject has been notably increased by several American experimenters.[27]

It seems remarkable that the addition of several of the expanders, in the aggregate amounting to only 1 or 2 per cent, can increase the capacity of the negative plates by as much as several hundred per cent. Expanders, therefore, fill a very important role in the operation of the finished battery. The three common types of expanders may be used separately or in combination. Each has its characteristic effect in the battery with the result that they supplement each other.

Lampblack has less effect on the capacity of the plates than the others, but it fills a useful purpose in helping to clear the negative plates on formation. In carefully adjusted amounts it has a beneficial effect at low temperatures, increasing slightly the capacity of the plates, and it tends to counteract the tendency to high final charge voltages when the other expanders are present. The amount of lampblack used is about 0.15 per cent, seldom exceeding 0.2 per cent.

Barium sulfate continues as in the past to be a valuable expander. It is used in amounts varying from a fractional percentage to as much as 3 per cent. Willihnganz found that precipitation of barium sulfate in the pores of the negative plates immediately and greatly increased the capacity of the plates. Its beneficial effects become apparent during discharge. Since it is extremely insoluble in the electrolyte, it is helpful in the solid state. This is explained by its orthorhombic crystal structure, which is almost identical with that of lead sulfate. The two sulfates have practically the same unit cell size. In the operation of the battery, therefore, the crystals of barium sulfate provide nuclei for the precipitation of lead sulfate as it forms in preference to depositing on the lead itself. Polarization is thereby diminished and the apparent capacity of the active material extended. Strontium sulfate

---

[26] O. Scarpa, Function of inert substances in lead accumulators, *Chimie & industrie, 17,* 408 (1927), also *Elettrotecnica, 6,* 176 (1919).

[27] Eugene Willihnganz, Nature of the action of expanders in the storage battery, Natl. Lead Co., *Pub. 63* (1942), and Storage-battery addition agents, *Trans. Electrochem. Soc., 92,* 281 (1947). E. J. Ritchie, Addition agents for negative plates of lead-acid storage batteries, I, Introduction, *Trans. Electrochem. Soc., 92,* 229 (1947), and II, Pure organic compounds, *J. Electrochem. Soc., 100,* 53 (1953). A. C. Zachlin, Functions and behavior of the components of expanders for the negative plates of lead-acid storage batteries, *J. Electrochem. Soc., 98,* 325 (1951).

likewise has the same crystal structure and could be used, although there appears to be no immediate reason for doing so.

Interest in the use of organic expanders began before 1920. The substitution of rubber separators for wood in early attempts to make dry-charged batteries had not proved successful, and the difficulty was attributed to the absence of wood. Finely ground wood was incorporated with the paste in making negative plates. It soon became apparent that this improved the capacity of the batteries at high rates and low temperatures. From these early experiments have sprung many investigations on the effect of organic expanders and the methods of preparing them.

Lignin is a major constituent of wood, amounting to about 30 per cent in the coniferous varieties. Other major constituents are cellulose and pentosans. Separation of these constituents is an important part of processes for making paper and other products. In the course of these separations the lignin is found in the waste sulfite liquors. Lignin is soluble in alkalies but not in water or most mineral acids, including sulfuric acid. These waste liquors are an abundant source of ligninsulfonic acid, which can be concentrated and used by the battery industry. By another process, the so-called soda process, a heavy black liquor is obtained which precipitates sodium lignates by decreasing its alkalinity. These lignates are then treated with acid to liberate the lignin, which is then dried and used as a powder. Lignin can also be obtained from wood by hydrolyzing the wood strongly with acid. The soluble portion then contains the sugars, and the lignin remains in the insoluble residue. The preparation of organic expanders for battery use is described by Ritchie.[28]

Organic expanders are used in quantities of less than 1 per cent. They are usually combined with barium sulfate and lampblack. In the working cell the organic attaches itself to the spongy lead surface, helping it to remain active and free from an otherwise impervious coating of lead sulfate. Polarization is thereby reduced and the capacity of the battery increased. Too much organic may increase the difficulty in clearing the negative plates when they are on formation. The organics do very definitely increase the final charging voltage of the batteries by about 0.2 volt per cell. A so-called "non-lignin paste" would have about 2.5 per cent $BaSO_4$ and 0.2 per cent lampblack as the expander blended with the lead oxide.

A special lead-nickel glass used with expanders has been described[29]

[28] *Loc. cit.*

[29] F. J. Williams and J. A. Orsino, Lead-nickel glass of controlled chemical durability for storage batteries, *J. Am. Ceram. Soc.*, *29*, 313 (1946).

as a means of reducing the final charging voltage of lead-acid batteries. Traces of nickel are released gradually to the electrolyte by the very slight solubility of the glass. It was known previously that nickel has the property of decreasing the charging voltage, but only limited use has been made of this property.

PASTED PLATES

The Grids

The grids (Fig. 9) serve as supports for the active material of the plates and conduct the electric current. The grids also have an important function in maintaining a uniform current distribution throughout the mass of the active material. If the current distribution is uneven, the changes in volume of the plates, during their charge and discharge, will be uneven, resulting in a tendency of the active material of the plate to buckle or crumble. Light grids are, in general, used in batteries designed for heavy discharges of short duration, but in batteries designed for long life for which the discharge is intermittent or extended over a long period of time heavier grids are employed. The grids are cast, for the most part, of an alloy of lead and antimony, and frequently have designs by which the manufacturer can be identified. The grids most commonly used at the present time have cross bars which pass either straight or diagonally across the plate and are designed to lock the active material in place.

Grids for positive and negative plates are frequently of the same design, composition, and weight; but it is possible to make the negative grid lighter, because this grid is less subject to corrosion than the grid of the positive plate.

Casting the Grids. The molds ordinarily consist of two parts. They are made of cast iron. Grooves are cut in the opposite faces of the mold, according to the design of the particular plates. Small grids are cast double and later cut apart. Large grids for stationary batteries or submarines are cast singly.

The molds should be evenly heated to 135 to 180° C (275 to 356° F) to make the metal run freely, and they must be provided with suitable vents to let out the air, which would otherwise be trapped in the mold when the molten metal is poured in. A slight head of the molten metal is required to make it run into all the recesses of the mold, and the molten metal must be at a high enough temperature, when it is poured into the mold, to prevent premature solidification. The range of temperature is from about 425° C (800° F) to 525° C (975° F). Excessive temperatures, 975° F and above, should be avoided since oxidation

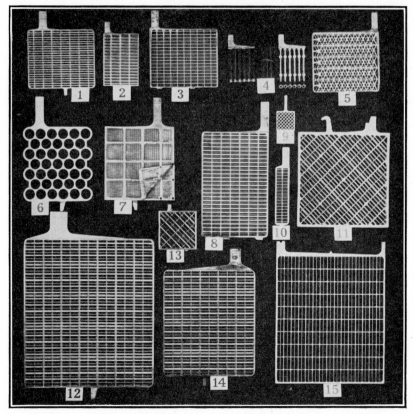

Fig. 9. Various types of grids. Nos. 1, 2, and 5 are grids for small glass-jar batteries; 3, grid for automotive battery; 4, smallest-sized grid of Ironclad construction, assembled and unassembled; 6, grid of the Manchester positive plate; 7, grid of the box-negative plate; 8, grid of a motive-power battery; 9 and 10, grids for small experimental batteries; 11 and 13, Floté reinforced grids for stationary batteries; 12 and 14, Tytex reinforced grids for stationary batteries; 15, lead-calcium reinforced grid for stationary batteries.

may change the composition of the alloy and affect the speed and quality of the casting.

The present tendency is to use lower percentages of antimony in the alloy than formerly. This makes casting somewhat more difficult and requires higher casting temperatures. Dross is likely to become more troublesome at these high temperatures; it should be avoided as far as possible because in the finished grid it is a cause of corrosion. In an effort to improve service conditions, Willihnganz[30] has described

[30] Eugene Willihnganz, New grid casting improves battery life, *Iron Age, 163,* 62 (June 9, 1949).

a new grid design and an improved casting technique to avoid cavities or "pipes" caused by shrinkage of the alloy and also to regulate the time of freezing of the lead matrix in the mold with reference to the freezing of the eutectic.

Molds are prepared for the casting process by smoking, spraying, or dusting the faces after preliminary heating. This is necessary for producing good castings which can be removed readily from the mold. The practice of dusting the mold frequently with pumice or other powder has been largely superseded by the application of a smoky acetylene flame or by spraying the mold with one of the several mold coatings that are now available. Spraying is done with an air gun: the nozzle is held 15 to 20 inches from the mold, and the spray is directed from several directions. The mold must be hot enough to evaporate the water constituent instantly. A thin uniform coating is desired. With occasional "touching up" this should suffice for several hours. If the mold becomes too hot for the molten metal to solidify promptly, a little water may be used at the gate. The molds are sometimes filled from a ladle, by hand, but usually they are filled directly from the melting pot. Heavy grids, such as those for the Manchester plates, are cast under pressure, the molten metal being blown into the mold by compressed air.

The melting pot containing the molten alloy is usually fired by gas and contains several hundred pounds of the alloy. During the casting of the grids, the molten metal is stirred from time to time and the dross skimmed off. In the best practice the trimmings from the castings and scrap alloy, unless clean and free from sweepings, are not thrown back into the melting pot, since there is danger of contaminating it with impurities. Ventilation is necessary to carry off the fumes which may sometimes be seen over the melting pot. For this a forced suction through a hood is desirable, since the natural ventilation of the room does not provide satisfactory protection to the workers from poisonous fumes.

The most common flaws that are observed in grids are due to dross, to premature solidification of the metal in the mold, to webs which appear when the two faces of the mold do not fit well together, and to warping, which may be caused by mechanical injury in taking the newly cast grid from the mold. The grids are trimmed after casting, to remove rough edges and minor imperfections, but extensive trimming should be unnecessary. Grids for small batteries, such as those used on automobiles, are pasted while still double. After this they are cut apart, and the lugs for connecting the grids to the connecting straps are cut to the proper length and brightened by a scratch brush before the plates are burned to the straps. Large and costly grids containing minor defects are sometimes repaired by burning in new ribs. Before being pasted, the grids must be free from grease or dirt of any kind.

Cast grids are generally preferred to punched grids, but the latter have found some use. They are made by casting a continuous wide ribbon of the lead alloy which passes in turn through the punch press and pasting machine.

With the object of making the battery much lighter, substitutes for the customary lead-antimony alloy grids have at various times been proposed. These include grids of Celluloid and ebonite, proposed many years ago, and more recently grids of light metals which are lead-plated. Rods of carbon can be plated with lead dioxide. None of these, however, has found extended use.

The Pasting Process

General Characteristics of the Pastes. The pastes now commonly used in making the familiar pasted-plate batteries are prepared by mixing some particular lead oxide or a blend of oxides with a dilute solution of sulfuric acid. Reactions occur that result in the formation of basic lead sulfate and the liberation of considerable heat. The temperature of the mixture rises to a maximum, which must be passed to avoid premature solidification before the paste can be applied to the grids. The lead sulfate is the cementing material which makes a firm plate that can be handled in the processes to follow. The lead sulfate also expands the paste ("bulking"), and this has an important effect on the subsequent operating characteristics of the finished battery. Too little expansion results in hard, dense plates and needless limitation of the ampere-hour capacity of the battery. They may fail in service by buckling. On the other hand, too great expansion may result in shedding of the active material and thereby shorten the useful life of the battery.[31]

Manufacturers individually have their specifications for the consistency of the paste and its "gram weight." This refers to the grams per cubic inch or as it is sometimes called the "cube weight." Naturally this varies with the processes employed, but several published statements indicate 64 to 67 grams per cubic inch for positive-plate pastes and 69 to 71 grams per cubic inch for negative pastes as being usual. In general the pastes when ready for application to the grids have the consistency of a fairly stiff mortar. There are several

---

[31] O. W. Brown, R. L. Shelley, and E. W. Kanning, Expansion as a controlling factor in positive-plate composition for lead storage batteries, *Trans. Electrochem. Soc.*, *64*, **355** (1933).

methods of measuring the consistency for purposes of manufacturing control. These include the use of a cone penetrometer, or allowing a cylinder of the paste to fall on a plate from a specified height and measuring the resulting deformation.

Many variations in the conditions of preparing and applying the paste are possible. Experience is an important factor in perfecting the proper methods and technique. Variations in the physical and chemical characteristics of the oxides, the percentage of true red lead which is present, the temperature and strength of the acid solution, the time of mixing, the treatment of the plate during and after the pasting process, and even the atmospheric humidity are among the conditions that affect the finished product. By attention to such details, control may be obtained of the bulk of the paste, the time of setting, the hardness of the plate, the time required for formation, the initial capacity when put in service, and, to a limited extent, the life of the plate.

Litharge has a valuable cementing action when mixed with any one of a variety of solutions of acids, bases, and salts. With glycerin it forms a hard, strong mass that is too limited in porosity for present-day battery uses but is well known as a cement for other purposes. With other solutions, including those of sulfuric acid or ammonium or magnesium sulfates, porous masses are prepared which are much more suitable for use as active material in batteries. The ammonium sulfate process which was a favorite years ago, like the others, has given way to the acid pastes of today. Red lead, Pb_3O_4, is much less reactive than litharge with these solutions, but battery red lead usually contains about 25 per cent of litharge. Mixed with sulfuric acid the paste darkens as lead dioxide and water are formed along with the sulfates. Some types of plates today are made with the red lead exclusively, but for the type of battery most familiar to the public, the automotive battery, red lead is blended with the uncalcined oxides to the extent of about 20 per cent. It has some advantages in regulating the time needed for formation, and it provides a battery whose capacity quickly reaches its maximum value. In too large amounts, however, red lead may shorten the useful life of the battery. Red lead and litharge are still important materials, but the trend began about 25 years ago toward the use of high-metallic uncalcined oxides, such as the oxides made by attrition of lead balls agitated in air or the Barton oxides, both of which contain 20 to 50 per cent pulverized lead. These are highly reactive with the sulfuric acid solutions, and they oxidize rapidly in the presence of moisture.

Formulas for the pastes are strictly the manufacturer's prerogative, and they vary with the type of battery to be made and the operating characteristics that he wishes to obtain. Merely as an indication of what to expect of pastes for automotive batteries, Ritchie[32] says that many formulations of paste for negative plates will be close to the following composition: blanc fixe 0.50 per cent; lampblack 0.20 per cent; organic 0.30 per cent, and leady materials 99 per cent. Pastes for positive plates also contain a preponderating amount of the un-calcined oxides blended with perhaps 20 per cent of the red lead, Pb_3O_4.

Mixing the Ingredients. The dry materials, having been carefully weighed and blended in a mixing machine, are made into a paste with a solution which is usually dilute sulfuric acid. If this has a specific gravity of 1.100 or less, the solution can be added directly to the dry mixture of the oxides and expanders, provided that they have been thoroughly blended. The more usual procedure, however, is to add a considerable portion of water to the oxides before adding a some-what stronger solution of the acid (sp. gr. 1.200 to 1.400). This has the advantage of eliminating dust and prevents the formation of a gritty paste which would otherwise result from using a solution of such strength. The acid must be added slowly while mixing is continued. A final portion of water can then be added as required to bring the paste to the proper consistency.

The amount of H_2SO_4 in solution, regardless of its exact concentra-tion, determines the amount of basic lead sulfate formed and conse-quently the expansion or bulking of the paste. If not carried too far, the capacity of the positive plates will increase as the expansion is increased. For factory control, the expansion of the paste may be measured by determining the weight of a specified volume of the paste, as was done in determining the apparent density of the dry oxides.

Mixing must be continued until the paste has become uniform and of the proper consistency to be applied to the grids. Consistency of the paste is controlled largely by the amount of water. The finished paste should be reasonably stiff when it is applied to the grids. Too soft a paste produces soft plates and is liable to form slick surfaces which usually blister when the plates are formed. Consistency is usually expressed by some arbitrary scale of numbers, which, properly interpreted, indicates whether the desired uniformity of the paste is being attained. Otherwise these numbers have little physical sig-nificance.

In order to obtain thorough mixing within 5 to 15 minutes, mechani-

[32] J. Ritchie, *loc. cit.*

cal mixers are commonly employed. These must not contaminate the paste with impurities that would be injurious to the battery. A mixer which can be cleaned easily is desirable.

Applying the Paste. The paste is applied to the grids by hand labor in many of the smaller manufacturing plants and by machine pasting equipment in most of the larger plants. When the work is done by hand, the paste is spread upon the grids with a wooden spatula or a smoothing trowel. A sheet of paper may be interposed between the table and each grid. Paper is used also to cover the plate after pasting. The paper keeps the paste from adhering to the table and also takes up some of the moisture. It is more usual now to rack the plates immediately after pasting them. Sufficient pressure must be applied to force the paste into intimate contact with the cross bars of the grid. The grids must be free from grease and dirt before pasting is begun. Sometimes they are washed and dipped in a dilute solution of sulfuric acid before being pasted.

Machines of several types have been developed for pasting the plates. In the machine operation, the grids pass under a hopper from which approximately the right amount of paste is received. This is pressed into the grids as they pass along, and the excess of paste is removed. Sometimes the plates are partially dried also before leaving the machine. The papers used in the hand-pasting process are not required in machine pasting.

Basic Sulfates in the Paste. The presence of basic sulfates in the paste has been recognized for a long time, but now it is possible to identify these basic sulfates by the use of X-rays. Mrgudich[33] says that, as sulfuric acid is added to lead oxide of the tetragonal variety containing 10 to 20 per cent of finely divided unreacted lead, the first reaction is assimilation of the sulfate groups into the oxide lattice without the precipitation of any known stoichiometric compounds. These groups occupy unstable positions in the lattice, straining it and imparting to it an excess of potential energy which he believes accounts for the properties of adhesion and cohesion. Others have questioned the validity of his conclusion. The paste which he prepared in various size batches maintained a fixed ratio of acid to oxide. Its composition was 51 ml of 1.400 sp.gr. sulfuric acid to 1 kilogram of the oxide. In the various samples he observed small amounts of tetrabasic sulfate, $4PbO \cdot PbSO_4$, which increased to 70 per cent when the pasted plates were dried in superheated steam. These plates were said to be satisfactory, but others in which the basic sulfate initially had amounted

[33] J. N. Mrgudich, X-ray studies of storage-battery pastes, *Trans. Electrochem. Soc., 81,* **165** (1942).

to 60 per cent or more as a result of too high a temperature were definitely unsatisfactory. In these pastes the PbO lattice was destroyed and replaced by lattices characteristic of the basic sulfate and lead. As the paste stability increased, the activity decreased.

His next experiments were to determine the effect of changing the ratio of acid to oxide. Increasing amounts of acid, up to double that used above, provided progressively increasing modifications in the PbO pattern without any new compound appearing until rather suddenly the basic sulfate changed to $3PbO \cdot PbSO_4$, and this continued until a maximum was reached when 154 ml of the 1.400 sp.gr. acid had been used.

Determinations by Lander[34] support the formulas for basic sulfates. He experimented with compounds prepared both by fusion and in the "wet way." By X-rays he found the tetrabasic sulfate, $4PbO \cdot PbSO_4$, and the monobasic sulfate, $PbO \cdot PbSO_4$. Between these is a hydrous compound, as he calls it, the tribasic sulfate, $3PbO \cdot PbSO_4 \cdot H_2O$, which is most important. The tetrabasic sulfate occurs at high temperatures.

Exposure of dried unformed plates for several months may produce surface films of basic carbonates of which several are known: $2PbCO_3 \cdot Pb(OH)_2$ and $4PbCO_3 \cdot 2Pb(OH)_2 \cdot PbO$ are recognized together with the ordinary lead carbonate, $PbCO_3$. The formation of these carbonates imparts additional hardness and strength to the plates.

Curing of Pasted Plates. With the modern trend toward the use of uncalcined litharge containing a considerable percentage of finely divided metallic lead have come significant changes in the methods of handling plates immediately after being pasted. The need for such changes[35] arises from the peculiar behavior of the pulverized lead in the presence of air and moisture. In a perfectly dry condition the lead does not oxidize, and this is equally true in an abundance of water. The latter condition is beneficial in operations up to the actual pasting of the plates, but after the paste is applied to the grids the residual moisture, the atmospheric humidity, and the ambient temperature all become significant. The problem is then to control these three factors so that oxidation to litharge occurs and the free lead content is reduced from about 25 or 30 per cent to about 5 per cent. Positive plates properly treated become strong and are still quite porous. Negative plates are less sensitive to these conditions.

[34] J. J. Lander, Compound formation between lead monoxide and lead sulfate, *Naval Research Lab. Rept. C–3262* (1948).

[35] R. H. Greenburg, F. B. Finan, and B. Agruss, The curing of lead storage-battery plates, *J. Electrochem. Soc., 98,* 474 (1951).

With residual moisture between 7 and 8 per cent, the rate of oxidation of the lead is a minimum, but further decrease in the moisture content by only 2 per cent, that is, to between 5 and 6 per cent, increases the rate of oxidation more than tenfold. The reaction is exothermic, and the heat liberated helps to dry the plates.

Practical operations with the uncalcined litharge having a high metallic lead content follow the usual procedures: first mixing with water, followed by the gradual addition of sulfuric acid solution (about 1.400 sp. gr.), and continued mixing until the temperature has passed its maximum and fallen 30° F or more. Too rapid addition of the acid should be avoided, as otherwise the mix temperature is likely to exceed 140° F, which is high enough. The pasted plates are passed through a flash drier to remove excess moisture and to permit stacking for 3 to 4 days, during which the changes outlined above occur. The plates will become warm in the stack as oxidation of the free lead occurs, but they cool down as the net lead content approaches 5 per cent. Aside from the improved strength of the dried plates, an excess of free lead over 5 per cent is undesirable as it is likely to cause washing or scaling of positives in the forming baths or in service life. Negative plates can be made directly from these materials with the addition of the expanders. In general the practical procedures are much the same as for the more critical positive plates, but the cube weight of the paste is slightly higher.

Calcined litharge and its blends with red lead are used less than formerly in making pastes but are still important. The red lead forms quickly and can be used to adjust the relative time of formation of positive and negative plates when formed together. The curing of freshly pasted plates has followed different processes. Probably the more common practice at the present time is to cure them in a tunnel drier under carefully controlled conditions of temperature, time, and humidity. The manufacturer has the choice of completely drying the plates before formation or transferring them to soaking baths or forming tanks while they are still in the moist condition. Freshly pasted plates require rapid handling, because partial drying is likely to lead to difficulties when the plates are formed.

Plates which are to be dried are often dipped for a few seconds in a solution of sulfuric acid (1.100 to 1.125 sp. gr.). Before this is done, however, the papers, if used, should be stripped from them. After being dipped, the plates are drained, racked, and allowed to dry. Dipping increases slightly the amount of sulfation and lessens the tendency for the paste to crack or check. Drying at ordinary temperatures and humidities requires 2 to 4 days, but at 100 to 150° F (38 to 65° C)

Storage Batteries

they may be dried in 12 to 15 hours. Ovens providing controlled humidity at elevated temperatures can be used to advantage. The humidity is initially high, usually at the saturation point, but this is reduced gradually and the temperature is raised as drying proceeds. Thorough drying is necessary to avoid "popping" (loosening of the pellets) when the plates are formed. Abrupt changes in temperature and pressure during the drying process should be avoided. In no event should the plates be subjected to freezing temperatures.

Plates which are to be treated by the "wet" process must be protected by damp cloths or equivalent means after being pasted until they are transferred to the soaking or forming tanks. Soaking, which differs from dipping mainly in the longer time that the plates are immersed, is usually done in special tanks. The time required varies from a few hours to a day or more, depending on the strength of the solution. The specific gravity of the solution falls rapidly at the beginning, but the rate decreases as time goes on. This decrease in specific gravity serves as an indication of the amount of lead sulfate

TABLE 4. WEIGHT OF ACID CONSUMED DURING SOAKING PROCESS

(Weight is given in grams for sp. gr. 1.250 to 1.050, 27° C [80° F])

Time, hr	1.250 sp. gr.	1.200 sp. gr.	1.150 sp. gr.	1.100 sp. gr.	1.050 sp. gr.
2	1.99	2.50	2.27	1.57	0.96
5	2.39	2.82	3.27	2.59	1.50
21	3.12	3.46	4.28	4.83	4.50
45	3.23	3.48	4.92	5.22	5.31

formed in the plate. The initial strength and volume of the solution being known, reasonably accurate calculations can be made, if correction is made for dilution of the acid by moisture in the plates and for evaporation. The time is reduced if the acid is relatively strong, but more lead sulfate is formed when weaker solutions are used, provided that sufficient time is allowed for the acid to penetrate the plates. Table 4 gives the results of experiments on small plates soaked in solutions of several strengths. A specific gravity of 1.250 is obviously too high and 1.050 too low. The time for soaking varies with the thickness of the plates. If the soaking process is omitted, the plates may go from the pasting room to the forming tanks.

Other Materials Used in Pasting the Plates. A lead compound consisting of 20 per cent lead oxide and 80 per cent lead sulfate chemically combined, known as "superite," has been produced for use with the ordinary lead oxides in preparing the paste with the object of increasing the bulk of the paste. From 5 to 10 per cent of this material may be mixed with the lead oxides and subsequently formed into the active material of the plates. Precipitated lead sulfate is also used similarly.

Hardeners and binders were occasionally added in the preparation of the paste, to increase the coherence and hardness of the plate. A great variety of substances has been used for this purpose. Glycerin and carbolic acid are probably the best known of these materials for making hard plates, but the present practice is to control the plate hardness by proper processing of oxides and solutions rather than by resorting to the use of hardeners.

Another class of materials, called porosity agents, was sometimes employed in the preparation of the paste. These are substances that can be dissolved out of the paste when the setting process is complete. They increase the bulk of the paste. Magnesium sulfate is the best known of these, but other sulfates and even sugar have been used. A magnesium sulfate paste has a good cementing action because of the crystalline structure formed as the paste dries, but a reaction such as that with ammonium sulfate seems to be lacking.

Other Kinds of Paste. In addition to the pastes that are commonly used as described above, there are a number of others that deserve brief notice. One of these is the so-called lead sulfate paste, which is made from lead sulfate mixed with ammonium hydroxide, NH_4OH. This paste is made into a thick dough and hardens when dried. The plate is formed in a bath of ammonium sulfate containing a small percentage of free ammonia. In this bath the sulfate can be reduced to the form of lead sponge, and positive plates are prepared by oxidation from the negatives. Glycerin paste consists of approximately 75 parts of litharge mixed with 25 parts of glycerin. This paste forms an unusually hard plate. Lead chloride paste is usually cast in the form required with 10 per cent of zinc chloride. Lead carbonate paste is made from basic or normal lead carbonate and lead oxide, and formed into a paste in an alkaline solution. Negative plates are prepared from this paste by reduction, and the positive plates are prepared by oxidizing the negative plates. Alkali pastes are prepared from litharge mixed with a solution of caustic potash having a specific gravity of about 1.10.

Formation of Pasted Plates

The plates are electrolytically oxidized and reduced in dilute sulfuric acid or a sulfate solution. The plates that are to become the positives are made the anode in the forming tank, container, or jar, and the plates for the negatives are made the cathode. The word "formation" applies primarily to Planté's process for increasing the capacity of the plates in his cell. This is a process quite different from that used for developing the pasted plate batteries, but the word "formation" is now in common parlance applied to the pasted plates as well. Formation, as applied to the Planté plates, means the creation of a layer of sponge lead on the surface of the negative plates and of lead peroxide on the positives to constitute the active materials of the cell. Formation of pasted plates, on the other hand, means the oxidation or reduction of the lead oxides or other materials which have been applied to the grids.

Positive plates, alternating with negatives, are mounted in tanks or other containers which provide for proper spacing of the plates and insulation between those of opposite polarity. The tanks may be of vitreous material, hard rubber, or wood with lead lining. Some manufacturers, however, prefer to assemble the plates in groups or complete elements before formation, which is then done in tanks or in the containers that are to serve for the finished battery. In any event it is presumed that the forming time for positives and negatives is about the same. This can be regulated by the composition of the paste and to some extent by the strength of the forming acid. Positive plates, more than negative plates, are subject to harmful effects of overformation. It is preferable, therefore, that the negatives should finish first. Some variations in procedure of forming plates are indicated below, but experience in forming plates, prepared in a particular way, is the safest guide.

The strength and amount of sulfuric acid used will depend upon the pervious treatment of the plates. Usually it is in the range 1.050 to 1.150 sp. gr. Increasing the strength of the acid increases the time required for forming positive plates.

Dry plates and those which are still wet from the pasting process will sulfate when they are immersed in the solution, and its specific gravity will decrease rapidly. The wet plates will cause further dilution of the solution, because they contain a considerable amount of water. Such plates are often allowed to soak, partially at least in the forming solution. Plates which have been dried must be allowed to remain in the solution long enough for the acid to penetrate the pores of the plate before beginning the forming charge. The pores at this time are not as open as they will be later. When a large group of tanks is filled with plates, those plates in the tanks filled first are necessarily immersed longer than those in the tanks filled last. Some manufacturers, therefore, make a practice of allowing an hour or more to elapse after the last tank is filled as a "time of set" before starting the charging current.

Plates that have been soaked, on the other hand, have reacted with the sulfuric acid and they contain a relatively large amount of sulfate. They will, therefore, have less effect on the forming solution initially, but in the later stages of the forming process they will greatly increase the specific gravity, because of lead sulfate derived from the plates. For this reason the initial specific gravity of forming solutions used with these plates is usually low. The charging current can be started promptly after the plates are immersed.

Formation undoubtedly starts in the region where the poorly conducting paste is in contact with the better conducting grid bars. For this reason, formed material, particularly of soaked plates, is observed around the edges of the pellets while the center portion still has a hard white core. It may be necessary to interrupt the charge and to make a discharge before resuming formation. This helps to counteract the tendency to overformation and softening of the positive active material. Other corrective measures include decreasing the strength of the forming solution and decreasing the current density.

Completion of formation is indicated by (1) the color of the active materials, that is, the plates have "cleared" and are uniform in color; (2) plates gassing normally; (3) cadmium readings are constant and of normal values. The exact values of the cadmium readings will depend on various conditions, but they are usually in the range 2.30 to 2.45 volts for the positive plates and -0.05 to -0.20 volt for the negative plates. The cadmium readings on positives may reach constant values several hours prior to complete clearing, but this is usually not true of the negatives.

The proper charging current will depend on the thickness and type of plates and on the temperature. For the more common sizes of automotive plates the current is usually specified as 0.75 to 1.50 amperes per positive plate. At 1 ampere the forming time is about 45 to 50 hours. Foaming and excessive gassing must be avoided. When either of these occurs the current should be reduced. It is customary in forming plates to provide a rest period of 2 to 4 hours when the current is cut off altogether or greatly reduced. This is necessary for best results on positive plates and avoids soft edges of the pellets. A second rest period is sometimes provided also. The current is used more effectively at lower rates, and these are desirable for thin plates. For plates in general a current density of 0.2 to 0.5 ampere per dm^2 (2 to 5 amperes per square foot) is reasonable. The area is reckoned as the apparent surface of both sides of the plate.

Some modifications must be made when the plates are formed in small tanks. These relate chiefly to the strength of the forming solution and the available volume of solution. This is important when the plates are formed in containers which are to serve for the finished cells. In that event when the elements are assembled with moist separators they carry enough water to dilute the forming solution appreciably. Assembled elements are usually allowed to stand in the forming solution several hours before beginning the charge. After formation is completed, it is advisable to pour out the electrolyte and replace it with acid of a strength that will finish at a specific gravity of 1.260 to 1.280 when the battery is fully charged. The initial charge, distinguished from the forming charge, is usually done at a current rate of 0.75 to 1.00 ampere per positive plate of the automotive size.

Groups of plates which have been formed in tanks or slotted rubber cases may be drained at the end of the forming process. Positive plates are sometimes rinsed and dried. Negative plates will become hot during the drying, because of oxidation of the sponge lead in the air. They may be cooled somewhat by forced circulation of air. Such plates must be regarded as being in a discharged state, although they contain little lead sulfate. When they are assembled into batteries a prolonged initial charge is necessary.

After the plates are formed some of them are dried by processes known and approved by the individual manufacturer. The oldest and simplest is air drying, which requires several days. Special methods for drying plates to be used in dry-charged batteries are described below. Other plates are assembled into elements while moist. It should be remembered that plates containing an appreciable amount of sulfuric acid will not dry completely in the air because sulfuric acid is hygroscopic. Negative plates must be cooled.

Although tank formation is regarded by many storage-battery engineers as the best process, many of the smaller types of batteries are container- or jar-formed. Either of two processes is involved: (1) Two-step formation, which means that, after the pasted plates have been conditioned and properly aged for a period of about 4 days, they are assembled into elements and immersed in a relatively low-specific-gravity solution of sulfuric acid in the containers that are to be parts of the finished batteries. The plates and separators are allowed to absorb electrolyte before charging begins. A rest period is provided, and the charge is resumed and continued (with perhaps a second rest period) until the plates have cleared. The electrolyte is then dumped, and the batteries are refilled with fresh electrolyte which will finish at the proper specific gravity at the conclusion of the charge.

(2) One-step or one-shot formation consists of immersing the processed plates assembled into an element in somewhat stronger solution of sulfuric acid. Reactions with the lead and oxides of the plates cause a marked rise in temperature. The maximum must be passed and the batteries cooled before charging is begun. If the modified constant potential charge is applied, the current will begin at a low rate because of the high internal resistance of the cell. The current will rise as formation progresses and ultimately fall to a low finishing rate as the counterelectromotive force of the cell increases to its maximum value. The electrolyte chosen for this one-step process of formation should have an initial specific gravity such that it will have the desired final specific gravity when the process of formation is completed. The relative merits of formation in tanks or in containers using the one- or two-step processes are the subject of controversy.

Plates for Dry-Charged Batteries. The positive plates present few difficulties. They may be dried after formation by customary methods. For best results they currently are washed in several changes of water to eliminate the acid. The plates then contain 80 to 90 per cent of lead dioxide, PbO_2, several per cent of lead sulfate, $PbSO_4$, and the remainder unconverted lead oxide, PbO. The dried plates can then be stored indefinitely.

The negative plates are more critical. Because they oxidize rapidly when charged and exposed to air in a moist condition, they must be exposed as little as possible and rinsed and dried rapidly. Various methods specify drying in inert atmospheres, such as carbon dioxide,[36] nitrogen, or a vacuum with applied heat. The use of infrared lamps[37] has been mentioned. One method described by the Eagle-Picher Lead Co.[38] involves the use of superheated steam at 360 to 370° F for drying, followed by cooling in saturated steam at a lower temperature. Cooling must be continued until the temperature is safely below 300° F, since moist plates will ignite at that temperature spontaneously. Further cooling and drying is accomplished in an air blast.

Effect of Temperature on Formation. Investigations on the effects of formation temperatures have shown the importance of knowing and controlling the temperatures that occur, and they illustrate the well-known fact that conditions favorable to one plate (e.g., the positives) are detrimental to plates of the other polarity, and vice versa. For-

[36] S. Makio, Drying and storage of secondary battery plates in the charged state, *Trans. Electrochem. Soc.*, *53*, 251 (1928).

[37] E. Jirik and A. J. McIlwraith, Battery plates dried with infrared lamps, *Elec. World*, *126*, 150 (Sept. 14, 1946).

[38] M. F. Chubb and P. F. Ebert, Performance characteristics of dry-charged batteries, No. 5 Eagle-Picher Lead Co., 1943.

mation temperature affects the structure of the active material. Greatly improved cold capacity with only minor decrease in life can be obtained by keeping the plates cool during formation, especially during the first half of the forming period.[39]

Formation at 80° F provides 50 per cent more cold capacity from pasted-plate automotive batteries than formation at 115° F. The negatives are better, but the life of the battery is likely to be less. On the other hand the positive plates formed at 115° F are better than those formed at 80° F. Lynes and his co-authors gave interesting data on the variability to be expected in routine laboratory testing, and they showed what pitfalls the uninitiated may encounter as a result of differing temperatures of formation, open-circuit stands, deep discharges, and conditions of cycling.

Other authors have found the low temperatures favorable for formation of negative plates. Hatfield and Brown[40] used negative plates formed at 20° F with positives formed at 90° F, and vice versa. Current rates for forming negative plates at 20° F are necessarily limited, and the plates do not clear as readily at 40° F as at higher temperatures.

Greenburg and Caldwell[41] showed that a decrease in forming temperature is accompanied by a decrease in the percentage of PbO_2 and an increase in an unidentified compound which they call "apparent PbO." They investigated the effect of temperatures between $+70$ and $-2°$ C (158 and 28° F). The composition of positive active material formed at 8° C showed 75.3 per cent PbO_2 and 4 per cent $PbSO_4$, the remainder being the apparent PbO. On the other hand formation at 71° C yielded 93.8 per cent PbO_2, 0.5 per cent $PbSO_4$, and 5 per cent PbO. They say that the PbO_2 is more reactive when formed at lower temperatures, and in this they agree with Hatfield and Brown, who say that it has greater initial capacity. The authors do differ, however; Hatfield and Brown say that positive plates cannot be satisfactorily formed at 150° F, but apparently the others did it. Lead sulfate forms more rapidly in newly formed active material of the positives than the negatives, as judged by the percentage of lead sulfate at the end of 14 days. This seems surprising. It is greater also in positive active material formed at lower temperatures.

[39] T. C. Lynes, F. Hovorka, and L. E. Wells, Storage-battery formation temperature and cold capacity, *Ind. Eng. Chem., 37,* 776 (1945).

[40] J. E. Hatfield and O. W. Brown, Influence of temperature of formation on the initial capacity and life of pasted SLI battery plates, *Trans. Electrochem. Soc., 72,* 361 (1937).

[41] R. H. Greenburg and B. P. Caldwell, Effect of forming temperature on lead storage-battery anodes, *Trans. Electrochem. Soc., 80,* 71 (1941).

Changes in Porosity. The lead sulfate which remains in the final composition of the plate is of great importance as a binder of the active material. This small percentage of sulfate remaining is not easily removed by excess charging.

Important changes in volume of the active material occur during the forming process. The lead sulfate is considerably less dense than the peroxide or sponge lead of the finished plates (Table 5). During the forming process, therefore, the pores of the plate open. This permits the electrolyte to come in contact with the material in the inner recesses of the plate. The porosity which the plate acquires during the forming process is of importance not only in the formation but also in the subsequent operation of the finished cell. In Table 5 are given the densities for the various materials found in the paste and in the finished plates.

TABLE 5. DENSITIES OF MATERIALS

Material	Formula	Density
Lead	Pb	11.3
Litharge	PbO	9.5
Minium (red lead)	Pb_3O_4	9.1
Lead dioxide	PbO_2	9.37
Lead sulfate	$PbSO_4$	6.3
Lead chloride	$PbCl_2$	5.8

Electrochemical Equivalents. It is of interest to consider the oxidation and reduction processes that take place during formation of the positive and negative plates, and to calculate the number of ampere-hours per kilogram of material that are required in accordance with Faraday's law. When litharge, PbO, is oxidized to lead dioxide, 1 atom of oxygen is added to the PbO molecule. The molecular weight of PbO is 223.2. The amount of oxygen to be added per kilogram of material is therefore

$$1000 \times 16/223.2 = 71.7 \text{ grams}$$

The equivalent of oxygen in ampere-hours per gram is 3.350. This value may be calculated from the following formula: 96,500 coulombs,[42] representing the number of coulombs required for the liberation of 1 gram equivalent of a substance, multiplied by 2, the valence, divided by 16, the atomic weight of oxygen, and divided by 3600, the number of seconds in 1 hour. The product of the equivalent of oxygen

[42] See discussion of Faraday's law in Chapter 4 for the most precise value of this constant.

and the number of grams, 3.350×71.7, equals 240.2 ampere-hours per kilogram. Since the reduction of litharge to lead involves the taking away of 1 atom of oxygen from the molecule PbO, exactly the same number of ampere-hours per kilogram is required for the reduction of 1 kilogram of material. In Table 6 is shown the number of ampere-hours per kilogram of material required for the oxidation to the dioxide state or the reduction to sponge lead of the various materials that are used for pasting storage-battery plates. These are the minimum figures and are exclusive of the energy that is wasted by gassing during the forming process. In calculating the values for minium, it is assumed that Pb_3O_4 represents the composition of the material.

TABLE 6. AMPERE-HOURS PER KILOGRAM REQUIRED FOR THE OXIDATION AND REDUCTION OF OXIDES, CHLORIDES, AND SULFATES OF LEAD

Material	Reduction to Lead, amp-hr	Oxidation to PbO_2, amp-hr
PbO	240	240
Pb_3O_4	313	156
$PbSO_4$	176	176
$PbCl_2$	193	193
Pb	. . .	514

Since the material of the plate at the beginning of the formation process consists of lead, one or more of the oxides, and a certain amount of lead sulfate, the following example is given to illustrate the use of the table in computing the theoretical number of ampere-hours required for the formation of a negative plate. In this example we shall assume that the plate, exclusive of the grid, weighs 300 grams and contains 20 per cent of lead sulfate and 80 per cent of lead oxide, PbO. From the table we compute at once that the number of ampere-hours required for the lead sulfate is $\dfrac{0.20 \times 300}{1000} \times 176 = 10.6$. Similarly, the number of ampere-hours required for the reduction of the lead oxide, PbO, is $\dfrac{0.8 \times 300}{1000} \times 240 = 57.6$. The total number of ampere-hours required, therefore, for both materials is 68.2. This example may be solved by another method if the electrochemical equivalent of lead is known. By this method the amount of lead in each of the constituents of the paste is calculated.

$$0.20 \times 300 = 60 \text{ grams of } PbSO_4 = 41 \text{ grams of Pb}$$
$$0.80 \times 300 = 240 \text{ grams of PbO} = 223 \text{ grams of Pb}$$
$$\text{Total weight} = 264 \text{ grams of lead}$$

The electrochemical equivalent of lead is 3.865 grams per ampere-hour. The number of ampere-hours is therefore 264 divided by 3.865 = 68.2 ampere-hours, which is the same result as was obtained in working this example by the first method.

The efficiency of the formation process depends largely on the amount of gas evolved. Ordinarily the efficiency of the forming process will not exceed 50 per cent. It is necessary, therefore, to double the answer obtained in the example, in order to obtain the number of ampere-hours required in ordinary practice. If the formation process is continued beyond the ordinary stopping point, the efficiency will be considerably lower. The elimination of the last traces of sulfate in the plates is difficult and not desirable, since this small amount of sulfate is in effect a cement to hold the active material of the plates together.

The potential relations of the plates during the forming process are of importance, since the potential that must be applied to them multiplied by the number of ampere-hours determines the energy and therefore one factor in the expense involved in the process. The potential of the negative plates during the forming process is measured against the positive plates or the dummies, which are essentially positive. The potential difference between the negative plates and the dummies increases toward the end of the forming process in accordance with changes in the ionic concentration. The lead sulfate of the plates dissolves to a slight extent in the electrolyte and in dissociating forms lead ions, Pb^{++}. These ions are deposited on the negative plates during the forming process. As they deposit, more lead sulfate dissolves, and when the lead sulfate is almost exhausted near the end of the process, the lead ions become scarce and the potential difference rises to the point where hydrogen ions, H^+, are liberated. The potential of the positive plates during formation is similarly measured against the negative plates or dummies in the forming bath. This potential difference is about 2.3 volts at the beginning. This voltage decreases to slightly less than 2 volts after the forming process begins, owing to the decrease in the resistance of the material of the plates. It rises again during the latter part of the formation period to 2.45 volts or higher.

Much gas is evolved during the formation process. Gassing begins normally before the time the theoretical number of ampere-hours for the reduction or oxidation of the material in the plates is completed. The conditions which determine gassing depend on the relative number of lead ions, Pb^{++}, and hydrogen ions, H^+, in the electrolyte when forming negative plates. Since a scarcity of the lead ions raises the potential of the negative plates, which, during this process, are the cathodes, to a point where hydrogen ions are liberated, it is possible to increase the efficiency of the formation process by maintaining a plentiful supply of the lead ions or keeping the hydrogen-ion concentration low. Some of the manufacturing companies in Europe have, therefore, made use of forming baths which are solutions of neutral sulfates of aluminum and magnesium, because the hydrogen-ion concentration in these baths is much less than in the sulfuric acid solutions. Heating the electrolyte is a way of increasing the concentration of the lead ions, but this may loosen the active material of the plates. Similar relations with respect to the oxygen evolution hold in the forming process for the positive plates. Neutral solutions have been used in Europe in forming these also.

The formation of plates pasted with materials other than the oxides ordinarily used is of much less importance, but it is interesting to note that plates have been successfully formed from lead sulfate and lead chloride pastes. The plates pasted with lead sulfate are formed very much as are the oxide plates. Since lead sulfate is a non-conductor of electricity, the plates have a very high resistance initially. Plates pasted with lead chloride are formed in quite a different way. This paste is soluble and cannot be peroxidized directly, since the lead chloride would go into solution and the lead would be precipitated on the cathode with the liberation of chlorine at the anode. The lead chloride paste is therefore reduced in a bath of zinc chloride with a plate of zinc which, with the lead chloride, forms the elements of a primary cell. During this process the lead chloride, $PbCl_2$, is reduced to lead, and the zinc becomes zinc chloride, $ZnCl_2$. Positives are formed electrochemically.

The positive plates, after formation, may be preserved in the dry state. Negative plates, however, if dried in air will become hot, owing to the formation of oxide, and will again require a prolonged charge before they can be made ready for use.

PLANTÉ PLATES

Manufacture of Planté Plates

The essential difference between Planté plates and pasted plates consists in the fact that the active materials of the former are derived from the body of the plate itself, whereas for the latter they are formed from oxides or other pastes applied to the plate mechanically. The active materials of the Planté plate are obtained by oxidizing the surface of the lead plate or reducing this material to sponge lead. A plate intermediate between the pasted and Planté varieties consists of a soft-lead grid pasted with the oxides and formed. The active material gradually falls out, but the capacity is maintained by the corrosion of the grid. In this way the plate becomes essentially a Planté plate. Planté plates serve quite a different purpose from the pasted plates. They are ordinarily much larger and heavier than the pasted plates and have a relatively smaller capacity. They are used chiefly for stationary batteries, in which considerations of space and weight are of less importance than durability.

The essential parts of the Planté plates are the underlying portion of lead and the developed surface which consists of a number of leaves designed to increase the surface of the plate and thereby increase the capacity. The effective surface of the Planté plates is 6 to 10 times greater than the apparent surface. The blanks, as the sheets of lead are called before the surface is developed, are prepared by casting pure lead in the form of flat ingots, which are then rolled to the required thickness. Soft lead of a very high degree of purity is required for this purpose. From the rolled sheets the blanks are cut or stamped in accordance with the size and design of the particular type of plate to be made.

Several methods have been used for increasing or developing the surface of these plates. One of these is the so-called plowing process. The blank plate is placed in a machine similar to the shaper used in a machine shop. The tool of the shaper is designed to produce a leaf of the proper shape and width. One leaf is produced at each stroke. Ribs to make the plate more rigid are made by jumping the tool at certain points.

A second method of developing the surface of Planté plates is commonly known as the swaging process. The master swaging block has the design for the plate cut in its surface, the ribs to produce recesses in the finished plate being in relief on the block. The swaging block is pressed against the surface of the soft-lead plate while it rocks back and forth, and the lead is pressed into the desired form.

A third process for the development of the surface of these plates is called the spinning process. The soft-lead plate or blank is held in a frame which reciprocates between revolving mandrels having a large number of steel disks which are pressed gradually into the plate on either side. The lead of the plate flows in between these steel disks, forming leaves with grooves between. The depth to which the steel disk may penetrate the surface of the plate is so regulated as to leave a thin web of supporting material in the center of the plates. Horizontal ribs are obtained in these plates by spinning the plate in sections. Vertical ribs are obtained by spacing washers placed between the disks.

Other plates with highly developed surfaces have been obtained by casting (Tudor plates), but this method is used more in Europe than in this country.

An important type of Planté plate is the Manchester positive. Heavy grids of lead and antimony are cast with a large number of round holes into which are pressed buttons of soft lead with corrugated surfaces. These buttons, or "rosettes," are prepared from lead ribbon which is extruded by a hydraulic press. The lead ribbon is passed through a crimping machine which crimps the surface, cuts the ribbon to the desired length, and then rolls the piece into the form of a rosette or button. The buttons are pressed into the holes of the lead-antimony grid by a hydraulic press. In order to lock the buttons in place, the holes are made with a slight bevel so that, as the lead button grows during the operation of the cell, it becomes more and more tightly locked in the supporting grid.

Formation of Planté Plates

The formation of Planté plates[43] is an electrochemical process that requires considerable time and the expenditure of a large amount of electrical energy. In the original process used by Planté and generally referred to at the present time as Planté formation, the plates were alternately charged and discharged, with occasional reversals of the direction of the charging current, until the plates had acquired sufficient capacity. The demand for a more efficient process of formation led to the use of forming agents, which were added to the sulfuric acid solution to hasten the formation by chemically attacking the lead of the plates. This is the method most used at the present time. The formation of Planté plates has also been accomplished by dipping the plates in certain solutions which have a strongly corrosive action on the lead, resulting in a layer of finely divided material that may then be reduced to sponge lead or oxidized to lead dioxide.

When two plates of lead are dipped in a solution of sulfuric acid and an electric current is passed through the cell, a very thin layer of lead dioxide is formed on the plate which is the anode, and the oxide of lead covering the surfaces of the cathode is reduced to a very thin layer of sponge lead. The evolution of oxygen gas at the anode and hydrogen at the cathode begins almost immediately. If the charging current is cut off, the lead dioxide on the surface of the anode forms a very

---

[43] G. Planté, Formation of secondary elements of lead plates, *Compt. rend.*, *95* (Aug. 28, 1882).

large number of little primary cells with the underlying lead, so that a vigorous local action begins. Lead sulfate is formed on the surface of the underlying lead, and in a few minutes the plate has lost its charge completely. The sponge lead on the surface of the plate which was the cathode has practically no potential difference from the underlying lead of the plate itself, and consequently no such vigorous local action takes place on this plate. If the charging current is again renewed, a larger amount of lead oxide is formed on the anode owing to the conversion of the lead sulfate previously formed as a result of the local action on this plate. Each time this process is repeated the amount of lead dioxide increases, but in order to obtain an increase in the amount of lead sponge on the surface of the negative plates it is necessary to reverse the current from time to time so that use may be made of the process that occurs at the positive plate. The strength of the sulfuric acid solution that is used has some influence on the amount of oxygen that is fixed on the surface of the anode. Experiments were made by Gladstone and Tribe[44] to determine the amount of oxygen fixed on the surface of electrodes which were 77 square centimeters in area by the action of 1 ampere flowing for 20 minutes. The solutions they used consisted of various strengths ranging from 1 part of acid to 5 parts of water down to 1 part of acid to 1000 parts of water. In Table 7 is given the amount of oxygen, expressed in milligrams, fixed by the action of the current for the different strengths of acid.

TABLE 7. OXYGEN FIXATION

Acid Strength, acid to water	Oxygen Fixed, mg
1/5	127
1/10	146
1/50	151
1/100	155
1/500	125
1/1000	Lower oxides and basic sulfates formed

The common method, at the present time, for the formation of Planté plates involves the use of forming agents which attack the lead of the plate. These agents are usually nitric acid or the salts of some acid such as nitric, although a number of other substances have been used, including chlorates and perchlorates, chlorides and fluorides,

44 J. H. Gladstone and A. Tribe, La chimie des accumulateurs, *Lumière élec.*, 9, 25 (1883).

bichromates, permanganates, formic acid, oxalic acid, alcohol, hydrox-ylamine, and sulfurous acid. The forming process with these addition agents is practically confined to the positive plates, which are the anodes in the forming bath. Negatives are obtained by a reversal of the positive plates and the consequent reduction of the lead dioxide to the sponge-lead state.

In general, the action of the forming agents is to retard the early formation of lead dioxide on the anode that would otherwise make a protective covering on which oxygen would be liberated. The anions depositing on the lead surface give rise to relatively soluble lead salts and increase the concentration of lead ions from which lead sulfate may be formed. This is oxidized ultimately to lead dioxide.

The amount and strength of the sulfuric acid solution and the amount of nitric acid or other forming agent are matters of importance both in determining the depth of the formation and in the finishing of the plate. The current density and temperature also affect the depth of formation. Since the nitric acid is reduced at the negative plate, which is the cathode in the forming bath, any factor, such as increasing current density or increasing temperature, which tends to increase the rate of reduction of the nitric acid, decreases the effective amount of it in the forming bath. During the forming process the amount of the forming agent present in the bath steadily decreases. It is necessary that at the end of the formation it should decrease to zero, in order that the finished plates may not contain traces of the agents that would cause subsequent growth and buckling of the plates in service. At the end of the forming period the dioxidizing effect of the charging current must predominate over any chemical action of the addition agents, in order that the plates may be satisfactorily "sealed off." This "sealing off" process consists in completely covering the underlying lead with a dioxide film which serves both as the active material of the plate and as a protective coating. The forming bath is ordinarily a solution of sulfuric acid ranging from 1.040 to 1.050 in specific gravity, to which the forming agent may be added from time to time as the formation progresses. At the end of the forming process the current density is ordinarily increased, and the plates, after being removed from the forming bath, are washed and sometimes given a further charge in a solution of pure sulfuric acid which is free from nitrogen compounds or other addition agents. Another method, seldom used, of finishing plates intended for positives is to reverse them in a pure solution of sulfuric acid and water followed by a second reversal to restore the dioxide condition. This process eliminates the forming agent.

The formation of these plates in solutions of sulfuric acid has been the common practice, but the use of neutral sulfate and alkaline baths is important and has been reported by Peters.[45]

In some cases plates have been dipped in strong solutions of nitric acid as a preliminary to the forming process. The corrosive action that takes place on the surface of the plates, particularly if the acid is concentrated, results in the formation of complex compounds of lead and nitrates and nitrites which are only slightly soluble. The lead plates, having a layer of this material on the surface, can then be formed into dioxide or sponge lead by ordinary forming processes. Another similar method has been to convert the surface of the lead plates into the carbonate.

SEPARATORS FOR LEAD-ACID CELLS

The development of thin, porous separators which are placed between the alternating positive and negative plates in storage cells has made possible the development of compact, portable batteries. Prior to the use of these porous separators, rods of glass or hard rubber or perforated and corrugated hard-rubber sheets were inserted between the plates of the cells to prevent possible short circuits through buckling of the plates. It was not possible, in the early types of batteries, to place the plates close together, because scaling of the active material or the formation of "trees" on the negative plates would make a metallic connection between plates of opposite polarity.

The primary object of the separators is to prevent metallic conduction between the plates of opposite polarity while freely permitting electrolytic conduction. Present types of separators include those made of wood veneer, perforated and slotted separators of hard rubber, microporous rubber, fibrous glass mats, and a surprising variety of new types, used or proposed, that became possible with the development of modern plastics. These include microporous plastics, fibrous materials impregnated with insoluble resins, regenerated cellulose films, layers of diatomaceous earth, fabrics of Vinyon, Saran, Dynel, woven glass, and porous vitreous materials.

Wood Separators

Certain kinds of wood have been found suitable for the manufacture of separators. In the past some of the woods employed were bass, poplar, fir, cedar, cypress, and redwood. Several of these are important today, but the almost universal choice was Port Orford white-cedar, which grows in a limited region of Oregon, taking its name from

---

[45] *Centralblatt f. Accumulatoren,* series of papers beginning Volume 2, p. 293, 1901, and continuing through Volume 5.

a seaport in the vicinity. With the great expansion of battery production for automotive purposes that occurred with the beginning of World War II, an acute shortage of Port Orford cedar arose that led to a search for substitutes. The result of valuable research at the

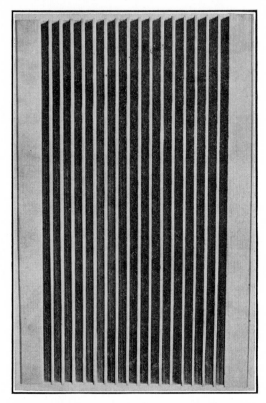

Fig. 10. A wood separator for a motive-power cell. The grooved side of the separator is placed next to the positive plate.

Forest Products Laboratory[46] showed that Douglas-fir, noble fir, Alaska yellow cedar, redwood, and bald-cypress when properly treated chemically are acceptable substitutes.

Production of wood separators for storage batteries has been estimated at about 1000 million per year at the present time. The separators are made from a knife-cut or sawn veneer in the form of a

[46] L. V. Peakes, Jr., R. A. Lloyd, V. S. Barnes, J. H. Berry, and G. J. Ritter, Substitute woods for Port Orford white-cedar for storage-battery separators. *Ind. Eng. Chem.*, *38*, 780 (1946), Rept. R1476.

long, narrow sheet of the required width. Grooves are cut in one side (Fig. 10), leaving a series of narrow ribs, between which are thin portions called the web. The method of cutting the separators affects the strength and electrical resistance of the finished product.

Rotary-Cut Separators. The veneer is cut from the log as a wide thin sheet while the log is rotated in a lathe. Rotary-cut separators are commonly called "flat grained," because the cutting edge of the knife is applied nearly tangentially to the rings of the log. The grain, however, is usually irregular and not entirely "flat," which would not be a desirable condition.

Sliced Separators. This method of cutting separators came into extensive use about 1931. Blocks of wood, after being steamed, are placed in the slicing machine, and the veneer is cut by a shearing motion of the knives. If the grain of the wood appears as parallel lines, it is called "vertical grain," but other positions of the block with reference to the knives result in "slash grain."

Sawn Separators. These are cut by a saw, the direction of the grain depending on the position of the block. Quarter-sawn separators show the grain as parallel lines extending through the separator, but the angle of the grain through the separator may vary as much as 45° and still present the appearance of parallel grain. As the angle approaches 90° the grain becomes nearly flat.

"Fibrite" Separators. Another variety of wood separator is made from finely divided bark of the redwood tree. This separator, known as "Fibrite," is available in flat sheets somewhat resembling blotting paper. Its porosity is said to be as much as 90 per cent. Because of the nature of the material it is not made as a grooved separator.

The most common defects observed in woods used for separators are knots, checks, and shakes. Knots, unless very small, are detrimental in separators. Checks are splits in the log, extending in a radial direction from the center, and are caused by stresses occurring during the seasoning process. A shake is a separation between two annual rings. As a general rule, it occurs in only one part of the ring. Any one of these defects appearing in the finished separator is sufficient cause for its rejection. After wood separators have been grooved and cut to size, they should be examined individually before a strong light in much the same way as eggs are candled. In this way defective separators are easily detected. Because of the frequent occurrence of knots, checks, and shakes, the wastage in preparing separators is considerable.

The principal constituent of wood tissues is cellulose, but with the cellulose are the so-called incrusting layers of lignin and the sap, which contains resinous material, tannins, coloring matter, and various mineral salts. There are a great many varieties of cellulose, but they have certain properties in common.

Celluloses, considered as a whole, are insoluble in the simple solvents and have a composition characteristic of carbohydrates. The empirical formula for them may be written $C_nH_{2m}O_m$. Celluloses resist in various degrees the processes of oxidation and hydrolysis. Generally speaking, however, they are much less subject to oxidation and hydrolysis than the other substances which are closely associated with them in the wood fiber. The woody tissues consisting of cellulose and the incrusting material are frequently called lignocelluloses. The non-cellulose part of these compounds is generally termed lignin. The lignin complex is easily decomposed by hydrolytic or oxidizing agents, yielding acetic acid, which is generally known to be deleterious to the life and performance of storage batteries. For example, wood immersed in dilute sulfuric acid at 60 to 100° C readily yields a considerable amount of acetic acid, and the same result is obtained at lower temperatures if an oxidizing agent, such as nitric acid, is also present. The amount of acetic acid that may be formed from the wood varies with the degree of hydrolysis or oxidation of the material. The discussion of the chemistry of cellulose is beyond the scope of this book, but the reader is referred to the books by Cross and Bevan, *Cellulose, an Outline of the Chemistry of the Structural Elements of Plants,* and by Hawley and Wise, *Chemistry of Wood.*

Treatment. The wood separators, after being cut to size, are subjected to a chemical treatment which has for its object the removal of soluble and easily hydrolyzed matter, and the expansion of the pores. Many different methods have been used, most of which are covered by patents. Some of these methods have been based on those used for the preparation of wood for paper-making. In general, the treatment consists of the removal of the mineral salts and other soluble matter contained in the sap, as well as the easily hydrolyzed parts of the lignocellulose, by the use of steam, boiling water, or dilute solutions of various chemical reagents, including caustic soda, sodium bisulfite, sodium sulfite, sodium sulfide, and sulfuric acid.

The more usual procedure is to pack the separators loosely in tanks that are then filled with a solution of caustic soda, 1 to 2 per cent, heated to the boiling point or slightly below. Insufficient treatment means high electrical resistance in the battery and incomplete elimination of deleterious substances; excessive hydrolysis, on the other hand, means an unnecessary loss in the mechanical strength of the separators.

The amount of material extracted from the wood depends on the severity of the treatment. The principal factors involved are: (1) the chemical agent used, as for example sodium hydroxide; (2) its concentration, usually within the limits 0.75 to 2 per cent; (3) the temperature, the time being greater as the temperature is less, 97 to 100° C is usually recommended; (4) the time, usually within the limits 10 to 12 hours; (5) agitation of the hot solution, usually accomplished by release of compressed air; (6) ratio of the solution to the wood, which is defined by Peakes[47] and his co-authors as the gallons of treating solution to 1.0 pound of dry separator material; (7) washing in hot and cold water, 12 hours.

The various woods differ in their resistance to treatment. Peakes and his associates recommended less drastic treatment for redwood, beginning with hot water and followed by a dilute solution of sodium carbonate. Bald-cypress, on the other hand, is quite resistant to sodium hydroxide, and the treating solution which they recommended was 5 per cent NaOH plus 5 per cent Na_2S for an aggregate time of 11 hours.

Rubber Retainers

Perforated or slotted separators of hard rubber, also called retainers or envelopes, are commonly found in various types of storage batteries. They assist in holding the active material of the positive plates in place and they protect other separators used with them from the strong oxidizing conditions of the lead dioxide. Perforated retainers are generally used in combination with other types of separators such as wood, porous rubber, or glass mats. Perforated-rubber retainers are placed in immediate contact with the positive plates, except when glass mats are used between the plates and the perforated rubber. The thickness of the rubber sheets is ordinarily about $\frac{1}{64}$ inch. Perforations may be round holes, elongated slots, or special shapes. The advantage claimed for the slots is that, because of their narrowness, particles of active material are more effectively retained on the surface of the positive plates, without loss of conductivity which is provided for by the length of the slot. Some of the special shapes of perforations are said to release bubbles of gas more readily than others. Whatever the shape of the perforation, the average percentage of area punched out is usually specified and called the "porosity." Thus, the round perforations, which may vary in size and number from 19 to 210 per square inch of surface, provide porosity from 13 to 46 per cent. The elongated slots, which vary from 0.012 to 0.020 inch in width and

---

[47] *Loc. cit.*

Fig. 11.   Sections of perforated-rubber retainers.

0.18 to 0.25 inch in length, number 75 to 125 per square inch.   Porosities up to about 40 per cent are available.   Various types of perforated and slotted retainers are shown in Fig. 11.   The percentage of material punched out in these slotted and perforated sheets is a matter of considerable importance in determining the performance of the battery. If too much material is punched out, the rubber is weak and liable to breakage, but if too little material is punched out the resistance of the cell is materially increased.

Porous-Rubber Separators

These are made from latex or smoked-sheet rubber.

Latex,[48] which is the secretion obtained principally from rubber trees, is a suspension of hydrocarbons in an aqueous serum containing also small quantities of proteins, resinous materials, mineral salts, and sugars.   Fresh latex is slightly alkaline and contains 30 to 38 per cent

[48] Latex is discussed in *Chimie et technologie du latex de caoutchouc* by Georges Génin, 1934; *Chemistry and Technology of Rubber Latex* by C. Falconer Flint, 1938.

of rubber hydrocarbon. The remainder is mostly water. On exposure to air the latex becomes slightly acid and coagulates. To prevent this, ammonia is added before it is shipped. Because of the large amount of water which it contains latex is sometimes partially concentrated by centrifuging or by evaporation to decrease costs of transportation. Latex is sensitive to traces of copper, cobalt, manganese, and certain other metals, contact with which should be avoided.

In 1914 Schedrowitz and Goldsborough found that coagulated latex could be vulcanized in steam and the water in its pores thus prevented from escaping during the process. After vulcanization, evaporation of the water left a highly porous mass. Beckmann[49] developed a process of making microporous sheets of rubber for battery separators and other purposes.

Beckmann's process involved vulcanizing a mixture of latex and sulfur in hot water or saturated steam. The pores in his finished product were estimated to be 2 to 3/1000 mm and to number 20 million in an area of 1 square centimeter. Later processes have altered the formula somewhat by the addition of various materials to benefit the negative plates, to increase porosity, to serve as protective colloids, to hold in check the rapidity of gellation, to strengthen the separators, etc. One process[50] starts with a 60 per cent of rubber latex and a suspension of sulfur to which is added zinc oxide and several organics. This was vulcanized in steam until 90 per cent of the sulfur was combined. The finished separators were lower in resistance than the corresponding wood separators, with which they were compared, and gave double the life in service according to the tests that were made. One notable advantage mentioned by the authors is the higher engine-cranking speed at low temperatures when using the microporous separators.

Another way of making microporous-rubber separators is by the smoked-sheet process. Several kinds prepared with the addition of silica gel in the manufacturing process are available.

A microporous-rubber separator (Peerless) of 60 per cent porosity and having a resistance of 0.036 ohm per square inch is shown in Fig. 12.

Threaded-rubber separators were formerly made as thin sheets of vulcanized rubber through which several hundred thousand threads

49 H. Beckmann, Das mikroporöse Gummidiaphragma für Akkumulatoren, *Elektrotech. Z.*, *51*, 1605 (1930).

50 H. W. Greenup and L. E. Olcott, Latex rubber separators, *Ind. Eng. Chem.*, *29*, 192 (1937).

passed as tiny wicks, and later were made as latex separators containing an open-weave fabric.

Porous-ebonite separators, such as the Wilderman type, were prepared from pulverized, partially vulcanized rubber, or completely vulcanized rubber to which some unvulcanized rubber was added.

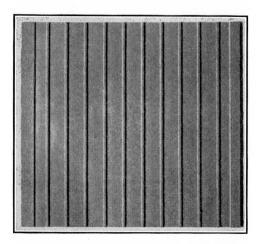

Fig. 12.   Peerless microporous-rubber separator.

Plastic Separators

The advent of modern plastics has made possible a variety of new types of separators. Thin films of uncoated, regenerated cellulose supported by glass mats are sufficiently conducting to receive consideration. Other plastics which are normally non-conducting can be made conducting by incorporating with the plastic a material which can be dissolved out, or if starch is used the starch can be liquefied by hydrolysis or by the action of an enzyme. There are many other types of separators, subject to experiment or actually in use, which consist of some fibrous material, woven or matted, and impregnated with a plastic. In sulfuric acid, Vinyon, Orlon, Dynel, Saran, Acrilan, and polystyrene have reasonably good resistance to the attack of acid.

Perforated and slotted separators and retainers are made of plastics such as polystyrene, and these are used for much the same purposes as similar separators and retainers of hard rubber.

Some of the better-known plastic separators that have found commercial use are known as "Poralite," alpha-cellulose impregnated with a resin; "Pormax," a microporous separator of polyvinylchloride; "Darak," which consists of two sheets of different kinds of paper resin-impregnated and cured in laminated form. A later form is a single corrugated sheet with reinforced ribs. This is made with or without glass retainer mats between the plies, or with glass mats over the ribs. "Revere" is another plastic separator offered as a substitute for wood. A microporous separator made of compressed polystyrene fiber, dimpled in lieu of ribs, appeared during the Second World War. Its characteristics were generally satisfactory, but its maximum safe temperature was stated to be 165° F. This is a limitation that may apply to other plastic separators also. Batteries are not intended to operate at such temperatures, but sometimes they do.

Fibrous-Glass Retainers

The use of glass mats as retainers or separators in storage batteries began in this country in 1925. Since then increasing use has been made of them. Increased cycle life of the batteries can be expected when they are employed, but the benefits to be derived from them are influenced by other details of construction.

Glass mats are placed next to the positive plates in order to retain the active materials in place and to provide a reservoir for the electrolyte. These porous-glass separators are always used in combination with other types of separators, such as perforated or porous-rubber or wood separators.

The diameter of the glass fibers is variously stated to be from 0.0005 inch to something less than 0.001 inch. Matted into a separator they are bound together by either a soluble bond, usually starch or gelatin, or an insoluble bond, such as furfural or one of the more common plastic resins. The mat thickness for automotive batteries is usually from 10 to 15 mils, and for industrial uses 20 to 30 mils. Thickness is measured under a pressure of 2.75 pounds per square inch.

A variation in the production of glass materials for battery use is found in the so-called "Sliver" or "Slyver" mats, which are made of very fine, parallel glass fibers massed vertically against the surfaces of the positive plates and passing around the bottom of the plates. The parallel-vertical arrangement serves as a retainer for the active material and provides minute vertical channels for the escape of gas. Sometimes Sliver is used in combination with heavier glass fibers in a jackstraw pattern. This mat holds the Sliver in place and provides for some expansion of the positive plate.

Since the glass mats are not sufficient protection against "treeing" in the working battery, the glass fibers or complete mats are combined with other types of separators, such as the Darak mentioned above. Another form is a combination of glass mat with a backing of diatomaceous earth.

Flat Types of Separators

In certain types of glass-jar batteries where the spacing of the plates is greater than in the more compact portable types, flat separators of porous rubber or wood veneer are used. These are usually reinforced by split-wood dowels or by strips of rubber.

The Exide-Ironclad batteries contain separators which are flat sheets of porous rubber without grooves. The construction of the positive plates provides the space required for the acid, and the rubber tubes serve the double purpose of a rubber separator and a retainer for the active material of the plate. The rubber tubes are slotted along the side to provide access for the electrolyte to the active material, but, as these slots are exceedingly narrow, the lead dioxide within cannot escape.

Some of the newer Willard types of small batteries in plastic cases are provided with Fibrite separators (see page 53). These separators are very porous and hold a relatively large amount of electrolyte. They are, therefore, flat separators making direct contact with both positive and negative plates.

Relation of Design to Cell Performance

The grooved side of the separator is always placed next to the positive plate. This is done for several reasons. The actual contact of the separator with the highly oxidizing material of the positive plate is minimized, and a greater volume of acid, for use by the positive plate during discharge, is provided. At high rates of discharge, the maximum capacity of the positive plate is attained only for moderately high concentrations of the acid. It is necessary, therefore, that the positive plate should have an ample supply of acid to maintain the concentration. The negative plate, on the other hand, reaches its maximum capacity at a relatively lower concentration.

Starting and lighting batteries for automobiles, which are required to deliver large currents, must necessarily have a low internal resistance. These require relatively thin separators, commonly $\frac{5}{64}$ inch in thickness. Other batteries, for which the rates of discharge are small, are provided with thick plates and a relatively wide separation.

Electrical Resistance

The resistance to the passage of the electric current through the separators is an important quantity, varying with the kind of material of which the separator is made, the treatment it has received, and the direction of the grain, if the separator is of wood. In general, it is known that woods such as basswood and poplar have a lower resistance when saturated with electrolyte than cedar or cypress. Tests are commonly made at high rates of discharge to determine the voltage characteristic of the battery, since this in turn depends on the internal resistance of the battery and so in part upon the separator resistance. It is desirable to have more direct measurements of the separator resistance, however, and several methods have been devised for measuring the resistance of individual separators.

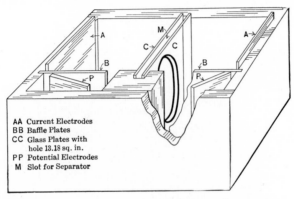

AA Current Electrodes
BB Baffle Plates
CC Glass Plates with
     hole 13.18 sq. in.
PP Potential Electrodes
 M Slot for Separator

Fig. 13. Cell for measuring resistance of separators.

One method which has found use in the industry consists in measuring the electrolytic resistance of a specially constructed cell, with and without the separator interposed across the path of the current, under otherwise comparable conditions. The difference of the two measurements gives the effective resistance of the separator. The cell shown in Fig. 13 contains two negative plates at the ends of the cell which serve as current terminals, and two other plates set at an angle are used to measure the potential drop across the separator that divides the cell at the middle. The separator is inserted between the glass plates, which have a fixed circular opening. A 60-cycle alternating current is used to avoid polarization effects. The electrical circuits are shown in Fig. 14. A transformer with two secondary windings of 5-ampere capacity is used to reduce the 110-volt circuit to about 10 volts. The current from one secondary is passed through the cell and maintained at 5 amperes. The current from the other secondary is passed through a calibrated slide wire and a non-inductive resistance of about 100 ohms. This circuit serves as a potentiometer in phase with the current through the cell. An alternating-current galvanometer can be used to determine the balance, care being taken to maintain

the field of the galvanometer in phase with the potentiometer. The
accuracy of the measurements is tested by measuring the value of a
known resistance. The resistance of individual separators, from the
same lot and similarly treated, may vary from 5 to 15 per cent, but
the precision of the measurements is about 1 per cent. Several weeks'

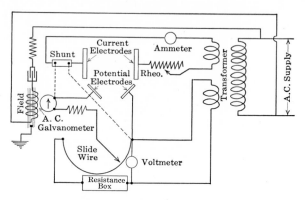

Fig. 14. Circuit for measuring resistance of separators.

immersion in sulfuric acid solutions is required prior to the measurements, in order that the resistance of the separators may reach a
constant value; but for practical purposes 24 hours is usually enough.

Another method which consisted of measuring eight or ten small
disks in a close-fitting glass tube was described[51] previously but has
been used less than the method described above. It is, therefore,
omitted here. Its chief fault was the possible by-passing of current
around the edges of the disks, or "fringe effect."

When measurements are made at different temperatures, it is found
that the percentage of increase in resistance of porous-rubber separators is approximately the same as the percentage of increase in resistivity of the acid solution for the same decrease in temperature.
This is not true of wood separators, which change in resistance more
than proportionally to the change in acid resistivity. This effect is
illustrated in Fig. 15.

Manufacturers ordinarily express the resistance of separators as
the resistance of a single separator, and then the resistance figure is
materially less than the resistance per square inch (see Table 8).
Each square inch is in parallel with every other square inch, and
therefore the resistance per square inch must be divided by the
number of square inches to get the resistance of a separator. As an

---

[51] This book, 3d ed., p. 56.

TABLE 8. MEASUREMENTS ON TREATED STORAGE-BATTERY SEPARATORS

Kind of Wood and Number of Samples	Breaking Strength (Wet) Cross Section 0.0386 Sq In.		Electrical Resistance per Sq In. Average Thickness 0.0520 In.		Volatile Acid per Gram of Separator, Calculated as Acetic Acid, g	Reducing Substances per Gram of Separator, Calculated for PbO_2, g
	Initial, in Water 24 Hours, lb	After 7 Weeks in 1.250 Sp. Gr. Sulfuric Acid, lb	Initial, in 1.250 Sp. Gr. Sulfuric Acid, at 25° C, ohms	After 25 Days in Acid, 1.250 Sp. Gr. at 25° C, ohms		
Port Orford cedar 10	273	123	0.052	0.045	0.0039	1.35
Fir 6	200	90	0.043	0.037	0.0051	1.60
Cypress 3	236	87	0.063	0.052	0.0042	1.38
Redwood 3	191	63	0.033	0.028	0.0044	1.23
Yellow cedar 2	262	102	0.046	0.041	......	1.40
Spruce 2	228	73	0.035	0.032	......	1.07

Notes on Table 8

1. Samples consisted of a considerable number of separators obtained from various sources. These had been treated according to the manufacturer's customary methods.

2. Breaking strength was measured on four strips cut from each sample. Each strip included two or three ribs and the web portion between, but the results have been calculated to a uniform cross section of 0.0386 square inch. The tensile strength machine in which the test pieces were broken was calibrated to read to 1 pound.

3. Electrical resistance measurements were made on ten disks cut from each sample. Diameter of disks was 1.395 inches, and diameter of tube was 1.41 inches. These were nominally $\frac{5}{64}$-inch separators, but the average thickness over ribs and web was 0.0520 inch, and all results have been calculated to the same basis.

4. Volatile acids were determined by extracting the materials in H_2SO_4 solutions followed by distillation in accordance with Craig's method described in his paper, *Bur. Standards J. Research*, **6**, 169 (1931). The results are expressed as grams of volatile acid, calculated as acetic acid, per gram of separator. Untreated separators yielded much higher amounts.

5. Reducing substances were measured by a permanganate method. The results are expressed as the amount of PbO_2 reduced per gram of separator. After separators have been in a battery for about 18 months, the amount of reducing substances that can be extracted is about half that from new separators.

approximation the resistance given in Table 8, after the specimens were soaked in acid for 25 days, divided by 28 square inches give values closely approximating those in the paper by Peakes and his associates (*loc. cit.*) for corresponding kinds of wood.   The resistance

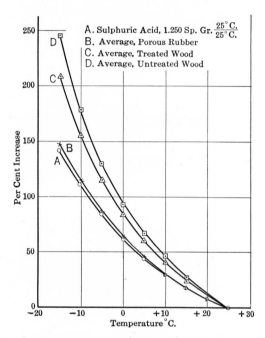

Fig. 15.   Effect of temperature on resistance of separators compared with the resistivity of sulfuric acid solutions.

of microporous-rubber separators and some of the newer plastic separators may be materially lower than the values for wood.   This depends on their porosity.   As an approximation the porosity may be assumed to be inversely proportional to the resistance.

Effect of Acid on the Wood

Sulfuric acid chars the wood, as may readily be observed.   The extent to which this action takes place depends very largely on the concentration and the temperature of the acid.   When the specific gravity of the acid is 1.250 or below, the charring action is relatively slight, even when the wood is immersed for a long period of time.   At 1.300 the charring action is noticeably greater, and it becomes serious at concentrations higher than this.   Samples were kept in the solutions at a constant temperature of 20° C (68° F) for the duration of the experiment, and they were then taken out and subjected to tensile-strength tests while still wet. The results of this experiment are shown in Table 8. Each value is the mean of four determinations. The effect of the acid in weakening the wood fiber is clearly shown.

Rubber Jars and Containers

Vulcanized rubber is commonly used for the containers of starting and lighting, radio, airplane, vehicle, marine, and submarine cells. The container for a single cell is called a jar, but others are divided

Fig. 16. Hard-rubber container for three cells of an automotive battery.

into several compartments for two or more cells. The degree of vulcanization varies with the proportion of sulfur and with the temperature and duration of the heating. The quality of the final product, as for example the battery jars, depends on these factors and also on the percentage of rubber in the compound. Present-day containers may be made from natural, synthetic, or reclaimed rubber, or a mixture of any of these in varying proportions. The formulas usually include also some fillers, accelerators, and carbon black. It is, therefore, not surprising that the quality should differ widely. Containers have superseded the use of the individual jars for automobile batteries. A hard-rubber container for three cells is shown in Fig. 16.

The covers are molded. They differ greatly in shape and arrangement of the openings. Covers for the smaller cells, such as for starting and lighting batteries and vehicle batteries, generally have three holes in the cover; two of these are for the protruding terminal posts and

Fig. 17. Cover of a cell of the automotive type, showing lead-insert bushing.

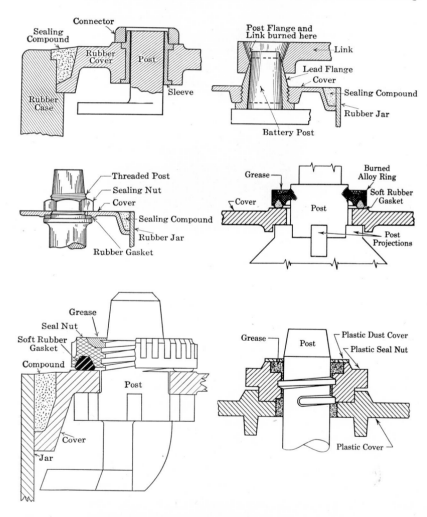

Fig. 18. Six methods of sealing the terminal posts.

the third for the combined vent and filling aperture. (See Fig. 17.)
The sealing compound is a blown oil asphalt with a melting point of about 200° F. The softer compounds melt below this temperature, and the harder ones above it. As the compounds contain volatile constituents which are driven off if melted for any considerable time, it is not desirable to use the compound a second time.

The method of sealing the terminal posts at the point where they pass through the cover is a matter of great importance. Unless the posts are satisfactorily sealed, they are likely to work loose in the cover and cause leakage of the electrolyte. Six of the familiar arrangements are shown in Fig. 18. The sealing nuts of rubber or lead alloy are screwed down on the post, thus pressing the cover against a soft-rubber gasket. In the figure, the first three types relate to automotive batteries, the last three show types used on power batteries and sealed jar batteries.

Composition Containers

Composition containers are molded products of thermoplastic, bituminous materials, usually containing asphalts, asphaltites, inert fillers, fibrous materials, lampblack, and sometimes mineral waxes and comminuted rubber. Many specific compositions of these materials have been patented.

Asphalt is a term applying to a species of bitumen and certain pyrogenous substances of dark color and variable hardness, which are composed of hydrocarbons that are comparatively non-volatile and free from oxygenated bodies. These include native asphalts and pyrogenous asphalts which are residues from the distillation or the blowing of petroleums. Asphaltite is a species of bitumen of which Gilsonite is the variety generally used in making battery cases. This class of materials consists of comparatively hard, non-volatile solids of a dark color.

Inert fillers which are used include infusorial earth, pumice, and other forms of siliceous material. The fibrous materials are usually cotton or asbestos, but mention is made in the literature of animal fibers. Of the native mineral waxes, Montan wax is sometimes employed. Petroleum fluxes are sometimes added also.

The cases are cast under pressure while the material is hot. As it flows into the mold, it is important that the material coming from one direction should form a satisfactory bond with the material coming from another direction. Cases are removed from the mold when sufficiently cooled.

Polyester plastic resins have been used for making automotive battery cases. They are relatively light, strong, and heat-resistant. The finished product must resist attack by sulfuric acid solutions of a strength commonly used in batteries; it must not become porous or deformed; it must not contaminate the electrolyte with impurities of which iron and particularly manganese are the most important.

Glass Jars

Glass jars are very commonly employed for stationary batteries except for the largest sizes. Desirable qualities in the glass jars are transparency, freedom from attack by the acid, the absence of blow holes, and the ability to withstand temperature variations. In the event that a glass jar is broken the entire element should be removed and immersed in water in a non-metallic vessel while a new jar is being obtained. If more than a week or two is needed, the positive plates should be removed and dried, but the negatives and separators should remain in water.

Plastic Containers

A wide variety of modern storage batteries is found in transparent or opaque plastic jars and cases. Use of these saves some space and an appreciable amount of weight over the corresponding sizes and types in glass jars. Plastic jars are less liable to breakage than glass and can be molded into desired sizes and shapes. Although several kinds of plastics are suitable for the purpose, the material most used is polystyrene. Covers for the cells or multiple-unit cases are of the same material.

Before the modern plastics were available Celluloid was employed to a limited extent. This is a plastic of good mechanical properties which can be machined or molded, but it has the serious drawback of being highly flammable. Two fires of Celluloid battery cases are within the author's experience. It is doubtful that Celluloid has any real advantages now that newer plastics can be used. Plastics have advantages in developing new types of cells, as the cases can easily be made in the laboratory by cementing together sheets of material as needed in experimental work. The chosen form is then translated into molded form for commercial production.

Ceramic Jars

These have found application for telephone, control, and standby batteries. The jars are pure white, highly glazed ceramic material that must be entirely free from absorption of water or the acid electrolyte.

Lead-Lined Tanks

These are used for large stationary batteries and are generally applicable where considerations of space and weight are not of importance. The wood casing must be of good quality, since the lead lining has little mechanical strength and is liable to deformation. The best wood available is a resinous long-leaf yellow pine. The upper rim of the lead lining is usually reinforced and is called the crown.

VENT PLUGS

All lead-acid storage cells are provided with some kind of a vent in the cover. This is necessary to provide for the escape of gases passing off from the cell and to provide an opening for the addition of water as needed to the electrolyte. Automotive and similar batteries are provided with vent plugs so designed as to prevent careless additions and the attendant overflow of electrolyte on the top of the battery. There are several ways of accomplishing this, but most of them are based on the principle that water can flow into the battery only to a prescribed height when the normal outlets for the escape of gas (or air) are closed. One device consists of a lead ring which tilts and closes the gas exit when the vent plug is removed, but replacing the vent plug restores the ring to its normal position, gas vent open. Another device limits the electrolyte height to the bottom of the filling tube by showing a bull's eye when looking into the battery when the level is correct. The bull's eye disappears if the electrolyte rises too high into the filling tube. Still another type of battery cover has as its gas vent a small nipple beside the vent plug which in this case is really a plug without any vent. To fill the cell the plug is placed over the nipple, closing it. Water is added until it rises into the filling tube, and when the plug is removed from the nipple the slight excess falls into the cell and the electrolyte is at the proper level. A fourth device consists of a prism which by appearing to the eye as clear or red indicates whether water is needed. Many patents have been issued on vent plugs and filling arrangements. It is quite beyond the scope of this book to give an extended description of them.

Vent plugs on the smaller types of batteries are usually of the bayonet or screw type. The "bayonet" plug is fastened in place by a partial turn, and the screw type as its name implies is threaded. Less commonly soft-rubber plugs which are pushed into place are used. Large batteries which are used in confined spaces as in the battery boxes of railroad cars are provided with hinged vent plugs that can be flipped open or closed by the tip of the automatic water filler. To prevent sloppage of electrolyte or the escape of spray, vent plugs should be provided with baffle plates.

Batteries for signal service or for use on aircraft as described in Chapter 10 often must be of the non-spill type. Various valves to prevent the escape of liquid when the batteries are inverted have been devised, but only two are commonly in use. The most familiar is the elongated plug of the double-chamber construction. The vent at its lower end is so placed as to be out of the liquid whatever position the battery may be in. The other type makes use of a tilting lead cone. These are more fully described in the section on aircraft batteries.

ASSEMBLY OF LEAD-ACID CELLS

The remainder of the process for the manufacture of lead storage batteries deals with problems which are of comparatively little interest to a student. They will, therefore, be described very briefly. The

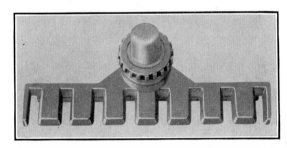

Fig. 19.　Connecting strap and pillar post with seal nut in place.

plates, after formation, are assembled into positive and negative groups by burning the plate lugs to the connecting straps or bars (Fig. 19). These groups are interleaved, the positive and negative plates alternating. The separators are then placed between the plates of opposite polarity, with the grooved side of the separator facing the positive plate. If slotted or perforated-rubber separators or fiber-glass mats are used, they are inserted between the positive plates and the wood separators. The assembly of positive and negative groups with separators is called the element of the storage cell. After assembly of the element is complete, it is placed in the jar or container, and the cover of the cell (Fig. 17) is sealed on.

Connectors between the individual cells may be of several different kinds. Pure lead connectors are used when high conductivity is required without severe mechanical conditions. The most common type of connector is made from a casting of lead-antimony alloy. These connectors may be burned to the projecting terminals of the cell and are more durable than the soft-lead connectors. Copper connectors, heavily lead-plated, are sometimes used in batteries for vehicle or locomotive work, since they combine high conductivity with flexibility, the latter property enabling them to absorb the strains that might otherwise crack the jars of the battery. The copper strips usually have a loop in the middle to increase the flexibility and have alloy castings on the ends which may be burned to the terminals of the cell. Intercell connectors are commonly referred to as burned or bolted, these terms indicating whether the connectors are attached to the cell post by the lead-burning process or are clamped by a threaded bolt. A combination of burned and bolted connectors is now used to some extent as a matter of convenience. The burned end of the connector is at the positive where corrosion is most likely to occur. The bolted connection is at the negative.

The resistance of the intercell connectors is a matter of considerable importance, since the energy that is wasted in the form of heat in these connectors varies as the square of the current. In Chapter 10 a discussion of the resistance of the intercell connectors is given for the automobile, vehicle, and airplane types of batteries.

TYPES OF LEAD-ACID CELLS

In the previous discussion, pasted and Planté plates have been described. They find many uses in a wide variety of cells, of which only a limited number can be mentioned here. Reference should be made to published catalogs of manufacturers for details of construction and features which make their batteries adapted to specific kinds of service.

Batteries with Flat Pasted Plates

Pasted plates are employed in many types of batteries. Without doubt the most familiar is the automotive battery which is made in greater numbers than any other. Pasted plates are used also in sealed glass-jar batteries for stationary service. In size, pasted plates range from small plates used in potential batteries to the largest sizes used in tank batteries. Typical pasted plates for automotive batteries are shown in Fig. 20. Distinguishing features are the grids and active material which extends through the plates. Positives have dark active material, which is lead dioxide, and the negatives have gray sponge lead as the active material. Assembled in a completed cell, these plates with intervening separators are shown in Fig. 21. Although the ordinary flat pasted plate is easily adapted to many services, other

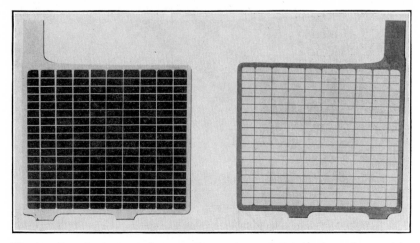

Fig. 20. Pasted plates of the type for starting and lighting batteries; positive left, negative right.

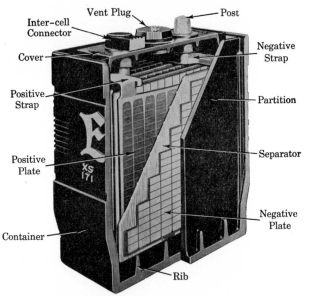

Fig. 21. Cut-away cell of the automotive type, showing details of construction.

types of plates and methods of constructing batteries may have distinct advantages for particular services. The choice of a battery for any service depends on features which are inseparable from operating conditions.

Batteries with Reinforced Grids

The first of these was developed primarily to provide a battery of longer life in floating service. Mechanical strength is one feature, but an equally important characteristic is the greater cross-sectional area of some of the grid bars to withstand corrosion, particularly in floating

Fig. 22.   Modern form of the Floté cell.

service. Reinforced, that is, heavier grid bars are provided at intervals, but the mass of the metal is within the active material of the plate and only the narrow edges of the bars are exposed. Lighter grid bars of the usual type are interspersed between the heavier members. Several grids of this type are shown in Fig. 9. The essential points in design relate to the distribution of the metal in positive grids. The grid weight is about 50 per cent of the weight of the finished dry plate. The outer frame with lug is from 21 to 30 per cent of the grid weight. The reinforcing members are about 40 per cent or more of the grid weight; the surface members account for only about 15 per cent of

the grid weight.   Positive grids usually exceed 0.3 inch in thickness.
A cell of this type is shown in Fig. 22.   The Floté element is provided
with corner clamps or tie rods at the bottom.

Batteries with Manchester Positives

These are cells of the Planté type familiarly known as the Exide-
Manchex type.   The grid of the Manchester plate is a casting of lead-

Fig. 23.   Exide-Manchex cell, type FME-17, showing Manchester positive plates
and pasted negative plates.

antimony (Fig. 9) providing openings for the active elements, which
are buttons of soft-lead ribbon.   These are corrugated and rolled into
spirals which are forced into the grids by hydraulic pressure.   The
buttons are electrolytically formed by the Planté process, and the
active material is therefore provided by the lead itself.   However,
initial formation penetrates only part of the lead ribbon, the remainder
being available for gradual conversion into active material to replace
other active material that may become dislodged and deposited as
sediment in the bottom of the cell.

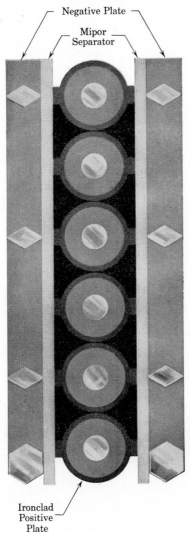

Negative Plate

Mipor
Separator

Ironclad
Positive
Plate

Fig. 24. Enlarged cross section of the Exide-Ironclad element.

The negative consists of a flat pasted plate. A fully assembled cell is shown in Fig. 23. These cells are made in a variety of sizes.

Batteries Containing Exide-Ironclad Plates

The distinguishing feature of this type of lead-acid battery is the positive plate which consists of vertical tubes of finely slotted rubber or plastic within which the active material is contained. The grid of this plate consists of a number of vertical spines cast integrally with the top bar and burned at the bottom to a bar after the tubes are filled with oxide. In later models the bottom bar is replaced by a molded plastic bar making a pressed fit with the end of each spine. Each spine forms the core of a pencil consisting of the active material and the enclosing tube. The slots in the tubes provide access of electrolyte to the active material but prevent the active material from being washed out. The tubes at the outside edges of the plates are reinforced by leaving the exposed edges solid. Each tube has two outer vertical ribs projecting on opposite sides at right angles to the face of the plate. These ribs not only reinforce the tubes but also serve as insulating spacers, taking the place of the ribs on wood separators. A cross section of the element is shown in Fig. 24. The negative plate is a flat pasted plate. The separators are flat sheets of microporous rubber, called "Mipor." One of these cells is shown in Fig. 25. These batteries find use in motive power, marine, and other services. They are made in five sizes of plates, for which operating data are given in the section on truck and tractor batteries in Chapter 10.

Fig. 25. Exide-Ironclad cell, type TH-19.

Kathanode Batteries

Glass mats are placed on both sides of pasted positive plates. They are made of filaments of spun glass, matted together, and laminated to form a porous retainer of approximately 0.10 inch in thickness. Outside the glass mats there is a perforated-rubber envelope of about

Fig. 26. Gould Kathanode construction with the "Z" plate. Positive grid has vertical and horizontal members of equal cross section, spun-glass retainer, perforated envelope. Microporous-rubber separators are added to make triple insulation.

44 per cent porosity which holds the glass mat in contact with the surfaces of the positive plate. Unperforated at the edges, these envelopes serve to prevent treeing. The Kathanode units, Fig. 26, each consisting of one positive plate, two glass mats, and the envelope, are assembled with pasted negative plates to form the element. Wood separators between positive and negative plates are used in glass-jar cells for stationary service, and porous-rubber separators, called "Durapor," are used in rubber-jar cells for motive power or other purposes. Cells of this type are shown in Figs. 27 and 28.

Fig. 27. Gould Kathanode suspended-element cell in glass, type DK.

Planté Cells and Modifications

In addition to the Planté plate, described above and widely known as the Manchester positive plate, there are other varieties of these plates. Of historical interest is the Tudor positive plate, which is cast in cellular form. This plate is not now made in this country, so far as the author is aware. Spun plates described on page **47**, and sometimes known as C. P., or "chemically pure," plates, are now used in combination with pasted negative plates. The term "chemically pure" relates to formation of the plates in solutions which are without the usual forming agents. A cross section of a portion of such a plate is shown in Fig. 29. The

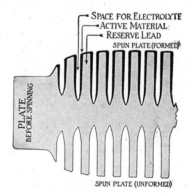

Fig. 28. Gould cell in rubber jar, type XLZ.

Fig. 29. Cross section of a spun plate.

layer of crystalline lead dioxide is rather thin initially, thus leaving space for expansion of the active material during discharge when an increase in volume occurs. Combinations of Planté plates with pasted negatives are shown in Fig. 30. The use of true Planté negatives, similar to the positives, has been declining. The cells are usually provided with wood separators. Planté cells find their chief applications in stationary service. They are in glass jars for capacities up to 800 ampere-hours at the 8-hour rate.

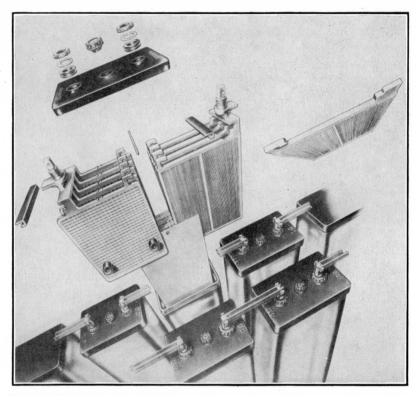

Fig. 30. Gould Planté cells. The positives are pure lead, grooved to produce evenly spaced ribs. The negatives are pasted plates.

Batteries for Low Rates

Low discharge rates imply discharge over long periods of time. A storage battery to be suitable for such a purpose must have very little local action or self discharge; otherwise a considerable portion of its capacity will be spent without doing useful work. The development of Willard's low-discharge battery[52] makes possible the use of lead-acid storage cells in circuits where the interval between charging may be a year or more, provided the current drain is not enough to exhaust the battery in a shorter time. The self discharge occurring in these cells has been reduced to less than 15 per cent per year by the use of pure lead grids without antimony. The cells are made in three sizes. Plates in the largest size are $\frac{15}{16}$ inch thick, 6 inches wide, and $8\frac{3}{4}$ inches high. Although designed for cycles of service of not less than 3

[52] C. C. Rose and A. C. Zachlin, Low-discharge cells, *Trans. Electrochem. Soc.,* *68,* 273 (1935).

months, they can deliver very appreciable currents for short periods of time in intermittent service.. The initial voltage is 2.12 volts per cell. With a high cutoff voltage of 1.95 volts per cell the average

Fig. 31. Willard cell for low-discharge rates. Local action in this cell has been reduced to a minimum.

voltage is 2.05 volts. These batteries find use in signal devices, alarms, light beacons, program systems, and laboratory installations. A cell of 600 ampere-hour capacity is shown in Fig. 31.

Storage Batteries in Plastic Cases

The rapid development of plastics has made possible their use for battery cases. In the smaller sizes plastics have largely superseded glass. Polystyrene is probably the most used. The containers are

Fig. 32. Three-cell battery in a plastic case.

molded and sealed. Figure 32 shows a three-cell battery in such a container, but a large variety of shapes and sizes is now available.

Closed-Tank Cells

The large open-tank cells of earlier days are largely replaced by the modern closed-tank cells in hard rubber cases. One of these is shown in Fig. 33.

Tank batteries present a problem because of their large size, ranging up to several thousand ampere-hours, and the possible accumulation of considerable amounts of gas unless forced ventilation is provided. The cell shown in Fig. 33 is provided with a ceramic cone through which the gases from within the cell can diffuse. Other explosion-proof types are illustrated in the section on telephone batteries in Chapter 10.

Approximate Sizes and Capacity of Plates

Although a great many different sizes and kinds of plates for lead batteries have been made from time to time, certain sizes have come into common use, and it is therefore worth while to give a table indicating the most common sizes for different types of service and the nominal capacity per positive plate of cells containing them. In the case of plates for stationary batteries, it will be noticed that there is a simple relation between the capacity of the various sizes, which is hardly suggested by the irregularity of the dimensions. In making use of Table 9, it should be remembered that the list of sizes is by no means complete and that the capacity will vary somewhat with the quality and thickness of the plates. The figures given here serve as a general guide to the size and performance of these plates.

HAZARDS IN THE PROCESS OF MANUFACTURE

The poisonous nature of lead and its compounds[53] gives rise to certain hazards which deserve attention in the process of manufacture. Governmental regulations with respect to the conditions of work, however, have greatly lessened or eliminated the number of cases of lead poisoning occuring among employees. Sources of poisoning are primarily the dust of lead and its oxides, and the fumes from the melting pots and lead burning. The compounds of lead are poisonous practically in proportion to their solubility.

Bluish fumes may often be observed over the melting pot when the alloy is ladled out or the dross skimmed off. The quantity of these fumes probably depends on the temperature of the molten metal, but they are generally pres-

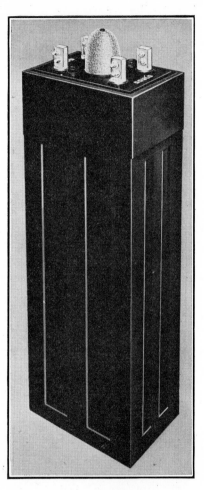

Fig. 33. Gould Tank cell, showing ceramic cone, one type of explosion-proof vent.

[53] Alice Hamilton, M.D., Lead poisoning in the manufacture of storage batteries, *Bureau of Labor Statistics Bulletin 165;* W. C. Dressen, T. I. Edwards, W. H. Reinhardt, R. T. Pope, S. H. Webster, D. W. Armstrong, R. R. Sayers, Control of the health hazard in the storage-battery industry, *U. S. Public Health Service Bull., 262,* 1–82, 87–110, 123–134 (1941); G. S. Winn and C. Shroyer, Control of lead exposure in storage-battery manufacture, *J. Ind. Hyg., 29,* 351 (1947).

Storage Batteries

TABLE 9. PLATE SIZES, LEAD BATTERIES

The sizes listed are those most commonly used, but there are variations in dimensions and capacity. The capacities are calculated from the rated capacities of ordinary types.

	Width, in.	Height, in.	Nominal Capacity per Positive Plate, amp-hr	Rate of Discharge, hr
Starting and lighting	$5\frac{5}{8}$	5	16	20
Isolated plant	$5\frac{13}{16}$	$5\frac{7}{8}$	22	8
Isolated plant	$7\frac{3}{4}$	$7\frac{3}{4}$	44	8
Car-lighting plates	$7\frac{3}{4}$	$10\frac{1}{8}$	50	8
Vehicle plates	$5\frac{3}{4}$	$8\frac{5}{8}$	28	6
Stationary plates	3	3	5	8
Stationary plates	$4\frac{3}{8}$	4	10	8
Stationary plates	6	6	20	8
Stationary plates	$7\frac{3}{4}$	$7\frac{3}{4}$	40	8
Stationary plates	11	$10\frac{1}{2}$	80	8
Stationary plates	$15\frac{5}{16}$	$15\frac{5}{16}$	160	8
Stationary plates	$15\frac{5}{16}$	$30\frac{3}{4}$	320	8
Stationary plates	$18\frac{5}{8}$	$18\frac{5}{8}$	184	8

ent, and a hood with a forced-air suction is necessary. The dross should not be thrown on the floor. The process of burning the plates to the straps and the connectors to the posts gives rise to lead fumes and requires that good ventilation be provided. Some dust arises from trimming the grids and cleaning the lugs, but the greater quantity of dust is due to the oxides of lead used for pasting the plates. The material for the paste must be weighed and mixed dry. This is a dusty process that requires special attention. It should be carried on in a separate room provided with forced ventilation, and the operatives should wear respirators. There is less danger in the pasting process when the paste is wet, but dry scraps of paste produce dust. Pasting rooms should always be provided with adequate cleaning facilities and ventilation.

The forming rooms are often filled with fumes that are irritating to the throat of a person not accustomed to them. These fumes are caused by the acid being sprayed into the air by bursting bubbles of gas as they rise to the top of the forming tanks. Although the discomfort is greater in such a room as this, the hazard is less. Medical evidence is conflicting as to the possible bad effects of the fumes. It is, of course, necessary that the ventilation be sufficient to carry off the hydrogen generated, in order to avoid explosions, and the sulfuric acid content of the air must be kept low to preserve the insulation as well as to avoid discomfort to the workers.

THE EDISON CELL (ALKALINE TYPE)

Positive Plates

The positive plates of Edison alkaline batteries are filled with nickelous hydroxide which is converted to a higher oxide of nickel by the process of formation. Since this material is a non-conductor, additions of flake nickel or graphite are made to provide the necessary conductivity. An early form of Edison cell employed graphite for this purpose, but with the change in construction of Edison cells in 1908 the use of graphite was discontinued, flake nickel being used in its place. Some other alkaline batteries, however, use graphite.

A positive plate is shown in Fig. 34.

The electrolyte for all these alkaline cells is a solution of potassium hydroxide in water. To this a small amount of lithium hydroxide is always added for the Edison cells.

Preparation of Nickelous Hydroxide. Metallic nickel in finely divided form is dissolved in sulfuric acid. The hydrogen which is evolved during this process is collected and saved for use in connection with other processes in the manufacture of the cell. Certain impurities, such as copper, zinc, antimony, and iron, are usually present in the solution and must be removed. The solution of nickel sulfate is then sprayed into tanks containing a solution of hot sodium hydroxide. Nickelous hydroxide is precipitated as a result of the reaction that takes place. The precipitate is collected and dried, but it contains considerable amounts of sodium hydroxide, carbonates, and sulfates. These may be removed by leaching the mass with hot water. The nickelous hydroxide is then dried again and tested to determine its quality. If it is found to be satisfactory it is ground and sifted, the material used ranging between 30 and 190 mesh.

Fig. 34. Positive plate of the Edison cell, type A.

Preparation of the Flake Nickel. The flake nickel which is added to the nickelous hydroxide in making the plates is required to give the active material sufficient electrical conductivity. The flake nickel consists of small squares $\frac{1}{16}$ inch on the edge and 0.00004 inch in thickness. It is prepared by an interesting nickel-plating process. Ten revolving copper cylinders are carried by a crane and dipped alternately in copper- and nickel-plating baths for sufficient time to deposit the thin layers of metal which are required. The cylinders are washed by sprays of water between each immersion in the plating tank. This process of alternately plating copper and nickel is repeated 160 times so that the sheet of deposited metal on the cylinders consists of 160 layers each of copper and nickel. The copper-nickel sheet is then stripped from the cylinders and cut into squares $\frac{1}{16}$ inch on the edge. The copper is dissolved chemically, leaving the nickel as thin flakes, each little square yielding 160 of these flakes. The flake nickel is washed and centrifuged and dried over steam coils. In its final state a bushel of the material weighs only $4\frac{1}{2}$ pounds.

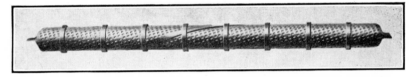

Fig. 35. Positive tube of Edison cell.

Tubes to Contain the Active Material. Since 1908, the active material for the positive plates of the Edison batteries has been contained in steel tubes (Fig. 35). Prior to this time the material was contained in pockets somewhat similar to those which are used for the negative plates. This earlier construction was abandoned because of the strains created by the swelling of the active material. The tubes are prepared from cold-rolled carbon-steel ribbon which is perforated by passing through special rolls that punch 560 holes to the square inch. The burrs are removed by emery wheels and the ribbon is cleaned by revolving wire brushes, to insure that the perforations are open so that the electrolyte may have free access to the active material of the plate. The steel ribbon is then nickel-plated. This is done as it passes continuously through a series of tanks containing the necessary solutions for cleaning, plating, and washing. The first of these tanks contains a solution of sodium carbonate to remove the grease from the steel ribbon. This is followed by a tank of hot water to wash off the alkali. The next tank contains a solution of nickel sulfate, in which the nickel plating is done. This is followed by two washing tanks containing hot and cold water, and finally the steel ribbon is passed through a tank containing a dilute solution of ammonium hydroxide. The time required for a given point on this ribbon to pass through the series of plating and other baths is about 8 minutes. After the nickel-plating process is finished, the ribbon is dried and annealed in an atmosphere of hydrogen. The purpose of this annealing is to fuse together the nickel plate and the underlying steel in order to prevent possible scaling off of the nickel plating. This is done in an atmosphere of hydrogen to prevent oxidation or discoloration that might otherwise take place.

The tubes which contain the active material of the positive plates are made from the nickel-plated steel ribbon, by winding it spirally. The seams are lapped and swaged flat. These tubes are made in rights and lefts, which are alternated in the assembly of the plate, thereby equalizing the strains which would tend to cause buckling. The tubes are made $\frac{1}{4}$ inch in diameter with a standard length of $4\frac{1}{2}$ inches.

Filling the Tubes. A cap is placed at the bottom of each tube, and a group of tubes is then put into a filling machine which tamps the active material into them. Eight tubes are mounted in molds below a row of weighted ramrods that fall by gravity, striking a blow of 2000 pounds per square inch. The filling machine is provided with two hoppers, one of which dumps a specified charge of nickelous hydroxide into each tube, and the other a certain quantity of the flake nickel. This is followed by a blow of the ramrod. The process is repeated 315 times, so that there are 630 layers of alternate nickelous hydroxide and flake nickel (Fig. 36). The layer of the nickel is somewhat thinner than the layer of the hydrate, the nickel constituting about 14 per cent of the contents of the tube. After the tubes are completely filled, the ends are pinched shut to form terminals which are suitable for clamping into steel grids. Before being placed in the grids, however, the tubes are reinforced by eight seamless nickel-plated steel rings. These are put on each tube to prevent possible bursting of the tube because of the swelling of the active material which takes place during the forming process. The active material of the positive plate is said to contract somewhat on discharge.

Positive Grids. These are steel punchings (Fig. 37) which are nickel-plated and annealed as has been described above. The pinched ends of the tubes are caught underneath ears projecting from the sides of the plate. The tubes are clamped in place by a hydraulic press.

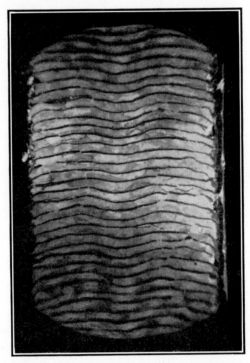

Fig. 36. Magnified section vertically through a positive tube, showing the alternate layers of nickelous hydroxide and flake nickel.

Fig. 37. Grids of the Edison cell: negative left, positive right.

Negative Plates

The negative plates of the Edison type are filled initially with a finely divided mixture of metallic iron, ferrous oxide, and mercuric oxide.

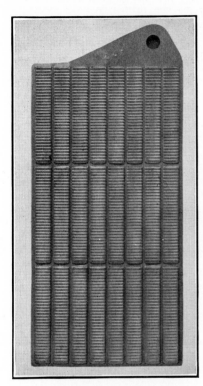

Negative plates of the Jungner cells contain a mixture of iron and cadmium. At one time cadmium alone was tried in these cells. The combination, however, proved to be better suited to the purpose.

A negative plate is shown in Fig. 38.

Active Material. Pure iron is dissolved in sulfuric acid. Hydrogen gas is evolved during this process, and this is saved for other uses in the manufacture of the batteries. The iron, in the form of ferrous sulfate, is recrystallized and then centrifuged to free it from mother liquor. It is dried at temperature of 500° C and roasted in an oxidizing atmosphere to the ferric state (Fe_2O_3). Traces of sulfate in this material are removed by leaching, and it is dried and reduced in an atmosphere of hydrogen in a muffle furnace. It is then partially oxidized to ferrous oxide, after which it is dried and ground.

Fig. 38. Negative plate of the Edison cell, type A.

Negative Pockets. The active material for the negative plates is contained in steel pockets (Fig. 39) with perforated sides. These are prepared from nickel-plated ribbon similar to that used for making

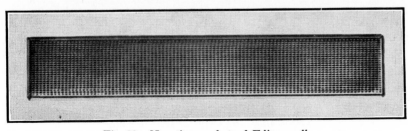

Fig. 39. Negative pocket of Edison cell.

positive tubes but more finely perforated. The ribbon is cut to the proper length and pressed into a form similar to a box, one end being left open for filling with active material. The pockets are fastened in the negative grid (Fig. 37), which is a nickel-plated steel punching having a series of parallel openings of the proper size to take the pockets of active material. The pockets are put in place, and a hydraulic press seals the boxes and crimps the sides. The dimensions of the pockets are given in Table 10.

TABLE 10. TYPES OF EDISON STORAGE-BATTERY PLATES

The types of plates are designated by letters; the types of cells by letters and numbers, the former indicating the type of plate and the latter the number of positive plates in the cell, except in type M. Ratings are based on 5-hour discharge for types A, B, C, D, M, N, F.

Type of Plate	Number of Tubes per Positive Plate	Dimensions of Tubes		Number of Pockets per Negative Plate	Dimensions of Pockets			Nominal Capacity per Positive Plate, amp-hr
		Diameter, in.	Length, in.		Length, in.	Width, in.	Thickness, in.	
A	30	$\frac{1}{4}$	$4\frac{1}{2}$	24	3	$\frac{1}{2}$	$\frac{1}{8}$	37.5
B	15	$\frac{1}{4}$	$4\frac{1}{2}$	16	$2\frac{5}{32}$	$\frac{1}{2}$	$\frac{1}{8}$	18.75
C	45	$\frac{1}{4}$	$4\frac{1}{2}$	40	*	$\frac{1}{2}$	$\frac{1}{8}$	56.25
D	60	$\frac{1}{4}$	$4\frac{1}{2}$	48	3	$\frac{1}{2}$	$\frac{1}{8}$	75.0
M	7†	$\frac{1}{4}$	$3\frac{1}{8}$	3	3	$\frac{1}{2}$	$\frac{1}{8}$	5.62
N	7	$\frac{1}{4}$	$3\frac{1}{8}$	3	3	$\frac{1}{2}$	$\frac{1}{8}$	5.62
F	4	$\frac{1}{4}$	$4\frac{1}{2}$	4	$2\frac{5}{32}$	$\frac{1}{2}$	$\frac{1}{8}$	5.0

* Twenty-four pockets are 3 inches and 16 pockets are $2\frac{5}{32}$ inches.

† The number for type M indicates the number of positive tubes in the battery instead of the number of positive plates.

Assembly

The necessary number of plates of each kind for any particular size of cell are mounted on steel rods which pass through the eyes of the grids at the top of the plates. The plates are separated from one another by means of washers fitting on this rod and are locked in place by a lock washer and nut on the end of the rod. Groups of positive and negative plates, corresponding to the positive and negative groups of the lead storage batteries, are assembled in this way. The rod which supports the positive or negative plates and connects them electrically is called the "connecting rod." The groups of positive and negative plates are intermeshed to form the element, and the plates of opposite polarity are separated by hard rubber pins. Side sepa-

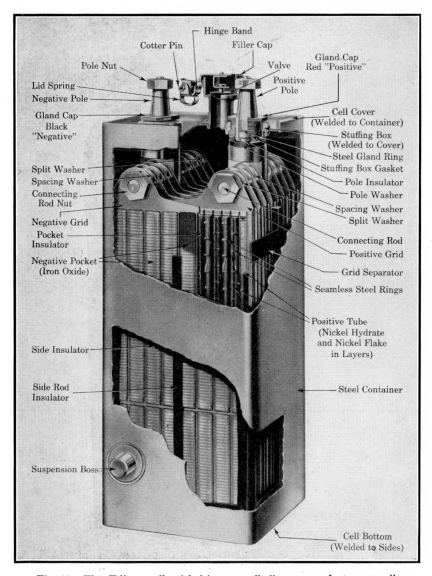

Fig. 40. The Edison cell, nickel-iron or alkaline type of storage cell.

rators, consisting of flat sheets of hard rubber, are used around the outside of the element to insulate it from the walls of the container. The element, when completely assembled, is slipped into the containing jar or can and rests upon a bridgework of hard rubber at the bottom. A complete cell is shown in Fig. 40. The can is made from sheet steel, nickel-plated. The newer cells are without corrugations as thicker metal provides the necessary strength. The metal is bent into shape around a form, and the side seams are welded by an oxyacetylene flame. The bottom of the cell is welded to the sides of the can, and after the element is in place the top seams are welded also. The container parts, such as cover, sides, and bottom, are nickel-plated before welding and after welding also. The terminal posts for the plate groups pass through openings in the top of the cell and are sealed into

Fig. 41. Gas valve and vent on the Edison cell.

place by a combination of hard- and soft-rubber gaskets and bushings. The bushing around the positive pole is red, while that around the negative pole is black. In this way a clear indication of the polarity of the cell is given. In the center of the cover is a combination valve and filler (Fig. 41). This consists of a hemispherical valve equipped with a spring that permits the valve to be opened for filling the cell with electrolyte or water but normally presses the valve against the seat. As the pressure of gas in the cell rises when it is on charge, the valve is displaced, allowing the gas to escape. This gives rise to the familiar popping sound which is often heard when batteries of this type are being charged. The satisfactory operation of the cell depends in considerable measure upon the gas valve, since the electrolyte, potassium hydroxide, has a strong tendency to absorb the carbon dioxide from the air. The valves may be easily opened, however, for filling the cells with water.

The terminal posts for the cells are of steel, tapered and ground to fit the lugs of the intercell or intertray connectors. The top of the post is threaded to take a lock nut. The lugs of the intercell connectors are drop forgings of steel, bored, reamed, and ground to fit the taper of the terminal posts. Both the terminal posts and the lugs are finished by nickel-plating and annealing in an atmosphere of hydrogen. This prevents the corrosion of the steel and provides a satisfactory contact surface in spite of the thin oxide film that exists on the nickel surfaces. The connecting links are copper swaged into the lugs, and the whole is nickel-plated. The lugs may easily be removed from the terminal posts by the use of a wrench and jack; these tools are necessary because the lug generally becomes "frozen" to the posts.

Electrolyte

The electrolyte for the alkaline cells is a solution of potassium hydroxide in water. For the Edison cells a small amount of lithium hydroxide is added because of its beneficial effect on the capacity of the cells. As a substitute for the potassium hydroxide, an electrolyte of sodium hydroxide has been tried, but the physical properties of this solution are not as good for the purpose as those of the potassium solution. The preparation of the electrolyte and the physical properties of it are described in Chapter 3.

Trays

The cells are usually mounted in trays which have open sides and in most cases are bottomless. Some of the larger locomotive batteries are now available in steel cradles or steel boxes. The cells are supported in these trays by bosses (Fig. 42) on the sides of the cans of the individual cells. The bosses fit into rubber buttons in the sides of the tray. These supports serve also to separate and insulate the cells, which is quite necessary because the containers for these cells are metallic conductors. The outside surface of the cells is coated with an insulating paint, and a film of rosin or liquid "Esbaline" is used to protect the top.

Types and Sizes of Edison Plates

The sizes of the Edison batteries are described in Chapter 10 in connection with the discussion of applications which are made of them. In this place, however, will be given a table analogous to that which has been given for the lead batteries, showing the different types of plates for Edison storage batteries, the number of tubes and pockets that they contain, with dimensions, and the nominal capacity of these plates at the normal rate of discharge. The nomenclature for the Edison cells is simple. The type of cell is designated by a letter followed by a number. The letter indicates the type of plate which the cell contains, and the number indicates the number of positive plates. An exception is made for the smaller sizes of the Edison batteries; for

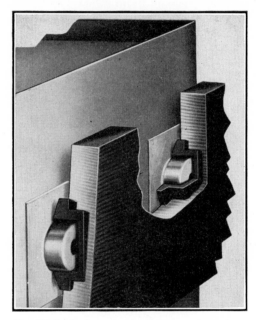

Fig. 42.   Method of supporting Edison cells in the tray.

these the number indicates the number of positive tubes instead of the number of positive plates.

The standard plate for the Edison battery is called the A plate. This is designed for general purposes, electric vehicles, trucks, locomotives, isolated farm lighting plants, and railway train lighting. The next most common plate is known as the B plate. It is used for lighting service, ignition, fire- and burglar-alarm systems, and clocks. B plates have half the capacity of A plates. C plates are equal to an A plate plus a B plate. D plates have double the capacity of A plates. The M and N types contain positive tubes $3\frac{1}{8}$ inches long and seven tubes to the plate. These are small plates designed for use with mine lamps, clock circuits, and time recorders. In addition to the types named above, there is also the F plate, used in mine-lamp batteries.

Some types of cells are designated by additional letters H or HW. These signify variations in the container: A6H is an A6 cell in a container that has extra height, and A6HW has extra height and width. The additional space allows more electrolyte to be used so that flushing is required less often.

NICKEL-CADMIUM CELLS (ALKALINE TYPE)

Although nickel-cadmium cells were little known in this country until recently, their production in Europe increased greatly after 1930. They are used for emergency lighting, control and switch-gear operation, marine services, railroad signaling, Diesel starting, and auxiliary power systems. Cadmium is said to be less subject to self discharge than iron and is relatively free from passivity at low temperatures. The average voltage of a cell on discharge is about 1.2 volts.

Nickel-cadmium cells now are known in two distinct forms. The older is the Jungner type of cell developed first in Sweden about 1900, and the other is an impregnated sintered-plate type which apparently originated in Germany during World War II. Both are now being made in this country. The sintered-plate type at the present time is used more for military purposes than the Jungner type.

The Jungner Type of Cell

The positive and negative plates are usually similar in construction, consisting of perforated pockets which contain the active materials. However, the positive plates in some European types of nickel-cadmium cells are of tubular construction and resemble closely those found in Edison cells. The pockets for both positive and negative plates are made from perforated steel ribbon which has been nickel-plated and annealed in hydrogen. Provision must be made for some expansion of the positive active materials within the pockets. Progressive swelling of positive plates in American-made "Nicad" batteries was corrected by the use of nickelic (black) hydroxide instead of the nickelous hydroxide (green) as the filling for the pockets. To that is added highly purified natural graphite to increase the conductivity.

Pockets of the negative plates are filled initially with cadmium oxide, CdO, or cadmium hydroxide, $Cd(OH)_2$, either of which is reduced to metallic cadmium in a spongy form on the first charge. Most manufacturers of these cells add iron (5 to 30 per cent) to the cadmium in order to obtain the required degree of fineness of the cadmium. The beneficial effect of the iron seems to be a more or less empirically determined fact, and the real function of the iron is not entirely clear. Some have thought that it forms an alloy with the cadmium; others have regarded it as an expander; and still others think that the sole function of the iron is that of a disperser to bring about a finely divided state of the electrolytically precipitated cadmium.

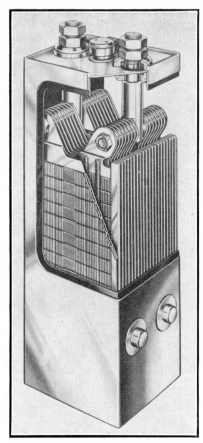

Iron in the metallic state undoubtedly increases the conductivity of the active material, but cadmium oxide itself is a reasonably good conductor. Whether iron oxidizes during discharge is also subject to differences of opinion. Crennell and Lea[54] indicate that it may do so and contribute a small part of the discharge capacity, but Hauel[55] is of the opinion that the iron takes no part in the current-producing process.

The pockets for holding the active material of both positive and negative plates of Nicad batteries are standardized at ½ inch, but the length may be varied. These are pressed into grids, usually called frames, made of nickel-plated steel. The plates are assembled into elements, rods of polystyrene being used as separators between plates of opposite polarity and sheet hard-rubber separators between the sides of the element and the inside of the steel container of the cell. The container is welded along all seams and at the top and bottom. It should not be grounded. Terminal posts are sealed through rubber glands. The positive is marked and further identified by a red bushing. The negative has a black bushing. Vent caps are spring-loaded and should be kept closed. The containers or cans are provided with bosses to support the cells in trays in much the same manner as for Edison cells.

Fig. 43. Nicad, nickel-cadmium cell.

[54] J. T. Crennell and F. M. Lea, *Alkaline Accumulators,* p. 120, Longmans, Green and Co., London, 1928.

[55] Anna P. Hauel, The cadmium-nickel storage battery, *Trans. Electrochem. Soc., 76,* 435 (1939).

TABLE 11. TYPE "S" STANDARD RANGE NORMAL-DUTY BATTERIES

Over-All Dimensions of Battery Trays in Inches

Cell Type No.	Capacity Amp-Hr, at 8-Hr Rate	Charge Current at 7-Hr Rate, amps	Weight per Cell,* lb	Width of Tray	Height†		Length of Tray Containing								
					Vent Cap Closed	Vent Cap Open	2 Cells	3 Cells	4 Cells	5 Cells 6 Volts	6 Cells	7 Cells	8 Cells	9 Cells	10 Cells 12 Volts
SYW7H	13.5	2.7	3.1	$4\frac{1}{2}$	$8\frac{1}{8}$	$8\frac{3}{4}$	5	$6\frac{7}{8}$	$8\frac{3}{4}$	$10\frac{5}{8}$	$12\frac{7}{8}$	$14\frac{7}{8}$	$16\frac{3}{4}$	$18\frac{5}{8}$	$20\frac{1}{2}$
SYW9H	18.0	3.6	3.9	$4\frac{1}{2}$	$8\frac{1}{8}$	$8\frac{3}{4}$	$5\frac{3}{8}$	$8\frac{1}{8}$	$10\frac{3}{8}$	$12\frac{3}{8}$	$15\frac{3}{8}$	$17\frac{5}{8}$	20	$22\frac{1}{4}$	$24\frac{5}{8}$
SYW11H	22.5	4.5	4.6	$4\frac{1}{2}$	$8\frac{1}{8}$	$8\frac{3}{4}$	$6\frac{5}{8}$	$9\frac{1}{4}$	12	$14\frac{3}{4}$	$17\frac{7}{8}$	$20\frac{1}{2}$	$23\frac{1}{4}$	$25\frac{5}{8}$	$28\frac{5}{8}$
SMC7H	30	6.0	6.3	$5\frac{1}{2}$	$12\frac{1}{8}$	$12\frac{3}{4}$	$5\frac{1}{8}$	7	$8\frac{7}{8}$	$10\frac{3}{4}$	$12\frac{5}{8}$	$15\frac{1}{8}$	17	$18\frac{7}{8}$	$20\frac{3}{4}$
SMC9H	40	8.0	7.9	$5\frac{1}{2}$	$12\frac{1}{8}$	$12\frac{3}{4}$	$5\frac{7}{8}$	$8\frac{1}{4}$	$10\frac{1}{4}$	$12\frac{7}{8}$	$15\frac{5}{8}$	$17\frac{7}{8}$	$20\frac{1}{4}$	$22\frac{1}{4}$	$24\frac{7}{8}$
SMC11H	50	10.0	9.4	$5\frac{1}{2}$	$12\frac{1}{8}$	$12\frac{3}{4}$	$6\frac{7}{8}$	$9\frac{1}{4}$	$12\frac{1}{8}$	$14\frac{5}{8}$	$18\frac{1}{8}$	$20\frac{1}{4}$	$23\frac{7}{8}$	$26\frac{5}{8}$	$28\frac{5}{8}$
SMC13H	60	12.0	11.0	$5\frac{1}{2}$	$12\frac{1}{8}$	$12\frac{3}{4}$	$7\frac{1}{2}$	$10\frac{5}{8}$	$13\frac{3}{4}$	$17\frac{7}{8}$	$20\frac{1}{2}$	$23\frac{5}{8}$	$26\frac{3}{4}$	$30\frac{5}{8}$	
SHR11H	75	15.0	15.7	$6\frac{1}{2}$	16	$16\frac{5}{8}$	$7\frac{1}{8}$	10	$12\frac{3}{8}$	$15\frac{5}{8}$	19	$21\frac{7}{8}$	$24\frac{5}{8}$	$27\frac{3}{4}$	$30\frac{1}{4}$
SHR13H	90	18.0	18.0	$6\frac{1}{2}$	16	$16\frac{5}{8}$	8	$11\frac{1}{8}$	$14\frac{3}{8}$	$17\frac{5}{8}$	$21\frac{1}{8}$	$24\frac{5}{8}$	$27\frac{7}{8}$	31	$34\frac{1}{4}$
SHR17H	120	24.0	23.0	$6\frac{1}{2}$	16	$16\frac{5}{8}$	$9\frac{5}{8}$	$13\frac{5}{8}$	$17\frac{5}{8}$	$22\frac{1}{8}$	$26\frac{5}{8}$	$30\frac{5}{8}$			
SHR21H	150	30.0	28.0	$6\frac{1}{2}$	16	$16\frac{5}{8}$	$11\frac{1}{4}$	16	$21\frac{1}{8}$	$26\frac{5}{8}$	$31\frac{5}{8}$	$36\frac{5}{8}$			
SHR25H	180	36.0	33.0	$6\frac{1}{2}$	16	$16\frac{5}{8}$	$12\frac{7}{8}$	$19\frac{1}{8}$	$24\frac{1}{4}$	31	$36\frac{5}{8}$				
SUO19H	216	43.2	38.6	$7\frac{1}{2}$	19	$20\frac{1}{8}$	$10\frac{1}{2}$	15	$20\frac{1}{8}$	$24\frac{5}{8}$	$29\frac{1}{8}$				
SUO23H	264	52.8	44.5	$7\frac{1}{2}$	19	$20\frac{1}{8}$	$12\frac{1}{8}$	$17\frac{1}{2}$	$23\frac{3}{8}$	$28\frac{5}{8}$	34				
SUO27H	312	62.4	50.7	$7\frac{1}{2}$	19	$20\frac{1}{8}$	$13\frac{3}{4}$	$20\frac{1}{2}$	$26\frac{5}{8}$	$33\frac{5}{8}$					
SL/25H	384	76.8	60.3	$8\frac{1}{2}$	$22\frac{7}{8}$	24	$13\frac{1}{4}$	$20\frac{1}{8}$	26	$32\frac{5}{8}$					
SL/31H	480	96.0	73.3	$8\frac{1}{2}$	$22\frac{7}{8}$	24	16	$23\frac{3}{4}$	$30\frac{7}{8}$						
SL/37H	576	115.2	86.0	$8\frac{1}{2}$	$22\frac{7}{8}$	24	$18\frac{3}{8}$	$27\frac{1}{2}$	$35\frac{1}{4}$						

* Includes electrolyte and portion of wood tray.
† The cells have extra large electrolyte space above the top of the plates.

A complete cell, cut away to show the construction, is illustrated in Fig. 43.

Of the Nicad batteries which are made in this country, type S is intended for normal duty. This, however, does not include cycling the cells. Type L is a variation intended for light services, and these cells have capacity ratings similar to the S cells. Type THR cells are designed for heavy duty, such as starting Diesel engines.

Table 11 gives data on the type S cells. The capacity in ampere-hours is specified at the 8-hour rate. The normal charging period is 7 hours. In the type designations the number represents the whole number of plates per cell, and the letter H, if used, indicates the "high" type of cell which provides extra space for electrolyte.

Some cells of the type S are intended for stand-by service and may be subject to heavy drains incident to emergency operation. These are listed in Table 12. Capacities are listed for the 8- and 3-hour rates and also for the 1-minute rate. In many instances these batteries will be floated on the line. Data for floating voltages are given in the table.

Sintered-Plate Cadmium Batteries

This type of cadmium battery differs materially in construction and performance from the more familiar type described above. Often known as the Durac[56] type, it was developed in Germany during World War II.

The plates consist of a highly porous structure of nickel impregnated with the active materials nickel oxide and cadmium. Fleischer distinguishes between plaques and plates. The plaques are the sintered metal bodies before being impregnated; after that they are called plates.

To obtain plaques of 60 to 90 per cent porosity it is necessary to use a nickel powder of low apparent density. This is prepared by thermal decomposition of nickel carbonyl vapor.[57] The apparent density of nickel carbonyl can be as low as 0.6 g/cc, but in battery construction, material of 1 g/cc is employed. It sinters in a protecting

[56] G. D. Chu, The Durac nickel-cadmium alkaline battery, Natl. Research Council Can., *Elec. Eng. Radio Br. Rept. ERB 172* (1947). Arthur Fleischer, Sintered-plates for nickel-cadmium batteries, *J. Electrochem. Soc., 94,* 289 (1948).

[57] S. E. Buckley, et al., Carbonyl nickel and carbonyl iron powders, their production and properties, *BIOS Final Rept. 1575* (1947). N. V. Sidgwick, in his *Chemical Elements and Their Compounds,* Clarendon Press, Oxford, says on p. 1451 of Volume 2, that the chief of the nickel carbonyls is $Ni(CO)_4$, a highly poisonous material made by passing carbon monoxide gas over finely divided nickel at a temperature below 100° C.

TABLE 12. TYPE "S" NICAD BATTERIES FOR SWITCHGEAR OPERATION, SUPERVISORY CONTROL, AND EMERGENCY LIGHTING

Cell Type No.	Capacities: Amp-Hr 8-Hr Rate to 1.10 Volts per Cell	Capacities In Amperes: 8-Hr Rate to 1.10 Volts per Cell	3-Hr Rate to 1.10 Volts per Cell	1-Hr Rate to 1.10 Volts per Cell	1-Min. Rate to 1.10 Volts per Cell	1-Min. Rate to 0.95 Volts per Cell	Nominal Battery Voltages 48 V; Floating Voltages 53–55; No. of Cells per Battery 38	120 V; 133–138; 95	240 V; 266–272; 190	Over-All Dimensions of 1 Tray, in.: Width	Length	Height	Charge Current, amps: 7-Hr Rate	14-Hr Rate	Approx. Weight per Cell, lb: Net*	Shipping
							No. of Trays per Battery									
SMC9H	40	5.00	12.4	23	66	132	4	10	19	$5\frac{1}{2}$	$24\frac{7}{8}$	$12\frac{1}{8}$	8.0	4.0	7.9	9.3
SMC11H	50	6.25	15.5	29	83	166	8	19	38	$5\frac{1}{2}$	$14\frac{7}{8}$	$12\frac{1}{8}$	10.0	5.0	9.4	11
SMC13H	60	7.50	18.5	35	100	200	8	19	38	$5\frac{1}{2}$	$17\frac{3}{8}$	$12\frac{1}{8}$	12.0	6.0	11	13
SHV17H	80	10.0	23.2	46	120	240	8	19	38	$6\frac{1}{2}$	$22\frac{1}{4}$	$12\frac{5}{8}$	16.0	8.0	17	21
SHV21H	95	11.8	27.6	54	142	284	8	19	38	$6\frac{1}{2}$	$26\frac{3}{8}$	$12\frac{5}{8}$	19.0	9.5	20	24
SHV27H	123	15.4	35.7	71	185	370	8	19	38	$6\frac{1}{2}$	33	$12\frac{5}{8}$	24.6	12.3	27	33
SHN28H	141	17.6	41.0	81	212	424	8	19	38	$6\frac{1}{2}$	$34\frac{1}{8}$	$13\frac{1}{8}$	28.2	14.1	30	36
SHT28H	168	21.0	48.7	96	252	504	8	19	38	$6\frac{1}{2}$	$34\frac{1}{8}$	$14\frac{1}{4}$	33.6	16.8	33	40
SHT30H	181	22.6	52.5	104	271	542	8	19	38	$6\frac{1}{2}$	$36\frac{1}{8}$	$14\frac{1}{4}$	36.2	18.1	34	41
SUK25H	204	25.5	59.7	117	306	612	8	19	38	$7\frac{1}{2}$	$31\frac{3}{8}$	$15\frac{3}{8}$	40.8	20.4	38	46
SUK27H	221	27.6	64.0	127	331	662	8	19	38	$7\frac{1}{2}$	$33\frac{3}{8}$	$15\frac{3}{8}$	44.2	22.1	41	49
SUK30H	246	30.7	71.3	141	369	738	8	19	38	$7\frac{1}{2}$	$36\frac{3}{8}$	$15\frac{3}{8}$	49.2	24.6	45	54
SLK29H	280	35.0	81.2	161	420	840	8	19	38	$8\frac{1}{2}$	$36\frac{5}{8}$	$16\frac{5}{8}$	56.0	28.0	56	67

* Includes electrolyte and portion of wood tray.

atmosphere without the need for compacting at temperatures as low as 500° C, but higher temperatures are generally used. Chu gives the sintering temperature of his nickel powder as 900 to 960° C.

Grids are usually a coarsely woven wire cloth of about 20 mesh and may have a frame around the wire. A thin nickel strip is welded to the cloth to provide the lug. The grids are usually of nickel, but nickel-plated iron has been used.

In making the plaques, the form of the plate is cut in a block of graphite about 1 inch thick. The carbonyl nickel is sifted into the opening to about half depth, and then the grid is laid on it and covered with the other half of the powder. A graphite cover closes the mold. The sintering furnace is provided with a cooling chamber into which the graphite block passes after about 10 minutes' heating. In the protective atmosphere of nitrogen and dissociated ammonia the plaque is finished, and it then has about 80 per cent porosity.

Impregnation is accomplished in solutions of nickel or cadmium salts, preferably the nitrates. Electrolysis is done in an electrolyte of 25 per cent sodium or potassium hydroxide heated to nearly 100° C. The current density is high and maintained for about 20 minutes while vigorous gassing occurs. The washing that follows in flowing tap water is critical and may last 3 hours until the effluent water has a pH of 9. The plates are dried at 80° C. This cycle of impregnation is repeated 4 or 5 times, each succeeding cycle making the washing more difficult.

The finished plates are assembled into cells, using plastic rods as separators. The initial voltage is about 1.4 volts. These cells are still more or less in the experimental stage, and few operating data are available. On life tests they are said to have passed 300 cycles of charge and discharge.

SILVER OXIDE BATTERIES

A third type of alkaline storage battery which has become available employs zinc and silver oxide as electrodes in a solution of potassium hydroxide saturated when in the discharged condition with zinc hydroxide. Interest in this type of cell became widespread during World War II because of the cell's large output per unit of space and weight. Many laboratories have been concerned with its development both as a primary battery and as a storage battery. Some of the previous investigations were reported in my book *Primary Batteries*.[58]

Silver oxide, Ag_2O, and the more highly oxidized state, silver

[58] G. W. Vinal, *Primary Batteries*, Chapter 9, John Wiley & Sons, New York, 1950. Data given there will not be repeated here.

peroxide, Ag_2O_2, are prepared by electrolytic oxidation. The charging voltage is limited to 2.1 volts per cell. Oxygen evolution occurs if the voltage reaches about 2.3 volts.

The chemical reactions of charge and discharge as given by the Yardney Electric Corporation are as follows:

$$Ag + Zn(OH)_2 \rightleftharpoons AgO + Zn + H_2O$$
$$\underset{\text{discharged}}{} \qquad \underset{\text{charged}}{}$$

The preferred method of charging these cells is by constant current at a rate approximating completion in 10 to 15 hours. The end of

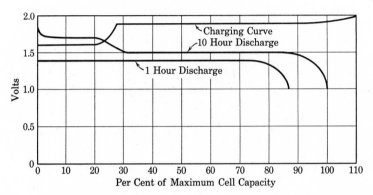

Fig. 44. Typical curves of charge and discharge of the Yardney Silvercel. Note the flat voltage characteristic.

charge is indicated when the terminal voltage reaches 2.0 or at most 2.1 volts per cell while current is flowing at the specified rate for the type of cell. Undercharging is not harmful except as the capacity of the cell may be limited, but overcharging is definitely to be avoided. The ampere-hour efficiency is about 80 per cent.

Charging may be done also by the modified constant-potential method from a d-c source of 2.2 to 2.5 volts with a series resistance in the circuit to limit the initial current to that corresponding to the 1-hour rate. This charging method is limited to charging a single cell.

As a third method of charging the stepwise decreasing current may be used. For each type there is a specified maximum initial charging current, and reductions to lower values are made each time the terminal voltage reaches the specified maximum, not exceeding 2.1 volts.

The silver cells can be operated in series or parallel provided they are of the same type and size, but some preliminaries are necessary. Before assembly into a battery they should be completely discharged

to zero volts at a rate equal to the 1-hour rate or less. Series-connected cells are charged as outlined above, but parallel-connected cells should not be charged while so connected.

Distilled water may be added to fully charged cells when they appear to be dry. Normally only a small excess of electrolyte will appear above the top of the element. Cells intended for storage up

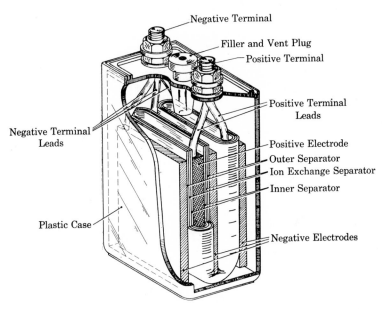

Fig. 45. Yardney Silvercel. Cut-away view of a cell of small size.

to 2 months should be discharged to zero volts. Storage may be at normal temperatures, but cold storage even as low as 0° F is preferred.

Operating characteristics of silver oxide cells are interesting because of the large output in proportion to the weight and the space that they occupy. Typical curves of charge and discharge are shown in Fig. 44, and a cut-away view of a cell to show construction is in Fig. 45. Local action may be considerable, and the silver oxide is easily reduced by small amounts of organic matter.

Aside from military applications, the silver cells are finding use in motion picture and television cameras, radio and sound apparatus, model planes and boats, and battery-powered test instruments and toys.

3

The Electrolyte

The electrolyte for lead storage batteries is a solution of sulfuric acid. This has been used since the beginning of the industry. In this chapter a brief description of the preparation of sulfuric acid and the physical and chemical properties of its solutions, which are important in studying the subject of storage batteries, will be given.

Sulfuric acid is one of the most important of chemical compounds because of the great variety of its uses. Its manufacture antedates the beginnings of modern chemistry. It is marketed under a variety of names, depending on the strength and purity of the acid. *Chamber acid* is a dilute acid containing from 62 to 70 per cent sulfuric acid; *tower acid* contains 75 to 82 per cent acid; *oil of vitriol* contains 93 to 97 per cent acid; *monohydrate* contains 100 per cent acid; *fuming acid*, or *oleum*, contains sulfur trioxide dissolved in concentrated sulfuric acid. The strongest acid that can be prepared by evaporation is about 98.5 per cent. The significance of these names will appear in later sections. For the storage battery, the chemically pure oil of vitriol, diluted to the proper concentration, is of greatest importance. Since the term "oil of vitriol" is sometimes used to designate the more impure or technical grade of acid, including the brown oil of vitriol, it is better to adhere to the term "sulfuric acid," meaning thereby the chemically pure acid.

THE PRODUCTION OF SULFURIC ACID

Sulfuric acid in the free state occurs in small quantities in volcanic regions, and natural sulfates are abundant; but these sources are not utilized for producing the acid in large quantities. The commercial processes are based on the oxidation of sulfur dioxide to the trioxide state, which in combination with water yields sulfuric acid.

The combustion of sulfur in the air results in the formation of sulfur dioxide, SO_2, which is a colorless gas having a suffocating odor. This dioxide is not combustible but may be further oxidized to sulfur trioxide, SO_3, through the agency of nitrogen oxides or in the presence of certain catalysts. Trioxide of sulfur is also called sulfuric anhydride. It is a transparent crystalline solid that unites with water with explosive violence and with the evolution of a large amount of heat, forming sulfuric acid. The general reaction may be represented by the following equation:

$$SO_2 + O + H_2O = H_2SO_4$$

This reaction, however, is not as simple as the equation would indicate. The oxidation of the dioxide to the trioxide state is a slow reaction that is difficult to perform, and the sulfuric anhydride forms, with the water, a series of compounds that are called hydrates. These may be represented by the formula $m \cdot SO_3 \cdot n \cdot H_2O$, in which m and n are whole numbers indicating the number of gram-molecules of the anhydride and water, respectively. If m is less than n or equal to it, the acid is of the ordinary non-fuming variety. When m exceeds n, the formula represents fuming sulfuric acid. This fact enables us to distinguish the fuming from the non-fuming acid by the percentage content of the sulfuric anhydride. One gram-molecule of the anhydride weighs 80.06 grams, and a gram-molecule of water weighs 18.016 grams. The limiting percentage of the anhydride in the non-fuming acid is, therefore, the ratio of 80.06 to 98.076, which equals 81.63 per cent.

Since 98.076 is the molecular weight of sulfuric acid, according to the formula H_2SO_4 it is possible to compute from the limiting percentage of the anhydride the amount of it present in any solution when the percentage of sulfuric acid is known. For example: The acid of specific gravity 1.300, commonly used in certain types of batteries, contains 39.1 per cent of sulfuric acid and $39.1 \times 0.8163 = 31.9$ per cent of the sulfuric anhydride.

Two processes are used at the present time for the manufacture of sulfuric acid. These are commonly called the "chamber" and the "contact" processes. The difference between them consists in the method employed to oxidize the sulfur dioxide to the trioxide state. The contact process is supplanting the chamber process for the preparation of concentrated acid, but about half of the sulfuric acid produced in the United States is made by the chamber process.

As a preliminary to either of these processes, the production of sulfur dioxide is necessary. The purest sulfur dioxide is produced by the direct combustion of sulfur, but a common method is by the roasting of iron pyrites, which contain ordinarily from 40 to 50 per cent of sulfur. Other raw materials include zinc blende, galena, and the spent oxides of gas works. The oxide obtained by burning pure sulfur is free from arsenic, iron, copper, and zinc, which usually appear in the gas obtained by roasting the pyrites.

In the chamber process, the oxidation of the sulfur dioxide is effected by the aid of oxides of nitrogen in a series of lead-lined chambers, from which the process takes its name. The hot gases from the sulfur burners or the roasting ovens pass into a tower, called the Glover tower, filled with broken firebrick over which flows a stream of dilute sulfuric acid (part of the acid from the chambers), which cools the gases and effects a partial concentration of acid. Concentrated sulfuric acid, containing oxides of nitrogen, also flows in this tower. The oxides of nitrogen are liberated near the top of the tower by the dilution of the strong acid by the weak acid. The oxides of nitrogen and sulfur mix, starting the formation of the sulfur trioxide; this process is continued in the chambers to which the mixed gases next pass. Water and nitric acid are sprayed into these chambers. The water absorbs the trioxide as it forms. The oxides of nitrogen pass on to the Gay-Lussac tower, where they are reabsorbed by concentrated sulfuric acid and are used again. The principal product of the plant is the chamber acid drawn from the chambers. As this is relatively dilute it is concentrated by evaporation. Platinum, iron, silica, and glass serve as materials for the concentrating apparatus. They are not entirely inert. About 1 gram of platinum is lost per ton of acid that is concentrated in platinum stills or dishes. This represents about 1 part of platinum in 1 million parts of the acid, which would be sufficient to cause trouble in a storage battery. Platinum has become so expensive that it is rarely used now for this purpose.

The contact process is simpler. Sulfur dioxide may be oxidized to the trioxide state in the presence of sponge platinum and certain other catalysts, at a temperature of about 400 to 450° C. The process will proceed with evolution of much heat after being started. Other catalysts that are sometimes employed for this purpose are iron oxide and vanadium. The combination of pure water and pure sulfur trioxide results in the formation of pure sulfuric acid, which may be made of any desired degree of concentration. The contact process also requires two stages for the formation of sulfuric acid. These may be represented by the following formulas:

$$SO_2 + O = SO_3 \quad \text{and} \quad SO_3 + H_2O = H_2SO_4$$

Impurities, such as arsenic from the pyrite, must be removed, as they interfere with the action of the catalysts.

Storage batteries require a high purity of sulfuric acid. For this reason acid made from pure sulfur is preferable to acid made from pyrites. The source of the acid cannot always be ascertained, and it is important, therefore, to have specifications to insure an adequate degree of purity.

The conditions for storing and shipping sulfuric acid vary with the concentration. Acid up to about 75 per cent has little effect on lead; above this strength the acid attacks lead if hot. Concentrated acid readily dissolves lead at about 250° C. Monohydrate and fuming acid attack lead at ordinary temperatures. Strong acid, particularly the concentrated acid, has very little effect on iron. Cast iron is somewhat more resistant than wrought iron, but cast iron is not suitable for use with fuming acid. Alloys of iron and silicon are very resistant to the non-fuming acid and are extensively used for concentrating the acid up to 98 per cent. Dilute acid cannot be stored in iron containers, and glass carboys or lead-lined tanks are generally used for this purpose.

PROPERTIES OF SULFURIC ACID SOLUTIONS

Concentrated sulfuric acid is a clear, colorless, and odorless liquid having the consistency of a light oil. Its specific gravity is 1.84 at 15° C (59° F), and it contains about 95 per cent acid. It is miscible in all proportions with water, forming a series of hydrates, several of which are of interest in battery investigations. When the acid is mixed with water a large amount of heat is evolved. Sulfuric acid is a dehydrating agent having such a strong affinity for water that it can remove even chemically combined water from organic substances such as the separators in storage batteries. Wood becomes charred if immersed in moderately strong acid, and the carbon gives the wood a darkened appearance. The concentrated acid boils at 338° C (640° F). When boiled, it liberates a gas of the same composition, which appears as a dense white vapor because of the absorption of water vapor from the air.

Heat of Dilution

The amount of heat which is liberated when 1 gram-molecule of acid is diluted with n gram-molecules of water is shown in Table 13. When the amount of water added is so great that the solution may be considered to be infinitely dilute the total heat evolved is found to be 23.54 kilogram calories. In Table 13 the difference between amounts of heat liberated at any two different concentrations represents the amount of heat that would be liberated in diluting the more concentrated to the less concentrated solution. For example: By diluting

TABLE 13. RELATIVE HEAT CONTENT, HEAT OF DILUTION, AND
SPECIFIC HEAT OF SULFURIC ACID SOLUTIONS
(One mole of sulfuric acid is diluted with n moles of water.)

n Moles of H_2O	Per Cent of H_2SO_4	Specific Gravity at 25° C	Relative Molal Heat Content at 25° C, kg cal*	Heat† of Dilution at 25° C, kg cal	Specific Heat (Heat Capacity) at 25° C, cal/g
0	100	. . . . .	23.54	0	0.337
. . . . .	93.19	. . . . .	20.35	3.19	0.370
1	84.48	1.774	16.72	6.82	0.442
2	73.13	1.648	13.58	9.96	0.466
3	64.47	1.548	11.66	11.88	0.501
5	52.13	1.416	9.45	14.09	0.582
10	35.25	1.262	7.30	16.24	0.722
20	21.40	1.150	6.23	17.31	0.831
50	9.82	1.066	5.78	17.76	0.918
100	5.16	1.034	5.62	17.92	0.955
200	2.65	1.017	5.41	18.13	0.976
400	1.34	1.009	5.02	18.52	0.987
1,600	0.34	1.002	4.04	19.50	0.997
6,400	0.08	. . . . .	2.90	20.64	. . . . . .
25,600	0.02	. . . . .	1.54	22.00	. . . . . .
∞	0.00	1.000	0.00	23.54	0.9989

* F. D. Rossini, *Chemical Thermodynamics*, p. 280, John Wiley & Sons, New York, 1950.

† The heat liberated when the ordinary concentrated acid (93.19 per cent) is diluted can be calculated by subtracting the respective molal heat contents from 20.35 kilogram calories.

$H_2SO_4 \cdot 3H_2O$ to $H_2SO_4 \cdot 100H_2O$, the amount of heat evolved is $17.92 - 11.88 = 6.04$ kilogram calories. The kilogram calorie is the amount of heat required to raise the temperature of 1 kilogram (2.2 pounds) of water 1 degree Centigrade, the initial temperature being 15° C.

The relative heat content of sulfuric acid solutions is of importance in the theory of storage batteries because of the relation that exists between the voltage and the concentration of the acid. In Chapter 4 it will be shown that the voltage of a lead storage cell varies in a definite way with the concentration of the acid in the pores of the plates. It is possible to calculate the voltage corresponding to various concentrations of electrolyte by an application of the laws of thermodynamics.

Although the amount of heat evolved when the acid is diluted with a given portion of water is accurately known, the temperature to which the mixture rises will depend on the specific heat of the liquid (which varies somewhat with temperature) and the loss of heat from the solution. When a large amount of acid is prepared for use in storage batteries by dilution in a lead-lined tank, the loss of heat is proportionately less than when smaller amounts are mixed in containers having less heat insulation. In any event, if it is assumed that the loss of heat is negligible, it is possible to compute the maximum temperature to which the solution will rise from the heat evolved and the specific heat. For example: If 1 gram-molecule of acid (98 g) is mixed with 10 gram-molecules of water (180 g), the initial temperature being 15° C, the temperature rise will be

$$(t° - 15°) = \frac{16,240 \text{ cal}}{0.722 \times 278} = 81°$$

whence $t° = 81° + 15° = 96°$. By experiment with a glass beaker wrapped in a towel, the maximum temperature attained was 85°.

Storage-battery electrolyte is sometimes prepared from the acid of 1.400 sp. gr. Then the heat of dilution is 16.00 — 14.35 = 1.65 kilogram calories, which is only 10 per cent of the heat evolved if electrolyte of 1.280 sp. gr. is prepared from the concentrated acid.

Contraction of the Solution

When 1 volume of sulfuric acid is diluted with 1 volume of water, the volume of the solution is not 2 volumes but slightly less after the solution has cooled to the initial temperature. The same is true for any proportions of the acid and water; the sum of the original volumes exceeds the volume of the solution. The contraction is greatest for solutions having a specific gravity of 1.600. Table 14 shows the contraction of solutions of various concentrations.

TABLE 14.   CONTRACTION OF SOLUTIONS OF SULFURIC ACID

(The contraction is expressed as milliliters per kilogram of the solution, 18° C.)

Specific Gravity	Contraction*	Specific Gravity	Contraction*
1.000	0	1.500	60
1.100	25	1.600	62
1.200	42	1.700	60
1.300	51	1.800	48
1.400	57		

* Calculated from Pickering's results, *J. Chem. Soc.*, *57*, 148 (1890).

It is often convenient to dilute the sulfuric acid by measured volumes to obtain a required percentage strength or specific gravity. Allowance must be made for the contraction of the solution in estimating the required proportions of acid and water. To facilitate the preparation of such solutions, the percentage content of sulfuric acid solutions by both weight and volume is given in Table 21.

To illustrate the application of Tables 14 and 21 to the preparation of electrolytes of various concentrations for storage batteries, the following example is given: Required the percentage of sulfuric acid by volume in a solution of 39.1 per cent by weight (1.300 sp. gr.). One liter of the concentrated acid weighs 1840 grams, whence 1840/0.391 = 4706, the total weight of the solution containing 1840 grams of acid. Subtracting from this the weight of the acid, the weight of water is found to be 2866 grams. Since the specific gravity of the water is unity by definition, the sum of the volumes is 2866 + 1000 = 3866 ml. The percentage of H_2SO_4 by volume on this basis would be 1000/3866 = 25.9 per cent, which is not the correct result. Table 14 shows the contraction to be 51 × 4.706 = 240 ml. The true volume of the mixture is, therefore, 3866 − 240 = 3626 ml, and the true percentage of acid by volume is 1000/3626 = 27.6 per cent, as shown in Table 21.

Resistivity

The resistance to the passage of an electric current through the electrolyte varies with the concentration and the temperature. The resistivity is a property of the substance itself. It is the resistance of a specimen 1 centimeter in length and 1 square centimeter in cross section. The relation $R = \rho(l/s)$, where R is the resistance, ρ the resistivity, l the length, and s the cross section, indicates the unit for measuring the resistivity. This may be written $\rho = R(s/l)$. Since the area s has the dimensions of length squared, the dimensional formula for resistivity is the product of a resistance and a length. The unit of resistivity is, therefore, the "ohm-centimeter." It is sometimes spoken of as "ohms per centimeter cube."

The electrolytes used in storage batteries are within the range of lowest resistivities of sulfuric acid solutions. It has been known for many years that solutions of about 30 per cent (1.223 sp. gr. at 15° C) sulfuric acid have minimum resistivity, but only recently has it become known that the proportions of acid and water must be varied slightly to obtain minimum resistivity at other temperatures. Thus, at +30° C the solution of least resistance contains 31.5 per cent of the acid, but at −25° C the solution should contain 26.5 per cent. The resistivity of sulfuric acid solutions increases rapidly as the temperature is

decreased, particularly when the temperature is below 0° C (32° F).
Determinations[1] of the resistivity are given in Table 15.

TABLE 15.   RESISTIVITY OF SULFURIC ACID SOLUTIONS

Temperature		Resistivity in Ohm-Cm				For Minimum Resistivity	
° C	° F	15%	25%	35%	45%	Per Cent	Ohm-Cm
30	86	1.596	1.180	1.140	1.312	31.5	1.129
25	77	1.689	1.261	1.231	1.422	31.1	1.213
20	68	1.800	1.357	1.334	1.549	30.6	1.310
10	50	2.090	1.606	1.602	1.885	29.8	1.562
0	32	2.51	1.961	1.998	2.371	28.8	1.928
−10	14	.....	2.50	2.60	3.10	27.9	2.48
−20	−4	.....	3.35	3.57	4.31	26.9	3.34
−30	−22	.....	.....	5.29	6.35	....	.....
−40	−40	.....	.....	8.39	9.89	....	.....

The resistivity of the electrolyte is one of the most important factors
in determining the resistance of a storage cell. Unless the internal
resistance of the cell is small, a considerable portion of the useful
energy of the cell is expended within the cell itself.

Freezing Points

The freezing point of the electrolyte varies with its concentration.
It is, therefore, often said to vary with the state of charge of the
battery. The importance of knowing the freezing points has increased
because of the use of storage batteries on automobiles and airplanes,
which are often subjected to low temperatures in cold climates and at
high altitudes. The freezing points of sulfuric acid solutions were
carefully determined many years ago by Pickering.[2] On page 338 of
his paper, the whole freezing-point curve from dilute to concentrated
solutions is given. There are several maximums and minimums in the
curve, owing to the hydrates of sulfur trioxide that are formed at the
different concentrations.

In Table 16 the freezing points of solutions corresponding to the
specific gravities of the first column are given. These values are in
accord with the values adopted as standard by the Manufacturing
Chemists' Association in this country. The specific gravities are for

---

[1] G. W. Vinal and D. N. Craig, Resistivity of sulfuric acid solutions and its relation to viscosity and temperature, *J. Research Natl. Bur. Standards, 13,* 689
(1934).

[2] S. U. Pickering, The nature of solutions as elucidated by the freezing points
of sulfuric acid solutions, *J. Chem. Soc., 57,* 331 (1890). The complete freezing-
point curve is on page 338, and the tables are on page 363 and the following.

Table 16. Freezing Points of Solutions of Pure Sulfuric Acid

Specific Gravity at 15° C	Freezing Points Centigrade	Fahrenheit	Specific Gravity at 15° C	Freezing Points Centigrade	Fahrenheit
1.000	0	+32	1.450	−29	−20
1.050	− 3.3	+26	1.500	−29	−20
1.100	− 7.7	+18	1.550	−38	−36
1.150	−15	+ 5	1.600	*	*
1.200	−27	−17	1.650	*	*
1.250	−52	−61	1.700	−14	+ 6
1.300	−70	−95	1.750	+ 5	+40
1.350	−49	−56	1.800	+ 6	+42
1.400	−36	−33	1.835	−34	−29

* Freezing points indeterminate.

the solutions at 15° C and not at the freezing point. For certain concentrations, a considerable error may be made in estimating the freezing point if allowance is not made for the change in specific gravity at the low temperatures. The temperature coefficient of specific gravity for both the Centigrade and Fahrenheit scales, for all concentrations of electrolyte, is given in Table 21.

The freezing points of solutions of pure sulfuric acid are not exactly the same as the freezing points of electrolyte of the same specific gravity taken from storage batteries. The difference is small. For electrolytes of less than 1.290 sp. gr. the freezing points are slightly higher, and for greater specific gravities the freezing points are lower, than for solutions of pure acid. Figure 46 shows the freezing-point curves of both the pure acid solutions and the battery electrolyte.

In Fig. 46 there is also shown a series of curves in which the specific gravities of solutions are plotted at every 10 degrees between 0° and 60° C and extrapolated below 0° until they intersect the freezing-point curve. These curves were chosen so that specific gravities of 1.200, 1.225, 1.250, etc., fall on the 25° line. By means of this diagram, the freezing point of any electrolyte may be estimated with reasonable accuracy if the specific gravity and temperature are known. For example: The electrolyte in a battery at −20° C (−4° F) is found to be 1.256 sp. gr. This point is located on the diagram, and the nearest oblique line is followed to its intersection with the freezing-point curve of the electrolyte, for which the ordinate is −38° C (−36° F). The temperature can, therefore, fall 18° C (32° F) before the electrolyte will begin to freeze.

The lowest freezing point is for solutions having a specific gravity of 1.290. When solutions of lower specific gravity than this freeze, crystals of ordinary ice separate from the solution; for higher specific gravities the crystals are the tetrahydrate of the acid. Since the lowest freezing points are for concentrations of acid corresponding to the electrolyte of automobile batteries when fully charged, it is apparent that there is no danger of such a battery freezing even under

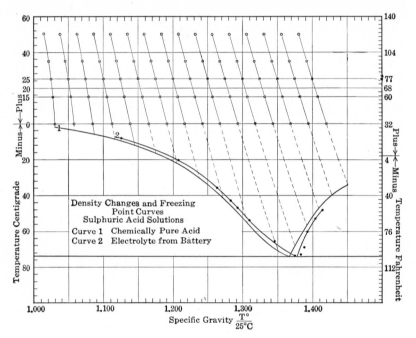

Fig. 46. Freezing-point curve for electrolyte and solutions of sulfuric acid.

the most severe conditions. If the battery is discharged, however, it will freeze at approximately $-10°$ C ($+14°$ F) or even higher. Electrolyte, or acid for preparing it, should be of the proper specific gravity when it is stored under severe conditions. Electrolyte of from 1.225 to 1.400 sp. gr. will not freeze under ordinary conditions, but nearly concentrated acid of 1.800 sp. gr. freezes at $+5°$ C ($+41°$ F). For concentrated acid the freezing point is much lower. At 1.835 it is $-34°$ C ($-29°$ F).

Vapor Pressure

Sulfuric acid is a powerful drying agent. Concentrated acid readily absorbs water vapor from the air in such quantities that a carboy left unstoppered will soon overflow if nearly filled with acid. Solutions of sulfuric acid have a definite vapor pressure that varies with the concentration and the temperature. The values of the vapor pressure are of theoretical interest in connection with storage batteries for the same reason as applied to the heat of dilution.

The electrolyte is often said to "evaporate"; most of the water lost from the cells, however, passes off as oxygen and hydrogen during charging. The electrolyte will absorb or give off water vapor according to the relative magnitude of its vapor pressure as compared with that of the surrounding air. If the vapor pressure of the electrolyte exceeds that of the water vapor in the air, evaporation will take place; if it is less, moisture will be absorbed. The simplest method of determining the pressure of water vapor in the air is by some form of dew-point apparatus, or a wet-and-dry bulb hygrometer. The temperatures obtained in such a measurement are referred to suitable tables[3] in which the corresponding vapor pressures, expressed in millimeters of mercury, are given. For example: On a humid and hot summer day, temperature 95° F (35° C) and relative humidity of 87 per cent, the vapor pressure would be 36.7 millimeters; on an average day, temperature 66° F (19° C) and relative humidity of 57 per cent, the vapor pressure would be 9.2 millimeters. These values should be compared with the vapor pressures for solutions of sulfuric acid given in Table 17

TABLE 17. VAPOR PRESSURE OF SOLUTIONS OF SULFURIC ACID

(The vapor pressures are expressed in millimeters of mercury.)

Specific Gravity at 15° C	Percentage of H_2SO_4	At 0° C, mm	At 15° C, mm	At 25° C, mm	At 35° C, mm
1.000	0.0	4.6	12.8	23.8	42.2
1.100	14.3	4.2	12.0	21.8	39.0
1.200	27.2	3.6	10.2	18.7	33.8
1.300	39.1	2.6	7.5	13.8	25.0
1.400	50.0	1.6	4.5	8.5	15.4
1.500	59.7	0.8	2.0	4.1	7.3
1.600	68.6	0.2	0.6	1.3	2.6
1.700	77.1	...	0.2	0.4	0.8

to determine whether evaporation or absorption of moisture will take place. Such a comparison shows that the electrolyte of specific gravity 1.300 will absorb water in the first case and evaporate in the second.

Under ordinary conditions, the vapor pressure of the electrolyte is not very different from that of the water vapor in the air. Stationary batteries, which have electrolyte of low specific gravity and open-top

[3] *Smithsonian Tables,* 8th ed. revised, p. 236, is a convenient table for use with a ventilated hygrometer.

cells, will lose water appreciably by evaporation. Automobile batteries, on the other hand, with a higher specific gravity of electrolyte
and closed tops, except for a small vent, will lose very little water by
evaporation in the strict sense of the word.

Because of the ability of sulfuric acid to absorb moisture, it is
evident that, if spilled, it will remain for long periods of time, or in
other words it will not "dry up." This makes it difficult to protect
painted surfaces, or other surfaces that are not acid-proof, against
corrosion. It is necessary to wash off, or better still to neutralize and
wash off, any electrolyte that may be spilled or become deposited as
a result of gassing upon any surface that is liable to injury.

Electrochemical Equivalent

According to Faraday's law the transformations taking place at the
plates of a storage cell are in the proportion of 1 gram equivalent of
the substance for the passage of 96,500 coulombs of electricity[4] through
the cell. This quantity of electricity is sometimes referred to as the
electrochemical constant, and it has been named the faraday.
The equation for the reaction within the cell is:

$$PbO_2 + Pb + 2H_2SO_4 \rightleftarrows 2PbSO_4 + 2H_2O$$

It is apparent that, for each gram equivalent of material transformed
at the positive and negative plates, 2 gram equivalents of acid are
decomposed and 2 gram equivalents of water are formed during the
process of discharge.

Since the gram-molecular weight of sulfuric acid is 98.076 grams
and the valence 2, the gram-equivalent weight is 49.038 grams. One
ampere-hour may be defined as 1 ampere flowing continuously for
3600 seconds and is equal, therefore, to 3600 coulombs. The number
of ampere-hours in a faraday is readily computed to be 26.80. From
these data the number of grams of acid decomposed by the discharge
of 1 ampere-hour is:

$$2 \times \frac{49.038}{26.80} = 3.660 \text{ g acid decomposed}$$

Similarly, the number of grams of water formed per ampere-hour of
discharge may be calculated. The molecular weight of water is 18.016

---

[4] The revised value, as of July 1, 1951, recommended for the faraday constant
by a subcommittee on Fundamental Constants of the Committee on Physical
Chemistry of the National Research Council is 96,493.1 absolute coulombs per
gram equivalent. The difference between this and the old value is immaterial
for the present discussion.

and its equivalent weight 9.008, whence

$$2 \times \frac{9.008}{26.80} = 0.672 \text{ g water formed}$$

In estimating the quantity of acid taking part in any reaction, it is necessary to draw a sharp distinction between the total quantity of acid reacting and the changes in weight taking place in the electrolyte as represented by the change in specific gravity. The reason for this is made apparent by the equation for the reaction given above. Since the decomposition of 2 equivalents of acid is accompanied by the formation of 2 equivalents of water, the net change for the passage of 1 faraday of electricity during discharge is a loss in weight corresponding to the sulfur trioxide, which may be expressed in grams as follows:

$$-98.076 + 18.016 = -80.060 \text{ g}$$
$$H_2SO_4 \quad H_2O \quad SO_3$$

The minus sign signifies the loss in weight which is manifested in the operation of the cell by the decrease in specific gravity of the electrolyte when discharge takes place. The change in sulfur trioxide content per ampere-hour is

$$2 \times \frac{40.030}{26.80} = 2.987 \text{ g}$$

When either charge or discharge takes place, there is a change in volume of the electrolyte as well as a change in the weight. An example completely solved will serve to make the process clear:

Example. A large stationary cell has a capacity of 4000 ampere-hours at the 8-hour rate of discharge. It contains 618 pounds of electrolyte of specific gravity 1.210 at 25° C when fully charged. Compute the changes in weight and volume and specific gravity of the electrolyte when 4000 ampere-hours are discharged.

In solving this example, it is convenient to convert the weight of electrolyte to the metric system, as this simplifies the later calculations. 618 pounds = 280.3 kilograms. Table 21 shows that this electrolyte contains 28.9 per cent by weight of H_2SO_4. The amounts of acid and water initially present are, therefore,

$$280.3 \times 0.289 = 81.0 \text{ kg of acid}$$
$$280.3 - 81.0 = 199.3 \text{ kg of water}$$

When 4000 ampere-hours are discharged, the amount of acid consumed is

$$3.660 \times 4000 = 14{,}640 \text{ g} \quad \text{or} \quad 14.64 \text{ kg}$$

The amount of water formed at the same time is

$$0.672 \times 4000 = 2688 \text{ g} \quad \text{or} \quad 2.69 \text{ kg}$$

Comparing the composition of the electrolyte at the beginning and at the end, we have

	At the beginning	At the end
Sulfuric acid	81.0 kg	66.36 kg
Water	199.3 kg	201.99 kg
Total	280.3 kg	268.35 kg

The net loss in weight of the electrolyte is therefore 11.95 kilograms. This corresponds to the sulfur trioxide removed, as may readily be seen by multiplying the grams of SO_3 equivalent to the passage of 1 ampere-hour as given above by the number of ampere-hours.

$$2.987 \times 4000 = 11,950 \text{ g} \quad \text{or} \quad 11.95 \text{ kg}$$

The next step is to compute the specific gravity of the electrolyte at the end of the discharge.

$$\frac{66.36}{268.35} = 24.7 \text{ per cent of acid in electrolyte at end}$$

Assuming the temperature to be 25°, this corresponds to a specific gravity 1.176 according to Table 21.

The change in the volume of the electrolyte which accompanies the discharge of this cell is computed from the initial and final weights and specific gravities.

$$\text{Initial volume} = \frac{280.3 \text{ kg}}{1.210 \text{ sp. gr.}} = 231.7 \text{ liters}$$

$$\text{Final volume} = \frac{268.35 \text{ kg}}{1.176 \text{ sp. gr.}} = 228.2 \text{ liters}$$

The decrease in volume is, therefore, 3.5 liters or about 0.88 ml per ampere-hour.

A table of the ampere-hour equivalents contained in unit volume of solution, for the various concentrations, affords the most convenient means of calculating the changes taking place in the operation of storage cells. This method is more direct than the use of formulas and diagrams which have been previously proposed for this purpose. In Table 18 are given the theoretical ampere-hours on the assumption that all of the acid could be utilized. This is an impracticable condition, since only part of the acid can be used, but the use of the table will appear from several examples. It is necessary to allow for the changes in volume of the solution which occur during charging and discharging. This is easy to do, because the increase in volume of the solution during charge and the decrease during discharge is practically 1 ml per ampere-hour. This quantity varies slightly with the concentration of the electrolyte. It is a little more than 1 ml per ampere-hour for acid of 1.300 sp. gr., and about 0.85 for acid of 1.200

Table 18. Electrochemical Equivalent of Solutions of
Sulfuric Acid

(Theoretical values per liter calculated for solutions at 25° C, assuming complete utilization of the acid in accordance with the reaction of a storage battery.)

Specific Gravity	Ampere-Hours per Liter	Specific Gravity	Ampere-Hours per Liter
1.040	17	1.220	100
1.060	26	1.240	110
1.080	35	1.260	120
1.100	44	1.280	130
1.120	53	1.300	141
1.140	62	1.320	151
1.160	71	1.340	162
1.180	81	1.360	173
1.200	90		

sp. gr. For most purposes, however, it is sufficient to assume that it is exactly 1 ml per ampere-hour.

Examples. 1. A large stationary cell has a capacity of 4000 ampere-hours at the 8-hour rate of discharge. It contains 618 pounds of electrolyte, specific gravity 1.210 at 25° C when fully charged. Find the coefficient of utilization of the acid at this rate of discharge.

618 pounds = 280.3 kilograms which, divided by the specific gravity, 1.210, = 231.6 liters. From the table, 1 liter = 95 ampere-hours. 231.6 × 95 = 22,000, the total number of ampere-hours. 4000/22,000 = 18.2 per cent, the coefficient of utilization.

2. A cell of the vehicle type, containing 6.75 pounds of electrolyte, 1.280 sp. gr. at 25° C, has a capacity of 220 ampere-hours at the 5-hour rate. What is the coefficient of utilization of the acid? *Answer,* 71 per cent.

3. A storage cell of a tractor battery gave 45 amperes for 5 hours and 42 minutes. The initial specific gravity at 25° C was 1.269, and the final specific gravity, corrected to 25° after allowing sufficient time for diffusion, was 1.099. How many liters of electrolyte were in the cell?

The total ampere-hours discharged were 45 × 5.7 = 256.5. Letting x represent the number of liters initially present, the number of liters after the discharge was $(x - 0.256)$. The ampere-hour equivalent of the electrolyte is, by Table 18,

Initial 124.5x,
Final 43.5$(x - 0.256)$.

Whence

$$124.5x - 43.5(x - 0.256) = 256.5$$
$$x = 3.02 \text{ liters. } Answer.$$

4. A locomotive cell which gives 90 amperes for 5 hours contains 15 pounds of electrolyte of specific gravity 1.280 at 25° C. What will be the final specific gravity after a 5-hour discharge, corrected to 25° after allowing sufficient time for diffusion to take place?

15 pounds = 6.80 kilograms of electrolyte, and this divided by the specific gravity, 1.280, = 5.31 liters.

The total ampere-hour equivalent is 5.31 × 130 = 690
The ampere-hours delivered are   5   × 90 = 450
Subtracting, the ampere-hours remaining are   240

During the discharge the volume of the electrolyte has decreased by 450 ml, so that the final volume of electrolyte is 5.31 − 0.45 = 4.86 liters.

240/4.86 = 49.4 ampere-hours equivalent per liter

The table shows this to correspond to 1.112 sp. gr. *Answer.*

5. A cell from a small starting and lighting battery, containing 580 ml of electrolyte, is rated to give 40 ampere-hours at the 5-hour rate. It is required that the specific gravity at the end of a full discharge shall not fall below 1.140 at 25° C. What must the specific gravity be at the beginning of the discharge?

Since the initial volume is 580 ml, the final volume after discharging 40 ampere-hours will be 540 ml. The ampere-hour equivalent of the electrolyte at the end of this discharge will be

$$0.540 \times 62 = 33.5$$

The ampere-hours initially present were

$$33.5 + 40 = 73.5$$

The ampere-hour equivalent per liter initially present was, therefore, 73.5/0.580 = 127

The table shows that this acid was specific gravity 1.274. *Answer.*

6. The electrolyte of a cell of a stationary battery of 280 ampere-hours' capacity has a specific gravity of 1.200 at 25° C at the conclusion of the charging period. The normal value of the specific gravity in this cell is 1.210. The amount of electrolyte is 40 pounds. How much lead sulfate remains on the plates?

40 pounds of electrolyte = 18.1 kilograms which divided by the specific gravity when fully charged 1.210 = 15.0 liters as the volume initially present. The ampere-hour equivalent for the 1.210 acid is

$$15.0 \times 95 = 1425$$

As a first approximation, it is necessary to assume the volume of electrolyte at 1.200 sp. gr. to be the same. The ampere-hour equivalent of this is

$$15.0 \times 90 = 1350$$

The difference is therefore 75 ampere-hours, and the volume of electrolyte at 1.200 is corrected to

$$15.0 - 0.075 = 14.9$$

for which the equivalent is

$$14.9 \times 90 = 1341$$

Since 1 ampere-hour decomposes 3.66 grams of acid, the deficit of acid is

$$(1425 - 1341) \times 3.66 = 308\,g$$

$$\frac{H_2SO_4}{PbSO_4} = \frac{98}{303} = \frac{308}{951}$$

Whence there are 951 grams of lead sulfate on the plates. *Answer.*

7. An aviation battery is rated to give 28 ampere-hours at the 5-hour rate. By actual test it was found to give only 24 ampere-hours. The battery contained 210 ml of electrolyte of specific gravity 1.280 at 25° C. Is this a sufficient amount?

$0.210 \times 130 = 27.3$ ampere-hours; that is, the battery could not give its rated capacity even if the electrolyte could be reduced to water.

Viscosity

A knowledge of the viscosity of sulfuric acid solutions is important, because the rate of diffusion of the acid through the pores of the plate is dependent upon it. In this respect the viscosity exerts a powerful influence upon the capacity, but it is seldom referred to in this connection. In Table 19 the absolute viscosities expressed in centipoises

TABLE 19. VISCOSITY OF SULFURIC ACID SOLUTIONS

Temperature		Viscosity in Centipoises				
° C	° F	10%	20%	30%	40%	50%
30	86	0.976	1.225	1.596	2.163	3.07
25	77	1.091	1.371	1.784	2.409	3.40
20	68	1.228	1.545	2.006	2.70	3.79
10	50	1.595	2.010	2.600	3.48	4.86
0	32	2.160	2.710	3.520	4.70	6.52
−10	14	.....	3.820	4.950	6.60	9.15
−20	− 4	.....	.....	7.490	9.89	13.60
−30	−22	.....	.....	12.200	16.00	21.70
−40	−40	.....	.....	.....	28.80	.....
−50	−58	.....	.....	.....	59.50	.....

are given. They have been computed for convenient percentages of sulfuric acid from measurements by Vinal and Craig.[5] The usual range of battery electrolytes lies between 20 and 40 per cent. The viscosity doubles between 25 and 0° C. Below zero the viscosity in-

[5] G. W. Vinal and D. N. Craig, The viscosity of sulfuric acid solutions used for battery electrolytes, *Bur. Standards J. Research, 10,* 781 (1933).

creases more rapidly, and it is at once apparent why the capacity of storage cells falls off so rapidly at low temperatures.

The relation of the viscosity to the rate of diffusion is discussed in Chapter 5.

MEASUREMENT OF SULFURIC ACID SOLUTIONS

Careful operation of storage batteries requires that the electrolyte contain the proper amount of sulfuric acid and water. Convenient and accurate means are necessary for measuring the concentration of these solutions. The method most generally used is to measure the specific gravity, or the density of the solution, at some definite temperature. Other methods include the use of arbitrary scales, such as the Baumé or the Twaddell scales. The concentration may also be expressed as the percentage, by weight or volume, of the sulfuric acid in the solution. This is convenient in preparing solutions of a definite strength. Chemical computations are often facilitated when based on the number of gram equivalents per liter. For storage or shipping purposes, the concentration as it affects the weight and bulk of the packages is important. This requires the pounds per cubic foot or the kilograms per liter.

Table 21 gives the comparative values of the concentration expressed by these different methods of measurement.

Meaning of Specific Gravity and Density

The term density is used to represent mass per unit volume and is usually expressed as grams per milliliter where liquids are concerned. Specific gravity is used to express the relative masses of equal volumes of the liquid and the water, each being at a specified temperature.

The specific gravity of sulfuric acid solutions used for the electrolyte varies appreaciably with the temperature. It is, therefore, customary to state the specific gravity thus: specific gravity $\dfrac{15° C}{15° C}$ or specific gravity $\dfrac{25° C}{25° C}$. The temperature above the line is the temperature of the solution, and the temperature below the line is the temperature of the water taken as the standard. A specific gravity referred to water at its maximum density (4° C) is identical with the density of the solution. For a solution at 25° C this would be written specific gravity $\dfrac{25° C}{4° C}$, or density $\dfrac{25° C}{4° C}$.

The difference between the specific gravities, as measured on the various scales in common use, is at most about 4 units in the third decimal place. For commercial work in storage batteries, this small difference can generally be ignored. Many of the small and cheap hydrometers used for testing automobile batteries do not indicate the basis of their scale. But research and laboratory work require a more exact knowledge. In Table 20, conversion factors to the fourth decimal

TABLE 20. CONVERSION OF SPECIFIC GRAVITIES*

(The table contains factors which are to be used to multiply the observed reading of specific gravity or density to obtain the difference between two scales of measurement. This difference is to be added to or subtracted from the observed reading, according to the plus and minus signs, to obtain the reading converted to the desired basis of measure.)

Converted from	Converted to						
	Density $\dfrac{25^\circ \text{ C}}{4^\circ \text{ C}}$	Density $\dfrac{20^\circ \text{ C}}{4^\circ \text{ C}}$	Density $\dfrac{15^\circ \text{ C}}{4^\circ \text{ C}}$	Sp. Gr. $\dfrac{15^\circ \text{ C}}{15^\circ \text{ C}}$	Sp. Gr. $\dfrac{60^\circ \text{ F}}{60^\circ \text{ F}}$	Sp. Gr. $\dfrac{20^\circ \text{ C}}{20^\circ \text{ C}}$	Sp. Gr. $\dfrac{25^\circ \text{ C}}{25^\circ \text{ C}}$
Density $\dfrac{25^\circ \text{ C}}{4^\circ \text{ C}}$	0	+0.0001	+0.0002	+0.0011	+0.0012	+0.0019	+0.0029
Density $\dfrac{20^\circ \text{ C}}{4^\circ \text{ C}}$	−0.0001	0	+0.0001	+0.0010	+0.0011	+0.0018	+0.0028
Density $\dfrac{15^\circ \text{ C}}{4^\circ \text{ C}}$	−0.0002	−0.0001	0	+0.0009	+0.0010	+0.0017	+0.0027
Sp. Gr. $\dfrac{15^\circ \text{ C}}{15^\circ \text{ C}}$	−0.0011	−0.0010	−0.0009	0	+0.0001	+0.0008	+0.0018
Sp. Gr. $\dfrac{60^\circ \text{ F}}{60^\circ \text{ F}}$	−0.0012	−0.0011	−0.0010	−0.0001	0	+0.0007	+0.0017
Sp. Gr. $\dfrac{20^\circ \text{ C}}{20^\circ \text{ C}}$	−0.0019	−0.0018	−0.0017	−0.0008	−0.0007	0	+0.0010
Sp. Gr. $\dfrac{25^\circ \text{ C}}{25^\circ \text{ C}}$	−0.0029	−0.0028	−0.0027	−0.0018	−0.0017	−0.0010	0

* Based on data contained in *Natl. Bur. Standards, Circ.* 19.

place are given for the scales in most frequent use. The method of using this table may be seen from the following example: A hydrometer reading density $\dfrac{20^\circ \text{ C}}{4^\circ \text{ C}}$ is used to measure a solution of which the specific gravity $\dfrac{25^\circ \text{ C}}{25^\circ \text{ C}}$ is desired. If the density reading is 1.280, what is the specific gravity?

$$\text{Specific gravity } \frac{25^\circ \text{ C}}{25^\circ \text{ C}} = \text{density } \frac{20^\circ \text{ C}}{4^\circ \text{ C}} + \left(\text{factor} \times \text{density} \frac{20^\circ \text{ C}}{4^\circ \text{ C}} \right)$$

$$= 1.280 + (0.0028 \times 1.280) = 1.284$$

Specific gravities of storage-battery electrolytes are usually expressed to the third decimal place, one unit in the last place being called a "point." Thus solutions of 1.285 and 1.270 sp. gr. are said to differ by 15 points. Colloquial expressions, such as "twelve-eighty acid" or "thirteen hundred acid," refer to solutions of 1.280 and 1.300 sp. gr.

The specific gravity of the electrolyte decreases with increasing temperature. The temperature coefficient is not a constant quantity but varies with the specific gravity. In Table 21 the values of this coefficient per degree Centigrade and per degree Fahrenheit are given for each value of the specific gravity. The coefficient applies strictly to specific gravities referred to water at 15° C, but for practical purposes it may be used for the 25° basis also. This coefficient does not, however, convert specific gravities at $\dfrac{15°}{15°}$ to the $\dfrac{25°}{25°}$ basis. The familiar rule that there is a change of 1 point for each 3° F difference from the standard temperature is correct only for solutions of 1.150 sp. gr. For greater concentrations the correction is larger. At 1.280 sp. gr. it is 2 points for 5° F, or 3 points for 4° C. When the temperature of the solution is above the standard temperature, the correction is to be added to, and when below subtracted from, the observed specific gravity to adjust this to the correct reading at the standard temperature.

Specific gravities for storage-battery purposes are most conveniently measured by means of hydrometers. These are of several varieties, but all consist of a float of some kind. For stationary batteries, the hydrometer is immersed in the liquid of the cell itself, and some have a recording attachment. Hydrometers for use with portable batteries are usually of the syringe type (Fig. 47), which permits drawing a portion of the electrolyte from the cell into a tube of glass in which the hydrometer is confined. After the reading is completed the solution is replaced in the cell from which it was drawn.

A well-designed hydrometer should be made of smooth, transparent glass, circular in cross section along the axis, with fixed ballast, and a scale on ledger paper with graduation that is correct to within one-half of the smallest division.[6] It is desirable that the basis of the hydrometer's scale be indicated, as, for example, $\dfrac{15°\ C}{15°\ C}$ or $\dfrac{60°\ F}{60°\ F}$. Flat hydrometers that will float between the plates and the jar of stationary cells are also used. The hydrometer should float vertically. In making readings the eye should be level with the surface of the liquid, disregarding the curvature due to the surface tension, as shown in Fig. 48.

Cells in glass or other transparent jars are often provided with

[6] *Natl. Bur. Standards, Circ. 16,* 5.

TABLE 21. MEASUREMENT OF SULFURIC ACID SOLUTIONS

Specific Gravity, 15° C / 15° C	Specific Gravity, 25° C / 25° C	Temperature Coefficient of Sp. Gr. 15° C Per ° C	Per ° F	Per Cent H_2SO_4 by Weight	Per Cent H_2SO_4 by Volume	Baumé Degrees	Twaddell Degrees	Pounds per Cubic Foot	Kilograms per Liter
1.000	1.000	.....	.....	0.0	0.0	0.0	0	62.4	
1.010	1.009	(.00018)	(.00010)	1.4	0.8	1.4	2	63.0	
1.020	1.019	(22)	(12)	2.9	1.6	2.8	4	63.6	
1.030	1.029	(26)	(14)	4.4	2.5	4.2	6	64.2	
1.040	1.039	(29)	(16)	5.9	3.3	5.6	8	64.8	
1.050	1.049	(33)	(18)	7.3	4.2	6.9	10	65.5	
1.060	1.058	36	20	8.7	5.0	8.2	12	66.1	
1.070	1.068	40	22	10.1	5.9	9.5	14	66.7	
1.080	1.078	43	24	11.5	6.7	10.7	16	67.4	
1.090	1.088	46	26	12.9	7.6	12.0	18	68.0	
1.100	1.097	48	27	14.3	8.5	13.2	20	68.6	
1.110	1.107	51	28	15.7	9.5	14.4	22	69.2	
1.120	1.117	53	29	17.0	10.3	15.5	24	69.8	
1.130	1.127	55	31	18.3	11.2	16.7	26	70.5	
1.140	1.137	58	32	19.6	12.1	17.8	28	71.1	
1.150	1.146	60	33	20.9	13.0	18.9	30	71.7	
1.160	1.156	62	34	22.1	13.9	20.0	32	72.4	
1.170	1.166	63	35	23.4	14.9	21.1	34	73.0	
1.180	1.176	65	36	24.7	15.8	22.1	36	73.6	
1.190	1.186	66	37	25.9	16.7	23.2	38	74.2	
1.200	1.196	68	38	27.2	17.7	24.2	40	74.8	
1.210	1.206	69	38	28.4	18.7	25.2	42	75.4	
1.220	1.216	70	39	29.6	19.6	26.1	44	76.1	
1.230	1.225	71	39	30.8	20.6	27.1	46	76.7	
1.240	1.235	72	40	32.0	21.6	28.1	48	77.3	
1.250	1.245	72	40	33.2	22.6	29.0	50	78.0	
1.260	1.255	73	40	34.4	23.6	29.9	52	78.6	
1.270	1.265	73	41	35.6	24.6	30.8	54	79.2	
1.280	1.275	74	41	36.8	25.6	31.7	56	79.8	
1.290	1.285	74	41	38.0	26.6	32.6	58	80.4	
1.300	1.295	75	42	39.1	27.6	33.5	60	81.0	
1.310	1.305	75	42	40.3	28.7	34.3	62	81.7	
1.320	1.315	76	42	41.4	29.7	35.2	64	82.3	
1.330	1.325	76	42	42.5	30.7	36.0	66	82.9	
1.340	1.335	76	42	43.6	31.8	36.8	68	83.6	

Approximately equal numerically to the specific gravity; for example, a solution of 1.300 sp. gr. will weigh approximately 1.300 kg per liter.

Table 21. Measurement of Sulfuric Acid Solutions (*Continued*)

Specific Gravity, 15° C / 15° C	Specific Gravity, 25° C / 25° C	Temperature Coefficient of Sp. Gr. 15° C — Per ° C	Per ° F	Per Cent H_2SO_4 by Weight	Per Cent H_2SO_4 by Volume	Baumé Degrees	Twaddell Degrees	Pounds per Cubic Foot	Kilograms per Liter
1.350	1.345	.00077	.00043	44.7	32.8	37.6	70	84.2	
1.360	1.355	77	43	45.8	33.9	38.4	72	84.8	
1.370	1.365	78	43	46.9	34.9	39.2	74	85.4	
1.380	1.375	78	43	47.9	35.9	39.9	76	86.1	
1.390	1.385	79	44	49.0	37.0	40.7	78	86.7	
1.400	1.395	79	44	50.0	38.0	41.4	80	87.3	
1.410	1.405	80	44	51.0	39.1	42.2	82	88.0	
1.420	1.415	80	45	52.0	40.1	42.9	84	88.6	
1.430	1.425	81	45	53.0	41.2	43.6	86	89.2	
1.440	1.435	81	45	54.0	42.2	44.3	88	89.8	
1.450	1.445	82	46	54.9	43.3	45.0	90	90.4	
1.460	1.455	83	46	55.9	44.4	45.7	92	91.0	
1.470	1.465	83	46	56.9	45.5	46.4	94	91.7	
1.480	1.475	84	47	57.8	46.5	47.0	96	92.3	
1.490	1.485	85	47	58.7	47.5	47.7	98	93.0	
1.500	1.495	85	47	59.7	48.7	48.3	100	93.6	
1.510	1.505	86	48	60.6	49.7	49.0	102	94.2	
1.520	1.515	87	48	61.5	50.8	49.6	104	94.8	
1.530	1.525	87	48	62.4	51.9	50.2	106	95.4	
1.540	1.535	88	49	63.3	53.0	50.8	108	96.0	
1.550	1.545	89	49	64.2	54.1	51.5	110	96.7	
1.560	1.554	89	49	65.1	55.2	52.1	112	97.3	
1.570	1.564	90	50	66.0	56.3	52.6	114	98.0	
1.580	1.574	91	50	66.8	57.4	53.2	116	98.6	
1.590	1.584	91	51	67.7	58.5	53.8	118	99.2	
1.600	1.594	92	51	68.6	59.7	54.4	120	99.8	
1.610	1.604	93	51	69.4	60.8	54.9	122	100.4	
1.620	1.614	93	52	70.3	61.9	55.5	124	101.0	
1.630	1.624	94	52	71.2	63.1	56.0	126	101.7	
1.640	1.634	95	53	72.0	64.2	56.6	128	102.3	
1.650	1.644	95	53	72.9	65.4	57.1	130	102.9	
1.660	1.654	96	53	73.7	66.5	57.7	132	103.6	
1.670	1.664	97	54	74.5	67.6	58.2	134	104.2	
1.680	1.674	98	54	75.4	68.8	58.7	136	104.8	
1.690	1.684	99	55	76.2	70.0	59.2	138	105.4	

Pounds per Cubic Foot will weigh approximately to the specific gravity; for example, a solution of 1.300 sp. gr. will weigh approximately 1.300 kg per liter. Kilograms per Liter Approximately equal numerically to the specific gravity.

TABLE 21. MEASUREMENT OF SULFURIC ACID SOLUTIONS (*Continued*)

Specific Gravity, 15° C / 15° C	Specific Gravity, 25° C / 25° C	Temperature Coefficient of Sp. Gr. 15° C — Per ° C	— Per ° F	Per Cent H_2SO_4 by Weight	Per Cent H_2SO_4 by Volume	Baumé Degrees	Twaddell Degrees	Pounds per Cubic Foot	Kilograms per Liter
1.700	1.694	.00100	.00055	77.1	71.2	59.7	140	106.0	
1.710	1.704	101	56	77.9	72.4	60.2	142	106.7	
1.720	1.713	102	57	78.8	73.6	60.7	144	107.3	
1.730	1.723	103	58	79.7	75.0	61.2	146	108.0	
1.740	1.733	105	59	80.6	76.2	61.7	148	108.6	
1.750	1.743	107	60	81.5	77.6	62.1	150	109.2	
1.760	1.753	109	60	82.4	78.8	62.6	152	109.8	
1.770	1.763	110	61	83.4	80.2	63.1	154	110.4	
1.780	1.773	110	61	84.4	81.7	63.5	156	111.0	
1.790	1.783	111	62	85.6	83.3	64.0	158	111.7	
1.800	1.793	110	61	86.7	84.8	64.4	160	112.3	
1.810	1.803	109	61	88.1	86.7	64.9	162	112.9	
1.820	1.813	108	60	89.8	88.9	65.3	164	113.5	
1.830	1.823	106	59	91.8	91.4	65.8	166	114.1	
1.835	.....	105	58	93.2	93.0	66.0	167	114.5	
1.840	1.834	.00103	.00057	94.8	94.8	....	168	114.8	

Right margin note, running vertically: Approximately equal numerically to the specific gravity; for example, a solution of 1.300 sp. gr. will weigh approximately 1.300 kg per liter.

EXPLANATION OF THE TABLE

Column 1. Specific gravities of solutions are given at $\dfrac{15° C}{15° C}$, which is approximately equal to $\dfrac{60° F}{60° F}$. Hydrometers used for measuring sulfuric acid solutions are frequently standardized at 60° F.

Column 2. Specific gravities of solutions corresponding to those given in Column 1 are given at $\dfrac{25° C}{25° C}\left(\dfrac{77° F}{77° F}\right)$.

Columns 3 and 4. Contain the temperature coefficients to calculate the specific gravity for any temperature $t°$ from the specific gravities given for 15° C. Values are given for both Centigrade and Fahrenheit degrees. They have been calculated from Domke's tables, *Wiss. Abh. Normal Eichungs Kom.*, 5, 75 (1904). Let the required specific gravity at $t°$ be represented by S_t, the specific gravity of the solution at 15° by S_{15}, and the temperature coefficient by α, then

$$S_t = S_{15} + \alpha(15° - t°)$$

The temperature coefficient for temperatures other than 15°, which may be taken as standard, varies slightly with the temperature. For solutions of the specific gravities most commonly used in storage batteries, the values given in the table may be assumed to be correct to within 3 per cent for a range of temperature from 0 to 45° C. Values in parentheses were obtained by extrapolation.

Column 5. Per cent by *weight* of H_2SO_4 solutions corresponding to those given in Column 1. These were computed from Domke's tables, pp. 131–148.

Column 6. Per cent by *volume* of H_2SO_4 solutions corresponding to those given in Column 1. These were computed from the formula:

$$\text{Per cent by volume} = \frac{d}{1.840} \times \text{Per cent by weight}$$

Values of d, the specific gravity of solutions for given percentages of H_2SO_4 by weight, were based on Domke's tables, and the computations were checked by comparison with Pickering's contraction table, *J. Chem. Soc.*, *57*, 148 (1890).

Column 7. Baumé degrees corresponding to the specific gravities given in Column 1 calculated from the formula:

$$\text{Degrees Baumé} = 145 - \frac{145}{\text{Specific gravity} \frac{60° F}{60° F}}$$

Column 8. Twaddell degrees calculated from the formula:

$$\text{Degrees Twaddell} = 200 \left(\text{specific gravity} \frac{60° F}{60° F} - 1 \right)$$

Column 9. Pounds per cubic foot of solutions having the specific gravities given in Column 1.

Column 10. The kilograms per liter are equal numerically to the specific gravity, if it is assumed that the difference between the specific gravity $\frac{15° C}{15° C}$ and the density $\frac{15° C}{4° C}$ is negligible. For example, the

$$\text{Specific gravity} \quad \frac{15° C}{15° C} \text{ for 38 per cent acid is } 1.29027$$

$$\text{Density} \quad \frac{15° C}{4° C} \text{ for 38 per cent acid is } 1.28915$$

$$\text{Difference } 0.00112$$

The assumption is correct, therefore, to a unit in the third decimal place.

built-in "charge" indicators. These are floats which rise or fall with changing specific gravity of the electrolyte. A familiar type comprises several brightly colored balls of wax, which are adjusted to the desired specific gravities corresponding to the respective colors by incorporating with the wax the necessary amounts of a heavy insoluble substance such as barium sulfate. Precautions must be taken to prevent bubbles of gas from clinging to the surface of the balls if they are to operate satisfactorily.

The Baumé and Twaddell Scales

The standard Baumé scale, for liquids heavier than water, as used in the United States, is related to the specific gravity by the following equation:

$$\text{Degrees Baumé} = 145 - \cfrac{145}{\text{Specific gravity} \dfrac{60° \text{ F}}{60° \text{ F}}}$$

The number 145 is called the modulus. Confusion has resulted in the past from the variety of values assigned to it.

The Baumé scale is now obsolete in this country for storage-battery work, but it is frequently referred to in foreign books on the subject. The Twaddell

Fig. 47. Use of the syringe hydrometer.

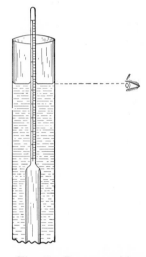

Fig. 48. Proper position of the eye for reading a hydrometer.

scale, extensively used in England, is related to the specific gravity in a more definite way. One degree Twaddell is equal to five units in the third decimal place of the specific gravity, from which it is apparent that the relation may be expressed by the equation:

$$\text{Degrees Twaddell} = \frac{\text{Specific gravity} - 1}{5} \times 1000$$

The use of specific gravities is preferable, in storage-battery work, to any of the arbitrary scales.

PREPARATION OF THE ELECTROLYTE

The electrolyte for lead batteries is prepared by diluting pure sulfuric acid with pure water. The acid is ordinarily sold as the concentrated acid of specific gravity 1.835 to 1.840 or partially diluted to 1.400. The latter is more bulky, which is a disadvantage in shipment, but it possesses a marked advantage in being easier to prepare for use. (See p. 108.) When the concentrated acid is diluted, the solution becomes very hot. The acid should always be poured into the water, never the water into the acid, because of the danger to the person making the mixture. Although the amount of heat evolved in either case is the same, the specific heats of water and of concentrated acid are quite different, as was shown in Table 13. A stream of water flowing into the concentrated acid causes the liberation of a great amount of heat, which, because of the low specific heat of the acid, causes a large local rise in temperature. Acid flowing into water cannot cause so great a rise in temperature, because the specific heat of the water is high. The solution should be stirred continually while the acid is being poured into the water, to prevent the heavier acid from flowing to the bottom of the vessel without mixing with the water. Suitable vessels for use in mixing or storing small quantities of the electrolyte are of china, vitreous earthenware, or glass. These are subject to cracking, and lead-lined tanks are preferred, especially for larger quantities. No metallic vessel other than lead should be used.

After the acid is diluted, it is necessary to wait until the mixture has cooled before it is poured into the battery, to avoid injury to the plates and separators. The cooling may be hastened by a jet of compressed air, but the air must be pure.

The great rise in temperature which occurs when the acid and water are mixed may be avoided altogether by using ice made from distilled water, instead of water. This is because the latent heat of fusion of the ice is approximately equal to the heat liberated by the dilution of the sulfuric acid. The ice, if drained of surplus water, may be added directly to the acid. In an experiment, 220 grams of ice were added to 98 grams of concentrated acid at room temperature. The temperature of the mixture was found to be $-2°$ C after all the ice had melted. The specific gravity of this mixture at 25° C was 1.225. Since the latent heat of fusion of the ice is 79.63 gram calories, the total heat absorbed by the ice was $220 \times 79.63 = 17,518$ gram calories. The total heat liberated by the dilution of the acid was, by interpolation of the values given in Table 13, 16,640 gram calories. The excess of heat absorbed indicates that the solution would be below 0° C, as was observed.

To facilitate the preparation of electrolytes of any required concentration, Fig. 49 shows the relative proportions of acid and water necessary. Manufacturers ordinarily furnish information as to the proper strength to be used in any particular battery.

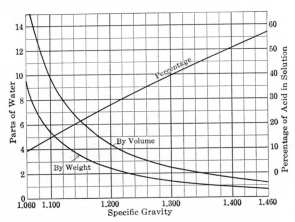

Fig. 49. Preparation of electrolyte of any specific gravity from concentrated sulfuric acid of 1.835 sp. gr.

It is often necessary to change the specific gravity of a solution of sulfuric acid by a given amount. The amount of water or acid that must be added to make the desired change can be calculated easily by using Table 21. For example: 290 pounds of electrolyte at 25° C are to be raised from 1.255 sp. gr. to 1.295 sp. gr. How much acid must be added?

By the table, 1.295 would contain 39.1 per cent acid or 60.9 per cent water, and 1.255 contains 34.4 per cent acid or 65.6 per cent water.

The excess of water is, by subtraction, 4.7 per cent; or

$$290 \times 4.7 \text{ per cent} = 13.63 \text{ pounds}$$

The amount of acid to be added is

$$\frac{100 \times 13.63}{60.9} = 22.4 \text{ pounds}$$

As another example, 340 pounds of electrolyte at 25° C of 1.325 sp. gr. are to be reduced to 1.285 sp. gr. How much water is to be added?

By the table, 1.325 contains 42.5 per cent acid, and 1.285 would contain 38.0 per cent acid.

The excess of acid is, by subtraction, 4.5 per cent; or

$$340 \times 4.5 \text{ per cent} = 15.3 \text{ pounds}$$

The amount of water to be added is

$$\frac{100 \times 15.3}{38.0} = 40.3 \text{ pounds}$$

CHOICE OF SPECIFIC GRAVITIES FOR BATTERY SERVICE

It is a matter of common observation that the specific gravity of the electrolyte in portable batteries is usually higher than in stationary batteries, but the choice does not depend entirely on space and weight considerations. Chemical reactions, temperature, character of service, and life of the battery are of importance in determining the proper specific gravity.

Enough acid in a given space within the cell must be provided to give the required output. The theoretical number of ampere-hours per liter has been calculated for various concentrations and is given in Table 18. It is not possible, however, to use all the acid in the cell. Space and weight requirements for portable cells do not permit of large volumes of electrolyte. In stationary cells space and weight are of less importance.

Chemical reactions, occurring within the cell during the time that it is at rest, place a limitation on the concentrations of acid that may be used. Local action (p. 134) increases rapidly when the concentration of the acid is increased. This is particularly true of the negative plate. Another chemical action taking place within the cell is that of the electrolyte on separators made of wood. These separators are destroyed by too strong acid. The action of 1.300 sp. gr. acid on the separator is much more marked than that of 1.250 and lower concentrations. The performance of batteries that are charged and discharged at frequent intervals, such as starting and lighting batteries or vehicle batteries, is not seriously affected by slight chemical actions resulting in the formation of lead sulfate. Batteries which are less frequently charged must be free from local action so far as possible.

The chemical reactions within the cell practically limit the higher concentrations to 1.300 sp. gr., and present tendencies are to use less acid.

The temperatures to which the battery is subjected in service have an important bearing on the specific gravity. Batteries exposed to low temperatures, such as automobile batteries in cold climates or airplane batteries, require a high density of acid to permit their capacity to be utilized without depleting their electrolyte to so low a specific gravity that freezing occurs. On the other hand, batteries for use in hot climates or on ships passing through the tropics require a lower specific gravity because of the increased chemical activity at the higher temperatures.

An extended investigation to determine the best concentration of the electrolyte for automotive batteries was made by Greenburg and Orsino,[7] as a result of which they recommended using electrolyte of 1.260 sp. gr.

The range of concentrations for different types of batteries when fully charged is approximately as follows:

Stationary batteries	1.200 to 1.225
Truck and tractor batteries	1.260 to 1.280
Starting and lighting batteries	1.260 to 1.280
Starting and lighting batteries in the tropics	1.200 to 1.230
Aviation batteries	1.260 to 1.285
Car-lighting batteries	1.210 to 1.230
Portable railway-signal batteries	1.220
Counter cells	1.210 to 1.250

PURITY OF THE ELECTROLYTE

Satisfactory service can only be obtained when the electrolyte is of a high degree of purity. It is necessary, therefore, that both the sulfuric acid and the water used in preparing the electrolyte should be pure. Distilled water is much to be preferred to natural water. The impurities contained in natural water may be small in amount, but their effect in the storage battery is cumulative because the evaporation which takes place results only in the loss of oxygen and hydrogen, which are the constituents of water, leaving the mineral and other impurities in the solution. Natural waters vary considerably in purity from one time of the year to another and in different localities. For Edison alkaline batteries, distilled water that has been kept in a closed container should be used exclusively to avoid carbonates. Rain water may be used in lead batteries if it is not collected from metallic roofs.

Limits for Impurities in Sulfuric Acid

Specifications for sulfuric acid, both in the concentrated form and when diluted for use in storage batteries, differ very greatly as to the amount of the various impurities considered allowable. In Table 22

[7] R. H. Greenburg and J. A. Orsino, Sulfuric acid and its effects on the storage battery, *Lab. Publ. 219–50,* National Lead Co. Research Laboratories.

are given figures taken from *Federal Specifications* OS801 and WB131e. If the amount of impurities allowed by specifications is made too small, a serious difficulty may arise in finding acid sufficiently pure to meet them. On the other hand, the specifications must limit the impurities to such amounts as are still within the limits for satisfactory operation of the batteries.

TABLE 22. PURITY OF SULFURIC ACID AND SOLUTIONS FOR
BATTERY USE

(Specific gravities are at 60° F.)

	Calculated as	Maximum Limits, Per Cent (Except as Noted)		
		Sp.Gr.1.835	49.50 to 50.5 Per Cent	Sp.Gr.1.280 (in new battery*)
Per cent H_2SO_4, not less than	....	93.19	50.0	36.8
Organic matter	....	†	†	†
Fixed residue	....	0.03	0.015	....
				Lead compounds only
Suspended matter	....	....	....	
Iron	Fe	0.005	0.003	0.012
Sulfurous acid	SO_2	0.004	0.002	....
Arsenic	As	0.0001	0.00005	0.00005
Antimony	Sb	0.0001	0.00005	0.0005
Manganese	Mn	0.00002	0.00001	0.00002
Nitrates	NO_3	0.00050	0.00030	....
Ammonia	NH_4	0.00100	0.00050	0.006
Chloride	Cl	0.00100	0.00050	0.012
Copper	Cu	0.00500	0.00300	0.005
Zinc	Zn	0.00400	0.00200	....
Selenium	Se	0.00200	0.00100	....
Platinum	Pt	†	†	†
Nickel	Ni	0.0001	0.00005	....

* Some manufacturers' specifications permit slightly more arsenic, antimony, and ammonia in new batteries. The upper limits are: As 0.0001; Sb 0.001; NH_4 0.012.

† To pass test, see Chapter 9.

It is assumed that, when the concentrated acid is diluted to a specific gravity of 1.400 (50 per cent) with pure distilled water, the percentage of impurities in the diluted solution will be approximately halved. There is no hard and fast rule for electrolytes in batteries. The limits given in Table 22 are about what may be expected for new batteries when fully charged. Some impurities tend to accumulate as the age of the battery increases, but not all of them remain in the electrolyte.

Platinum, silver, copper, arsenic, and antimony deposit on the negative plates; nitrates are reduced to ammonia; chlorides are eliminated as chlorine, in part at least, during charge; manganese is precipitated as manganese dioxide, but enough can remain in the electrolyte to do damage to organic materials; iron and ammonia remain in solution and are likely to increase with time.

Limits for Impurities in Water

Distilled water is to be preferred for use in storage batteries, but it is not always obtainable. The question then arises as to the maximum allowable impurities that natural water supplies may contain and still be safe for battery use. Table 23 is based on former Navy instructions.

TABLE 23. RECOMMENDED MAXIMUM ALLOWABLE IMPURITIES IN WATER
FOR STORAGE-BATTERY USE

Impurity	Calculated as	Parts per Million
Color		Clear and "white"
Suspended matter		Trace
Total solids		100
Calcium and magnesium oxides	CaO and MgO	40
Iron	Fe	5.0
Ammonia	NH_4	8.0
Organic and volatile matter	....	50
Nitrates	NO_3	10
Nitrites	NO_2	5.0
Chloride	Cl	5.0

The item that will exclude the greatest number of public supplies is probably chloride. The total solids and the oxides of calcium and magnesium are about in the proportions in which they occur in most waters, and therefore either one of them may be taken as the limiting factor. Iron is an important impurity, and the iron content of water as drawn from taps is not always shown by the analysis of the local water supply, because samples for the analysis are usually taken near the source, avoiding the mains and service lines. Another impurity, which is seldom mentioned, is manganese. This is perhaps rare in natural water supplies but is known to occur in river water which is acid, varying in certain localities from 0.2 to the highest observed, 10 parts per million. The limit on manganese for water to be used in batteries is about 0.00006 per cent.

Storage of water supplies for battery use in isolated communities needs consideration if harmful contamination is to be avoided. Preference is for unlined wood tanks of redwood, cypress, or cedar, in the order named. These should be filled with water and allowed to stand 1 week, after which they should be flushed and refilled.

Information about water supplies throughout the United States, including analyses of 670 public water supplies, has been published by the Geological Survey.[8]

Although the principle of ion exchange was discovered many years ago, recent years have witnessed a rapid development first with naturally occurring zeolites and later with synthetic resins for the purification of water and other solutions. The ion-exchange process is a method of removing positively or negatively charged ions by introducing large insoluble molecules that exchange like charges with the ions that are to be removed. The water is passed through a bed of the exchangers, and these in time require regeneration. With proper choice of the exchangers to be used, water can be highly purified and made entirely suitable for use in storage batteries as well as for other purposes.

Since natural waters usually contain bicarbonates, and purified or distilled water in contact with the air will come to equilibrium with the carbon dioxide present, the use of such water in storage batteries of the alkaline type is not recommended. For these, pure non-aerated water is preferred.

Local Action

Storage batteries of the lead-acid type are known to lose charge on standing as a result of local action at the plates. The amount should not be excessive. The rate of local action that may be expected of automotive batteries in good condition is approximately as follows:

Temperature	Loss per Day
100° F	0.003 sp. gr.
80° F	0.002 sp. gr.
50° F	0.0005 sp. gr.

Assuming that at most a loss of 75 points is permissible, this would require recharging the batteries at 100° F once in 25 days, at 80° F once in 37 days, and 50° F once in 5 months. If the specific gravity of the electrolyte in a battery is reduced to 1.170 after a full charge, less frequent charging will be required to maintain it in storage, perhaps once in 6 months at 70° F.

Cells in which local action has been reduced to a minimum were

[8] W. D. Collins, W. L. Lamar, and E. W. Lohr, The industrial utility of public water supplies in the United States, 1932, *Water Supply Paper 658*, Govt. Printing Office, 1934.

described by Rose and Zachlin.[9]   These cells were free from antimony in the grids and had thick, relatively soft plates designed for discharge at low rates over long periods of time.

Several extended investigations of local action and its causes have dealt with losses varying with temperature, composition of the grids, and the time of inactivity.   Zachlin[10] experimented with batteries in the range of -2 to $+47°$ C (28 to $117°$ F) and grids varying from 0.5 to 12 per cent antimony.   Local action increases with increasing temperature and antimony content.   Hoehne[11] investigated the behavior of batteries operated after periods of inactivity up to 400 days. He found lasting effects on both plates and separators, and some loss of capacity.

A certain amount of lead sulfate is formed as the result of local action whenever the plates are immersed in any sulfuric acid solution, even though it be the purest obtainable.   Detrimental impurities may (1) corrode the plate, (2) accelerate the formation of lead sulfate, or (3) be deposited in the pores of the plate.   In any event the weight of the plate changes, and this change affords the most sensitive and exact means we have for estimating the extent of the reaction.   This method was devised by Vinal and Ritchie[12] for measurements on positive and negative plates suspended in pure sulfuric acid solutions.

In order to obtain comparable results, it is necessary that the temperature be maintained at a constant value.   This was accomplished by immersing the glass jars containing the electrolytes in a large water bath thermostatically controlled to within about $0.01°$ C.   Two positive plates or two negative plates, suspended on glass hooks, were placed in each jar.   As a preliminary step the plates were given several cycles of charge and discharge, after which they were fully charged and then submerged in the electrolytes to be tested.   The electrolytes were saturated with lead sulfate because the previous work showed this to be necessary.

A sensitive balance mounted on a marble slab above the thermostat bath was used for weighing the plates while they were immersed.   Any

[9] C. C. Rose and A. C. Zachlin, Low-discharge cells, *Trans. Electrochem. Soc.*, *68*, 273 (1935).

[10] A. C. Zachlin, Effect of temperature on the rate of self discharge of lead-acid storage batteries, *Trans. Electrochem. Soc.*, *82*, 365 (1942); Self discharge of storage batteries, *ibid.*, *92*, 259 (1947).

[11] E. Hoehne, Das Verhalten gepasteter Bleisammler während u. nach längerer Nichtbenutzung, *Arch. Metallkunde*, *3*, 185 (1949).

[12] G. W. Vinal and L. M. Ritchie, A new method for determining the rate of sulfation of storage-battery plates, *Technol. Papers Bur. Standards*, *17*, 117 (1922), No. 225.

plate could be brought directly under the arm of the balance, as the jars containing them were carried on a revolving frame. The arrangement of the apparatus is shown in Fig. 50.

Fig. 50. Apparatus for determining the rate of sulfation of storage-battery plates.

Since the weighings were made of plates immersed in the electrolyte, a buoyancy correction was necessary. This correction was applied to the small difference of two weighings, and hence a slight error in the density of either the solutions or the active materials produced a negligible error in the final result.

Since the molecular weight of lead is 207.21 and that of lead sulfate 303.27, the gain in weight during the transformation of 1 mole of lead to lead sulfate is 96.06 grams; and the relative gain in weight is $\dfrac{96.06}{207.21}$

times the weight of lead acted upon. There is, of course, a large amount of lead that does not take part in the reaction and that decreases the sensibility of the weighings; but in spite of this fact and the heavy damping resulting from the viscosity of the electrolytes, the balance was sufficiently sensitive to permit an accuracy of a few tenths of a milligram to be obtained.

From the rate of change in weight of the plates, the equivalent loss in ampere-hour capacity can be computed. The valence is 2, and hence a gain in weight of the negative plate of 96.06 grams is equivalent to 2 × 96,500 coulombs. From this the change in weight is computed to be 1.79 grams per ampere-hour.

In Table 24 are given the measurements made on the negative plates of two manufacturers when the solutions were maintained at constant temperatures of 22, 25, and 30° C. The increase in the rate of sulfation from 22 to 25° C is proportionately greater than from 25 to 30° C.

TABLE 24. EFFECT OF TEMPERATURE ON THE RATE OF SULFATION OF NEGATIVE PLATES

Specific Gravity 25°	Gain in Weight, Grams per Hour, Saturated Solutions, for Plates of					
	Manufacturer B at			Manufacturer A at		
25°	22° C	25° C	30° C	22° C	25° C	30° C
1.150	0.0031	0.0048	0.0056	0.0012	0.0022	0.0022
1.200	0.0073	0.0118	0.0118	0.0016	0.0026	0.0027
1.250	0.0111	0.0178	0.0196	0.0028	0.0036	0.0051
1.300	0.0183	0.0283	0.0325	0.0046	0.0060	0.0082
1.350	0.0270	0.0385	0.0420	0.0122	0.0151	0.0196
1.400	0.0614	0.0906	0.1027	0.0350	0.0460	0.0544

Local action at the positive plate is essentially electrochemical, resulting from the difference in potential between the oxide and the lead-antimony alloy of the grid. Metallic impurities are not precipitated by the oxide, and Dolezalek and Finckh[13] have stated that no solubility of the dioxide in solutions of the specific gravities ordinarily used in storage batteries can be detected. The reaction at the positive plate in these experiments is therefore to be regarded as the general expression for the discharge of a cell, with the qualification that the negative electrode may be lead or lead-antimony alloy, depending on the construction of the plate.

$$PbO_2 + Pb + 2H_2SO_4 = 2PbSO_4 + 2H_2O$$

[13] F. Dolezalek and K. Finckh, Löslichkeit und Oxydations potential von Plumbisulfat und Plumbioxide, Z. anorg. Chem., 51, 320 (1906).

The gain in weight of the positive plates is not as simply related to the total amount of sulfate formed as it is with the negatives. The sulfate which forms on the grid or the underlying lead is equal to 96.06 grams for 2 × 96,500 coulombs, but the change in weight of the positive plates because of the formation of sulfate on the dioxide is 64.06 grams for the same quantity of electricity. The difference is accounted for by the loss of 1 mole of oxygen at the dioxide. This means that 160.12 grams of sulfate would be formed on the positive plate if the quantity of electricity flowing were 2 × 96,500 coulombs. The ampere-hour equivalent for the gain in weight resulting from local action is, therefore, 2.99 grams for each ampere-hour. The experimental results for the positive plates in solutions of varying concentration at 25° C are given in Table 25 for solutions saturated with lead sulfate.

TABLE 25. GAIN IN WEIGHT, GRAMS PER HOUR, OF POSITIVE PLATES IN SOLUTIONS AT 25° C SATURATED WITH LEAD SULFATE

Specific Gravity $\frac{25°}{25°}$	Gain in Weight for Plates of	
	Manufacturer A	Manufacturer B
1.150	0.0128	0.0077
1.200	0.0121	0.0077
1.250	0.0069	0.0047
1.300	0.0053	0.0050
1.350	0.0050	0.0048
1.400	0.0047	0.0049

It will be observed from the results given in Table 25 that the gain in weight of the positive plates decreases as the concentration of the solution increases. This suggests that the plates contained some soluble constituent, the solubility of which increased as the concentration increased, but the data available on the solubility of lead dioxide and lead sulfate fail to offer an explanation.

Effect of Various Impurities

The importance of obtaining exact information about the effect of impurities in storage-battery electrolytes arises from the detrimental effects which many of them produce on the operating characteristics and life of the storage battery, and such information is necessary as a basis for the preparation of specifications for sulfuric acid to be used in the batteries.

Determinations of the effect of many impurities have been made by the method of weighing the plates described above. The following data are taken from the papers by Vinal and Altrup,[14] and Vinal and Schramm.[15] A carefully measured quantity of the impurity was added to the electrolyte before the plates were immersed, and simultaneous measurements were made of the rate of sulfation of similar plates in pure solutions, the latter being designated as control experiments. The concentrations of the impurities are expressed as percentages by weight. An analysis of many control experiments makes possible the computation of the probable error of a single observation. For negative plates the probable error of a single observation varies from 0.15 gram at 50 hours to 0.60 gram at 500 hours; for positive plates the probable error is somewhat smaller.

Impurities Affecting the Negative Plates Only. These include metals (Table 26) that are deposited quickly in the metallic state and produce considerable gassing, and those chemical compounds that are reduced more slowly at the negative plates and result in little, if any, perceptible liberation of hydrogen. A closed circuit is formed between the lead of the plate and the impurity that is deposited upon it. Lead sulfate is formed in proportion to the quantity of electricity flowing and the plate gains in weight. Hydrogen is deposited on the surface of the metallic impurities. The potential required for the liberation of hydrogen on the various metals varies, but it is always in excess of the potential of the reversible hydrogen electrode. This excess is referred to as overvoltage. Hydrogen is liberated most easily on metals having a low overvoltage, such as platinum and the other metals of the platinum group. These are therefore the most harmful to the battery. Copper and tin having higher overvoltages are less harmful, and those metals having an overvoltage which places them above the discharge potential for hydrogen on lead, such as cadmium, zinc, and mercury, produce little or no effect. In some cases the deposition of hydrogen may result in the formation of other chemical compounds.

Platinum. Platinum has always been considered one of the most deleterious impurities, but it is not as common an impurity now as formerly, because sulfuric acid is no longer concentrated in platinum vessels. When negative plates are immersed in electrolyte containing even a very minute amount of platinum, gassing begins at once and the plate is rapidly discharged. The presence of 0.0001 per cent of platinum produced such violent gassing that the surface of the plate was

[14] G. W. Vinal and F. W. Altrup, Effect of certain impurities in storage-battery electrolytes, *Trans. Am. Inst. Elec. Eng.*, *43*, 709 (1924).

[15] G. W. Vinal and G. N. Schramm, *ibid.*, *44*, 288 (1925).

Storage Batteries

TABLE 26. LOCAL ACTION PRODUCED BY IMPURITIES AFFECTING
ONLY THE NEGATIVE PLATES

(Results are expressed as the gain in weight of a single plate in grams at
intervals from 50 to 500 hours.)

Impurity	Material Added	Per-centage Impurity	Time in Hours					
			50	100	200	300	400	500
None (Control experiments)		....	0.69	1.44	2.68	3.87	5.03	6.08
Platinum	PtCl₄	0.00001	0.7	1.4	3.0	4.8	6.8	8.4
		0.00003	13.2	19.3	26.1	29.9	32.4	34.2
		0.00005	27.4	28.1	28.8	29.2	29.3	29.5
Copper	CuSO₄	0.008	1.1	2.1	4.0	6.1	8.1	.....
Copper	CuSO₄	0.04	7.3	10.7	15.6	19.5	23.5	.....
Silver	Ag₂SO₄	0.1	13.5	18.6	24.0	.....	.....	.....
Tin	SnSO₄	0.1	4.6	7.0	9.4	11.0	12.5	13.9
Tungsten	WO₃	0.003	0.3	1.7	5.3	10.0	15.0	20.0
Bismuth	Bi₂O₃	0.2	4.5	5.8	8.2	.....	.....	.....
Sulfurous Acid	H₂SO₃	0.05	5.1	6.4	8.3	10.2	11.8	13.6
Sodium Bichromate	Na₂Cr₂O₇	0.05	3.3	5.2	8.4	11.4	14.1	.....
Arsenic*	As₂O₃	0.001	1.3	2.6	4.8	6.9	8.8	10.9
Arsenic	As₂O₃	0.10	0.8†	.....	.....	.....	.....	.....
Antimony	Sb₂(SO₄)₃	0.001	3.8	8.8	16.3	.....	.....	.....
Nitrates	HNO₃	0.001	1.3	2.0	3.6	.....	.....	.....
		0.004	3.1	4.0	5.2	.....	.....	.....
		0.008	5.3	6.4	7.7	.....	.....	.....
		0.035	23.0	25.3	27.3	.....	.....	.....

* These results are not as reliable as the others.
† At 55 minutes, plates gassing and solution turned brown, test abandoned.

apparently blasted off and much of the platinum was thereby removed,
which accounts for the fact that 0.00003 per cent acting more slowly
produced a greater discharge of the plates, as shown in Fig. 51. Negative plates contaminated with platinum are useless.

The curves of Fig. 51 make possible an estimate of the local action
in terms of the equivalent current that would be discharged by the
plate normally for the same rate of sulfation. The equivalent currents
are proportional to the slopes of the lines. The curve for plates in
pure acid shows the average equivalent current of the local action to
be 0.0059 ampere. With this as a basis, the equivalent currents during
the first part of the experiment for the other curves have been calculated to be as shown in the table on p. 141.

By this calculation, a physical meaning is given to the rather vague
term "local action."

Platinum Concentration, %	Current Equivalent of Local Action, amp
0.00001	0.0093
0.00002	0.0113
0.00003	0.107
0.00005	0.345
0.00010	1.71

Copper and Silver. Large amounts of copper and silver were added to the solutions as these metals do not produce such marked effects as

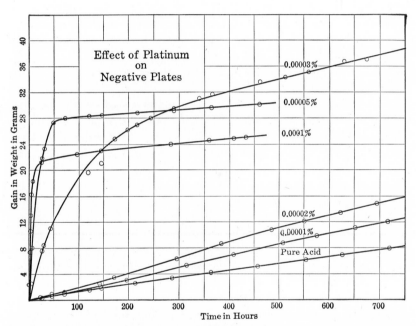

Fig. 51. Local action of plates in solutions contaminated with small amounts of platinum.

platinum. A considerable amount of the copper and silver was deposited on the plates as a spongy mass which afterward fell off. The gain in weight of the plates was therefore chiefly lead sulfate. The copper solutions, initially blue, lost color as the copper deposited, and an analysis of the electrolyte at the end of the experiment showed only a very small trace of copper remaining. Since the amount of solution to one plate in these experiments was approximately ten times that in a battery under operating conditions, it is apparent that a larger amount of copper or silver was deposited on a single plate than would be the case normally for the same concentration of these impurities. If more plates had been present, these impurities would probably have been deposited more uniformly and so would have been more effective. Such impurities as copper and silver cannot be eliminated by changing the electrolyte, but their effect may in some cases be mitigated, as will appear later.

Bismuth. Bismuth trioxide reacts with sulfuric acid to form bismuth sulfate, and this in turn is reduced at the negative plates to bismuth with the formation of an equivalent amount of lead sulfate. The bismuth is deposited as a brown powder on the plate. Twelve and a half grams of bismuth trioxide were added to the solution of sulfuric acid. This amount is equivalent to 24.4 grams of lead sulfate, but to this must be added the weight of the bismuth deposited in the pores of the plate, amounting to 11.2 grams, making the predicted increase in weight 35.6 grams. The two plates in the solution actually gained a total of 33.6 grams.

Arsenic and Antimony. Both these materials affect the negative plates, particularly antimony. The reactions of arsenic and antimony are probably analogous to those of bismuth, as the reduced material becomes visible after a short time. The presence of either antimony or arsenic in the electrolyte may be detrimental for other reasons also. Traces of arsine and stibine have occasionally been reported in the gas liberated from storage batteries when on charge.

Antimony, when present on the surface of negative plates, is a cause of local action which results in the formation of lead sulfate and the liberation of hydrogen. Jumau[16] recognized this in 1898, and two years later Strasser and Gahl[17] showed that hydrogen is liberated more easily on antimony than on lead, that is, the overvoltage for hydrogen on antimony is less. More recently the experiments of Crennell and Milligan,[18] Schubert,[19] and Vinal, Craig, and Snyder[20] have shown that corrosion of the positive grid in normal operation of a battery may supply enough antimony to the negative plates to affect their per-

[16] L. Jumau, Les accumulateurs électriques, p. 331, Dunod et Pinat, Paris, 1907.

[17] Strasser and Gahl, Über die Gaspolarisation im Bleiakkumulator, *Z. Elektrochem., 7,* 11 (1900).

[18] J. T. Crennell and A. G. Milligan, The use of antimonial lead for accumulator grids: A cause of self discharge of the negative plates, *Trans. Faraday Soc., 27,* 103 (1931).

[19] B. H. Schubert, Paper before Natl. Bat. Mfrs. Assn., Apr. 24, 1931.

[20] G. W. Vinal, D. N. Craig, and C. L. Snyder, Composition of grids for positive plates as a factor influencing the sulfation of negative plates, *Bur. Standards J. Research, 10,* 795 (1933), RP 567.

formance. As a result of each period of charge, antimony from the positive grids is deposited in minute amounts on the negative plates. This freshly deposited antimony is very active, but fortunately for the operation of the battery, successive charges and discharges of the cell cover the antimony previously deposited with lead or lead sulfate and thereby decrease its effect. Old plates contain more antimony in the active material than new plates and they are, therefore, subject to more local action, but the local action is usually not proportional to the amount of antimony carried in the active material. The maximum charging voltage of a cell without antimony (or other materials of low overvoltage for hydrogen) is normally more than for the ordinary cell, because polarization of the negative plates, which are cathodic during charge, is greater when antimony is absent. Maximum cell voltages on charge have been reported from 2.8 to 3.0 volts, but the value in any particular case will depend on current density, resistance, and other factors. This excessive voltage has been cited sometimes as a possible cause for corrosion and failure of positive grids which do not contain antimony. This cannot be the correct explanation, however, since the excess voltage is a result of higher polarization of the negative plates while the potentials of the positives are affected little if at all by antimony. This is illustrated in Table 27, which gives some typical conditions.

Studies of stibine, the hydride of antimony SbH_3, have extended our knowledge of this toxic gas. Sand, Grant, and Lloyd[21] made a study of the overpotential at antimony cathodes. Their work is without reference to storage batteries, but they obtained information of possible interest in battery operation. Using a commutator with 64 segments they could apply either direct or alternating current and found the surprising result that stibine is formed at lower current density with the alternating. Stibine does not form very readily, and fairly high current densities and high acid concentrations are necessary. Attempts to reproduce their results failed to confirm their conclusions.

Haring and Compton[22] made direct determinations of stibine produced during the charging of a battery. The amount of stibine formed increases abruptly when charging is completed and overcharging begins. It is produced coincidentally with hydrogen evolution. Requisites for the formation of stibine are the presence of metallic antimony and

[21] H. J. S. Sand, Julius Grant, and W. V. Lloyd, Overpotential at antimony cathodes and electrolytic stibine formation, *J. Chem. Soc.*, *130*, 378 (1927).

[22] H. E. Haring and K. G. Compton, The generation of stibine by storage batteries, *Trans. Electrochem. Soc.*, *68*, 283 (1935).

TABLE 27. EFFECT OF ANTIMONY ON PLATE POTENTIALS
AND CELL VOLTAGES

(Negative grids for these cells were all of the same commercial type.)

Description of Plates	Cycle Number	Per Cent Antimony Found in Active Material	Cell Voltage, v	Plate Potentials, Cadmium Readings	
				Positives, v	Negatives, v
Positive grids, pure lead	1	....	2.61	2.44	−0.15
	50	....	2.79	2.52	−0.24
	115	0.04	2.82	....	−0.25
Positive grids, 7% antimony	1	....	2.61	2.46	−0.13
	50	....	2.73	2.49	−0.22
	115	0.24	2.59	2.52	−0.06
Positive grids, 10% antimony	1	....	2.60	2.44	−0.13
	50	....	2.70	2.51	−0.16
	115	0.46	2.52	2.52	0.00
Positive grids, pure lead with new negative plates	...	....	2.64	2.49	−0.13
Same, 0.03% antimony added to the electrolyte	...	....	2.53	2.50	0.00
Positive grids, pure lead negatives after 2 years in service	...	....	2.51	2.49	0.00

nascent hydrogen. Plates having a deposit of antimony can be partially cleared by overcharging at a cell voltage of 2.55 volts or more. The gas then passes off as stibine.

The mechanism by which the negative active material becomes contaminated by antimony in the ordinary operation of a storage battery probably involves the formation of antimony trisulfate, $Sb_2(SO_4)_3$, directly or from the trioxide, Sb_2O_3, or from the trichloride, $SbCl_3$, if a trace of chlorine is present. At the negative plates reduction to metallic antimony occurs, and this deposits on the sponge lead.

Nitrates. Nitric acid added to the electrolyte was reduced at the negative plates and produced a marked increase in the rate of sulfation. Even so small a quantity as 0.001 per cent produced a measurable result. The reduction of nitric acid results in the formation of lead sulfate and ammonium sulfate. In the static tests, nitric acid was without effect on the positive plates, but this would not be the case in a battery in actual operation. The grids of the pasted plates would then be attacked. The use of nitrates as forming agents has been referred to previously.

Impurities Affecting the Positive Plates Only. These are chiefly organic compounds, such as starch, dextrose, sucrose, and extracts from wood separators.

The effects of acetic acid in storage batteries may appear to be contradictory, as they are not proportional to the amount of acetic acid present. Under some conditions, 1 per cent (or more, if the temperature is not too high) may be added to cells without deleterious effects, but under other conditions much smaller amounts will quickly destroy the positive plates. Experiments have shown that positive plates whose grids are adequately protected by a covering of lead dioxide are relatively immune to the action of acetic acid, but portions of the grid that are bare or covered only with lead sulfate are readily attacked. The difficulty in studying this problem lies in determining small amounts of acetic and other volatile organic acids in the presence of large amounts of sulfuric acid. Craig[23] has, however, developed a satisfactory method for determining acetic and formic acids in battery electrolytes.

Electrolytic effects of acetic acid and other materials yielding acetate ions occur during charging. Acetic acid, as an impurity in the electrolyte, has little effect on the rate of sulfation of plates standing in the electrolyte (see Table 28), but it has a very corroding effect on lead and lead alloys when these are made anodic.

Schreiner,[24] without discussing storage batteries, has shown that anodic oxidation of acetic acid in sulfuric acid solutions occurs as follows:

$$CH_3COOH \rightarrow CH_2OHCOOH \rightarrow CH(OH)_2COOH$$
(acetic acid)   (glycolic acid)   (glyoxylic acid)

$$\rightarrow COOHCOOH \rightarrow 2CO_2$$
(oxalic acid)   (carbon dioxide)

This represents the condition when the grids are fully protected by lead dioxide. Tests on cells fulfilling this condition showed that acetic acid, added to the electrolyte initially, is eliminated.

When grids are subject to attack, lead acetate is formed. This is soluble and increases the lead ions, with the result that considerable lead sulfate is deposited. The acetate ion, CH_3COO^-, migrates to the

[23] D. N. Craig, Determination of small quantities of volatile organic acids in sulfuric acid solutions, *Bur. Standards J. Research, 6*, 169 (1931).

[24] R. Schreiner, Anodische Oxydation der Essigsäure in Schwefelsäure Lösung, *Z. Elektrochem., 36*, 953 (1930).

(a)   (b)   (c)

Fig. 52. Effect of acetic acid on positive grid: (a) and (c) are grids destroyed by acetic acid; (b) is a magnified section of surface between grid and positive active material of a plate attacked by acetic acid, showing voids and sulfate spots.

TABLE 28. LOCAL ACTION PRODUCED BY IMPURITIES AFFECTING ONLY THE POSITIVE PLATES

(Results are expressed as the gain in weight of a single plate in grams at intervals from 50 to 500 hours.)

Impurity	Percentage of Impurity	Time in Hours					
		50	100	200	300	400	500
None (Control experiments)		0.34	0.51	0.66	0.86	1.05	1.15
Acetic acid	0.1	0.4	.....	.....	.....	.....	.....
	1.0	.....	0.9	1.6	.....	.....	.....
	3.0	.....	.....	.....	2.4	3.3	.....
Separator extracts (treated)	...	3.2	5.2	7.7	9.4	.....	.....
Separator extracts (untreated)	...	8.9	13.5	18.6	21.1	.....	.....
Dextrose	1.0	23.2	26.3	27.2	.....	.....	.....
Sucrose	1.0	23.6	26.7	27.6	.....	.....	.....
Invert sugar	2.0	23.6	26.4	26.8	.....	.....	.....
Starch	0.5	11.5	20.3	25.1	.....	.....	.....
Tannic acid	0.10	0.6	1.1	1.9	2.6	.....	.....

anode, which in this case is the positive plate of the battery, and attacks the lead repeatedly. Because the density of lead sulfate is less than that of lead, expansion occurs and the plate is said to "grow," while at the same time the grid loses its mechanical strength and ultimately disintegrates.

It is characteristic that the attack of acetic acid occurs in confined places, as between the grid and the active material. This is illustrated in Fig. 52. The acetic acid is less likely to attack the outer rim of the grid because this is covered, presumably, with lead dioxide. If the active material has pulled away from the grid, or if there are void spaces, or if formation is incomplete, areas are present which are subject to attack by acetic acid. When this occurs, the process is likely to continue.

Ethyl alcohol (C_2H_5OH) appears at first sight to be a most unlikely impurity to associate with the electrolyte of a storage battery, but a surprising number of inquiries have been made as to the effects that it produces. Alcohol finds its way into automobile batteries either as the result of a mistake when water should have been added or because of the impression that it will prevent freezing. Laboratory experiments have shown that alcohol, which is oxidized to acetic acid at the positive plates when the battery is charged, produces corroding effects similar to those of acetic acid.

Extracts from separators of cedar and cypress in both the treated and the untreated condition were made with sulfuric acid solutions of 1.250 specific gravity. The effects they produced led to experiments with other organic materials. The quantitative effects of materials affecting the positive plates are given in Table 28.

Impurities Affecting Both Positive and Negative Plates. These include several metals which may be present in the electrolyte in two states of oxidation, and chlorine.

Iron. Iron is perhaps the most common impurity. It is oxidized at the positive plates and reduced at the negative plate, resulting in the discharge of both. Experimental results on the effect of iron are given in Table 29.

When iron in the ferrous condition is added to the solution it is oxidized by the active material of the positive plates to ferric sulfate, accompanied by the formation of lead sulfate and water. The following equation for the reaction is assumed:

$$PbO_2 + 2FeSO_4 + 2H_2SO_4 \rightarrow PbSO_4 + Fe_2(SO_4)_3 + 2H_2O \quad (1)$$

The lead sulfate that is formed permits an accurate calculation to be made of the extent of the reaction from the gain in weight of the plates. The gain in weight of the positive plates must, however, be calculated as $PbSO_4 : (PbSO_4 - PbO_2)$, because the plate gains the sulfate radical SO_4 as the result of the reaction but loses simultaneously two oxygen atoms for each molecule of lead sulfate that is formed. The reaction expressed by equation (1) proceeds to completion if sufficient time is allowed. That is to say, all the ferrous sulfate is oxidized to the ferric condition, and beyond this point the rate of formation of lead sulfate is essentially the same as for plates in pure

Storage Batteries

TABLE 29. LOCAL ACTION PRODUCED BY IMPURITIES WHICH AFFECT
BOTH THE POSITIVE AND NEGATIVE PLATES
(Results are expressed as the gain in weight of a single plate in grams at
intervals from 50 to 500 hours.)

Impurity	Material Added	Percentage of Impurity	Time in Hours					
			50	100	200	300	400	500
(Positive Plates)								
None	(Control experiments)	.....	0.34	0.51	0.66	0.86	1.05	1.15
Iron	FeSO₄	0.012	0.7	0.9	1.1	1.3	.....	.....
		0.08	1.5	1.8	2.2	2.3	.....	.....
		0.4	5.6	6.8	7.2	7.5	.....	.....
Manganese	MnSO₄	0.08	2.9	3.7	4.4	5.0	.....	.....
		0.4	8.4	10.6	14.4	18.4	.....	.....
Chlorine	HCl	0.5	5.5	7.1	8.2	11.9	13.5	.....
	NaCl	1.00	23.7	25.4	26.0	26.2	.....	.....
(Negative Plates)								
None	(Control experiments)	.....	0.69	1.44	2.68	3.87	5.03	6.08
Iron	Fe₂(SO₄)₃	0.012	1.2	2.0	3.4	4.6	5.8	7.0
		0.08	6.3	8.0	9.5	10.8	12.2	13.6
Manganese	KMnO₄	0.04	2.2	3.0	4.3	5.6	.....	.....
		0.40	3.1	4.0	5.3	6.7	.....	.....
Chlorine	HCl	0.02	0.6	1.3	2.7	.....	.....	.....
	NaCl	1.00	22.0	27.1	30.2	32.7	.....	.....

acid solutions. The results of these experiments showed that the
curves representing data obtained from the iron solutions become
parallel to those for the pure acid solutions after about 150 hours.

TABLE 30. COMPARISON OF CALCULATED AND OBSERVED VALUES FOR
POSITIVE PLATES IN SOLUTIONS CONTAINING IRON

Amount of Iron Added		Equivalent Ferrous Sulfate,	Calculated Equivalent Lead Sulfate,	Observed Amount of Lead Sulfate,
Per Cent	Grams	g	g	g
0.4	22.5	61.2	61.2	60.2
0.08	4.5	12.2	12.2	11.8
0.012	0.675	1.8	1.8	3.4

We may, therefore, calculate the amount of lead sulfate that should
be formed and compare it with the amount determined by the weighings. Such a comparison is made in Table 30.
Since the reaction expressed in equation (1) came to a definite ter-

mination, this afforded an excellent opportunity to determine what the effect of introducing negative plates into the solution would be. This case represents the condition of a battery containing both positive and negative plates. One charged negative plate was immersed in each solution at the conclusion of 360 hours. These plates were not in elec-

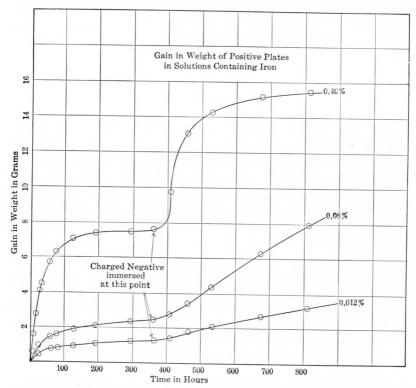

Fig. 53. Effect of iron in producing local action at the positive plate.

trical contact with the positive plates. The reduction of the iron to the ferrous condition began immediately, and the product in turn was reoxidized by the positive plates, accompanied by a further discharge. Curves showing the observations that were made are given in Fig. 53. The experiment was continued until 820 hours had elapsed. Iron has a destructive corroding action on the positive plates, particularly those of the Planté variety, because it increases the rate of sulfation.

The action of iron on the negative plates is much more pronounced than on the positives, and the local action produced is in excess of the amount that would be calculated from the reduction of the ferric sulfate. The effect is probably the result of two simultaneous reactions, which may be represented by the following equations:

$$Pb + Fe_2(SO_4)_3 \rightarrow PbSO_4 + 2FeSO_4 \qquad (2a)$$

$$Pb + H_2SO_4 \rightarrow PbSO_4 + H_2 \qquad (2b)$$

The amounts of iron added to the solutions were 4.5 grams and 0.675 gram. These are equivalent to 16.1 grams and 2.4 grams of ferric sulfate. On the basis of equation $(2a)$ these amounts will account for 12.2 and 1.8 grams of lead sulfate, respectively.

Table 29 shows the gain in weight of the plates in terms of the sulfate, SO_4, taken from the electrolyte. This gain in weight calculated to lead sulfate, $PbSO_4$, is in excess of what may be accounted for by the reduction of the iron salt from the ferric to the ferrous condition. This indicates that the presence of iron accelerates the reaction between sulfuric acid and lead, as represented by equation $(2b)$.

In the early stages of the experiment, when considerable ferric iron was present, this reaction was greatly accelerated. When 150 hours had elapsed, this accelerating effect of the iron seems to have died out. After 150 hours sulfate was being formed at a slightly greater rate on the plates in the solutions to which iron was added than in the pure acid. This effect is probably to be accounted for by the well-known slow spontaneous reoxidation of the ferrous sulfate by the air and its subsequent reduction by the negative plates during the long time that the experiment lasted.

Our knowledge of the reaction of iron in the storage battery has been extended by Lea and Crennell,[25] who have found that in addition to the discharge of the plates as a result of local action some permanent loss of capacity occurs which they attribute to adsorption of iron by the lead dioxide of the positive plates. They deny that iron accelerates the sulfation of negative plates as mentioned above, claiming that the self discharge of both positive and negative plates follows the simple theory of oxidation and reduction. However, Jumau[26] has stated that the effect of iron is greater on negative plates than on positives. His results support those of Vinal and Altrup. Lea and Crennell found that a definite tendency exists for iron to accumulate at the bottom of the cell. This may afford an explanation for the difference in condition occasionally observed between the upper and lower portions of positive plates.

[25] F. M. Lea and J. T. Crennell, Iron as an impurity in accumulators, *Trans. Faraday Soc.*, 23, 269 (1927).

[26] L. Jumau, Étude résumé des accumulateurs électriques, p. 93, 2d ed., Dunod, Paris, 1924.

Manganese. The experimental results obtained when negative plates were immersed in solutions containing manganese are shown in Table 29. It is at once apparent that the effects produced are not proportional to the amounts of manganese added. This is because of a reaction between the sulfuric acid and the potassium permanganate, which is independent of the reaction at the plates. The reaction between the permanganate and the 1.250 sp. gr. acid may be expressed by the equation:

$$4KMnO_4 + 6H_2SO_4 \rightarrow 2K_2SO_4 + 4MnSO_4 + 6H_2O + 5O_2 \qquad (3)$$

This is a slow reaction that may be demonstrated by a simple laboratory test, several hours being required to collect enough of the oxygen to make a satisfactory test. During the experiments the gas (oxygen) given off appeared in small amounts over the entire surface of the liquid. It was not localized at the plates.

The reactions that take place at the plates, in contradistinction to the above reaction, which occurs whether or not the plates are present, result in decolorizing the permanganate and in the formation of lead sulfate and manganese dioxide. The reactions are not fully understood, but the following equation, which is in accordance with the observed facts, is believed to represent the reaction:

$$2KMnO_4 + 4H_2SO_4 + 3Pb \rightarrow K_2SO_4 + 2MnO_2 + 3PbSO_4 + 4H_2O \quad (4)$$

A sludge which fell to the bottom of the jar was tested and found to be hydrated manganese dioxide. No gassing was visible at the plates.

Manganese was added to the solutions in which the positive plates were immersed, as manganous sulfate, $MnSO_4$. The solutions were initially colorless but began to show a purple coloration almost immediately after contact with the positive plates. This indicated the formation of permanganic acid, $HMnO_4$, probably according to the equation:

$$5PbO_2 + 2MnSO_4 + 3H_2SO_4 \rightarrow 2HMnO_4 + 5PbSO_4 + 2H_2O \qquad (5)$$

Although manganese may be thrown out of solution as relatively insoluble manganese dioxide, a small amount of manganese persisting in the solution as permanganate has a strong and destructive oxidizing action on organic matter, including both wood and rubber. The action occurs particularly at points where the organic matter is pressed into direct contact with lead dioxide of the positive plates. A few hundred thousandths of a per cent of manganese in the electrolyte has been found destructive.

Manganese causes wood separators to become bleached and thin;

the ribs are destroyed, and the web becomes perforated. If separators damaged in this way are dried, the side which was next to the positive plates usually has a characteristic silvery appearance. Rubber separators also are eaten away in a manner to shorten the useful life of the cell.

The characteristic pink to purplish coloration produced when the battery is on charge is a definite symptom of the trouble, but it cannot be readily observed when the cells are in opaque cases.

There are several possible sources of manganese in a storage battery. The sulfuric acid used in preparing the electrolyte may contain small amounts, but acid free from manganese in harmful quantities is obtainable. Material of the case is a more likely source. Sometimes the covers and bridges contain more manganese than the side walls. Iron and manganese often are found together in case material. Such material should be avoided.

Manganese occurs in some natural waters[27] which tend to be acid. Such water should never be used in storage batteries.

Chlorine and Chlorides. Hydrochloric is similar to nitric acid in its property of being a solvent for lead. Hydrochloric acid attacks both the sponge lead of the negative plate and the lead dioxide of the positive plate. In either case lead sulfate is produced. At the positive plate chlorine gas is liberated and this tends to mitigate the effects of chlorine in the battery, but many think that some chlorine is retained as a perchlorate when the battery is cycled. In testing battery electrolyte it is customary to use silver nitrate as an indicator for the presence of chlorine, but for total chlorine including both chlorides and perchlorates the Volhard test is necessary. Crennell and Milligan[28] made a study of the effect of chlorine as an impurity in storage batteries and found that it produced loss of capacity of both plates, rather more at the positive plates. The sludge was light-colored when chlorine was present. They did not detect the presence of perchlorates. In the same paper they reported that sodium ion was negligible in concentrations up to 1 per cent, but there is some evidence that it caused treeing. Salt water in a battery causes discharge with the formation of lead sulfate, sodium sulfate, and chlorine gas.

Shelley and Brown[29] reported that chlorine has a selective dis-

[27] W. D. Collins, W. L. Lamar, and E. W. Lohr, *loc. cit.*

[28] J. T. Crennell and A. G. Milligan, Effect of chlorine as an impurity, *Trans. Faraday Soc., 25,* 159 (1929).

[29] R. L. Shelley and O. W. Brown, Chlorine in the lead storage battery, *Proc. Indiana Acad. Sci., 42,* 123 (1933).

solving action on antimony of the positive grid. The antimony freed in this process deposits on the negative plates with well-known effects. Chlorine bleaches separators on the side next to the positive plates and shortens their life.

The Triad, Iron, Cobalt, and Nickel. The effects of iron in storage batteries have been discussed at some length previously, page 147. It affects both positive and negative plates. Cobalt reduces the polarization of the positive plates, lowering the required charging voltage, but it has a destructive effect on the separators. Nickel, on the other hand, reduces the polarization of the negative plates, without apparent harm to the separators. Nickel has been added to the battery electrolyte in some instances for the express purpose of lowering the required charging potential. Details of these effects are in a paper by Vinal, Craig, and Snyder.[30]

Ammonia. Ammonia may be present in the electrolyte as a result of absorption of ammonia gas by the electrolyte, or it may come from the use of ammonium sulfate in preparing the paste for pasted plates, or from the reduction of nitric acid. It exerts some forming action on the positive plate and has been said to be a cause of self discharge of both positive and negative plates. Tests have shown, however, that ammonia may be present in the electrolyte to the extent of even 50 parts in 100,000 without the battery's losing more than 8 per cent in capacity while standing during a period of 4 weeks after full charge.

Combinations of Impurities. Kugel[31] found that combinations of impurities such as tungsten and copper produced local action at the negative plates exceeding the effects of either of these materials singly. He advanced the theory that the polarization of the copper is decreased by the presence of the tungsten. Scarpa[32] has given a similar explanation, that the presence of the tungsten lowers the overvoltage, permitting the liberation of hydrogen on the surface of the copper. In either instance the local currents between the lead and the copper would be increased. Tungsten is an unusual impurity but furnishes an interesting example, as shown in Fig. 54. The more recent work of Vinal and Schramm[15] has revealed other combinations more likely to occur that result in the rapid discharge of the negative plates.

[30] G. W. Vinal, D. N. Craig, and C. L. Snyder, Note on the effect of cobalt and nickel in storage batteries, *J. Research Natl. Bur. Standards, 25,* 417 (1940).

[31] A. M. Kugel, Über die Selbstentladung negativer Akkumulatorplatten, *Elektrotech. Z., 13,* 8 and 19 (1892).

[32] O. Scarpa, Function of barium sulfate in accumulators, *Elettrotecnica, 5,* 371 (1918), and *6,* 176 (1919).

[15] *Loc. cit.*

TABLE 31. LOCAL ACTION PRODUCED BY COMBINATIONS OF IMPURITIES AFFECTING THE NEGATIVE PLATES

(Each experiment was started with the first named impurity, and then copper was added at the time shown.)

Combination	Material Added	Per-centage of Im-purity	Time of Adding Cu, hr	Time, hr					
				50	100	200	300	400	500
None (Control experiments)				0.69	1.44	2.68	3.87	5.03	6.08
⎰Tungsten	WO₃	0.003	246	1.1	2.4	5.4	23.1	25.6	26.7
⎱Copper	CuSO₄	0.05							
⎰Mercury	Hg₂SO₄	0.01	145	0.8	1.7	31.6	31.8	31.0	31.2
⎱Copper	CuSO₄	0.05							
⎰Molybdenum	MoO₃	0.01	145	1.1	2.0	30.5	.....	32.4	40.0
⎱Copper	CuSO₄	0.05							
⎰Zinc	ZnO	0.01	145	1.2	1.6	24.0	33.7	33.2	.....
⎱Copper	CuSO₄	0.05							
⎰Arsenic	As₂O₃	0.001	145	1.0	2.0	23.2	25.8	28.0	29.2
⎱Copper	CuSO₄	0.05							
⎰Antimony	Sb₂(SO₄)₃	0.001	145	3.6	8.8	28.0	38.5	38.0	36.0
⎱Copper	CuSO₄	0.05							

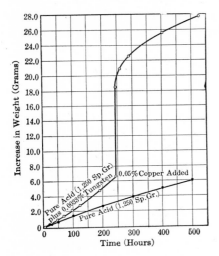

Fig. 54. Effect of a combination of copper and tungsten on negative plates.

These combinations all include copper as one constituent, the others being mercury, molybdenum, zinc, arsenic, and antimony, for which the detailed results are given in Table 31. The harmful effect of copper in a battery, therefore, depends more on the combinations with other impurities than on the effect produced by the copper alone.

Some local action is produced by the gases, oxygen and hydrogen, which are produced when the cells are on charge. The oxygen oxidizes the active materials of the negative plate, and the hydrogen reduces some of the lead dioxide of the positive plate. The extent of the reaction caused by these gases, when the electrolyte is saturated with them at the end of a charge, is small compared with the total capacity of the battery, but it probably accounts in part at

The Electrolyte 155

least for the fact that the most rapid rate of loss of charge is observed immediately after the charging period.

The Reduction of Sulfation

Frequent attempts have been made to reduce the sulfating action of sulfuric acid on the plates of storage batteries by the addition of various substances. Chief among these has been sodium sulfate, which is frequently sold under the name of Glauber salts. Schoop,[33] as far back as 1895, expressed the opinion that the use of sodium sulfate in the electrolyte was not beneficial. Its use appears to have originated with some experiments made in England by Preece. Neither is sodium sulfate a satisfactory substance for the removal of the sulfation from the plates. For this purpose Schoop suggested the use of sodium bisulfate, $NaHSO_4$, in place of the acid solution. When this is used, the acid electrolyte is poured out from the battery before the sodium bisulfate solution is added. It has been found possible to simplify the treatment for the removal of sulfation by merely pouring out the acid electrolyte and filling the cells with water. Lead sulfate is more soluble in water than in solutions of sulfuric acid as shown in Table 36. The cells are then put on charge at a low rate, provided they do not gas excessively, or at a current density that will maintain the voltage at the terminals of the cell at 2.3 volts, or less. A long time may elapse before a rise in specific gravity of the electrolyte is noted, but this should not be construed to mean that the battery is "not taking charge." Although the water treatment may not be the best in all cases, it has been found satisfactory, particularly for batteries of the starting and lighting type. Excessive sulfation and the water treatment are further discussed in Chapter 6.

Solid Electrolytes

Various attempts have been made to solidify the electrolyte in storage cells in order to eliminate spilling, and for other reasons. Materials that have been used in the past for making the electrolyte viscous have included albumen, starch, burnt clay, pumice, cellulose, soap, fatty acids, plaster of Paris, asbestos, sand, water glass, and fuller's earth. Of these various substances, water glass is perhaps the most suitable. The solidification of the electrolyte is brought about by the formation of silicic acid. The ratio of sodium oxide, Na_2O, to silica, SiO_2, varies somewhat, and the commercial silicate contains more silica than is indicated by the formula, Na_2SiO_3. The reaction for the sili-

[33] "Die Sekundär Elemente," *Encyklopädie der Elektrochemie*, IV, Part 2, p. 133, Wilhelm Knapp, Halle, 1895.

cate and sulfuric acid may be represented by the following formula:

$$Na_2SiO_3 + xH_2O + H_2SO_4 \rightarrow SiO_2 \cdot (x+1)H_2O + Na_2SO_4$$

It is necessary that both the acid and the water glass should be pure. Chlorides are perhaps the most common impurity occurring in the water glass. The time of setting before solidification takes place, and the stiffness of the jelly afterwards, are regulated by the proportions

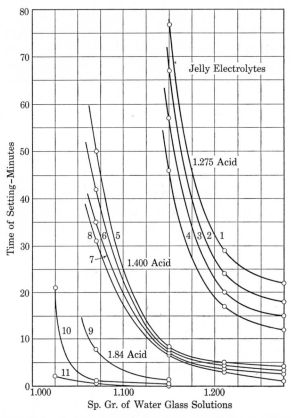

Fig. 55. Preparation of jelly electrolytes from solutions of sulfuric acid and water glass.

of the acid and the water glass. When thickening of the mixture begins, the final setting process occurs within a very few minutes. This mixture represents an interesting time reaction. If the mixture is made from dilute solutions, as, for example, sulfuric acid of specific gravity 1.275 and water glass of specific gravity 1.210, the greater the percentage of water glass in proportion to the acid, the more quickly the jelly sets and the more solid it becomes. The hard jellies are resonant.

It is possible to prepare the jelly as a clear, translucent, bluish mass which varies in consistency from a thick liquid to a fairly hard, resonant solid. The time of setting for various combinations is shown in Fig. 55. The curves are numbered from 1 to 11 and represent different proportions of the water glass and acid solutions measured by volume. These are as follows:

Curve 1, 5 parts 1.275 acid to 1 part of water glass
Curve 2, 4 parts 1.275 acid to 1 part of water glass
Curve 3, 3 parts 1.275 acid to 1 part of water glass
Curve 4, 2 parts 1.275 acid to 1 part of water glass

Curve 5, 5 parts 1.400 acid to 1 part of water glass
Curve 6, 4 parts 1.400 acid to 1 part of water glass
Curve 7, 3 parts 1.400 acid to 1 part of water glass
Curve 8, 2 parts 1.400 acid to 1 part of water glass

Curve 9, 1 part 1.84 acid to 4 parts of water glass
Curve 10, 1 part 1.84 acid to 3 parts of water glass
Curve 11, 1 part 1.84 acid to 2 parts of water glass

Batteries containing jelly electrolytes do not have as good electrical properties as those with the ordinary electrolyte. The internal resistance is higher and the capacity lower. They do not last well in service.

BATTERY ADDITIVES

Preparations in liquid and solid form sold to the public for rejuvenating worn-out, sulfated, or so-called "dead" batteries are frequently called battery additives. Since 1915 several hundred of these preparations have appeared with claims, more or less typical, that they extend battery life, reduce effects of sulfation, reduce operating temperatures, and some even claim to charge the batteries. A restriction is usually made that the batteries to be treated must be in sound mechanical condition. Beneficial results are also alleged if new or rebuilt batteries are similarly treated.

The most common additives consist of mixtures of sodium and magnesium sulfates in various states of hydration. These are added in small amounts to the acid electrolyte in each cell. Traces of other elements are often present as impurities in the major components or mixed with them.

Abundant test data are available to show that the performance of batteries treated with such materials is not significantly different from that of control batteries subjected to the same tests. The selection of batteries for test and the control batteries for comparison with them requires careful attention. Tests must be skillfully made according to well-recognized engineering principles.[33a] Simultaneous control tests are obligatory.

The ineffectiveness of small quantities of sodium hydrate, carbonate, or sulfate added to the usual sulfuric acid electrolyte was known to authorities more than fifty years ago. Wade,[34] in his well-known book published in 1902, said:

> Not only is it [the use of such material] supposed to hasten and make more perfect the recovery of badly sulfated cells, but it is sometimes said to retard sulfation in the first instance. Battery makers, however, seldom recommend its employment, and the opinion of many who have tried it is that it makes no difference whatever (this has been the Author's [i.e., Wade's] experience), nor are the specific examples of its success which have been put forward such as to carry much conviction. It would appear, therefore, that its effects, if any, are only felt under certain conditions which yet remain to be investigated.

The investigation which Wade foresaw as necessary to settle the controversial issues has now been made on a high level as a result of recent disputes. The Secretary of Commerce in 1953 requested the National Academy of Science to appoint a committee to appraise the work of the National Bureau of Standards on a battery additive of the sodium and magnesium sulfate type. After an extended investigation the committee reported[35] that the "material is without merit," confirming the Bureau's previous conclusion.

Other preparations have been essentially sulfates or phosphates of potassium, aluminum, ammonium, etc. In some additives, elements definitely known to be harmful to storage batteries have been found. These include copper, iron, nitrates, chlorides, etc. Various other elements in small amounts are sometimes present accidentally, and these often vary from one sample to another. Trace elements are sometimes alleged to be beneficial, but this is not supported by authoritative tests. A trace of manganese, for example, could be definitely harmful.

Other liquids and solutions which were a more common type of additive some years ago included solutions of sulfuric acid, with or without the addition of sulfates, and coloring matter. Even ordinary

---

[33a] Recognized tests are specified in the Standards of the Society of Automotive Engineers, p. 770 (1953), and in *Federal Specification, Batteries: Storage, Vehicular, Ignition, Lighting, and Starting*, W-B-131e (1953).

[34] E. J. Wade, *Secondary Batteries*, p. 337, published by the Electrician Printing and Publishing Co., London, 1902. (By permission.)

[35] *Report of the Committee on Battery Additives* of the National Academy of Sciences, National Research Council, Washington, Oct. 30, 1953.

water colored with a red dye was offered as a panacea for sulfated batteries. In principle its use was akin to the water treatment for sulfation mentioned in Chapter 6, but the price charged for the "water" was very high.

PROPERTIES OF ALKALINE ELECTROLYTES

The electrolyte for the alkaline storage batteries, in which the electrodes are nickel oxide and iron, is a solution of potassium hydroxide in water. To this solution, as used in the Edison batteries, is added a small amount of lithium hydroxide. The lithium hydroxide has a beneficial effect on the operation of the cells but is not necessary for the fundamental reactions that occur. The use of the lithium hydroxide is based on the results of experiment rather than on theoretical considerations. The lithia is considered essential for long life of the Edison battery, since a new battery without it has a falling life curve, but when the lithia is present the curve rises slowly for a considerable time before it begins to fall.

The concentration and chemical composition of the alkaline electrolytes, considered as a whole, do not change during the periods of charge and discharge. The measurement of the specific gravity is, therefore, of less importance than for sulfuric acid electrolyte used in the lead batteries. Occasional measurements of the specific gravity should be made, however, because there is a gradual weakening of the electrolyte, accompanied by a decrease in capacity of the battery to the point at which renewal is required.

The alkaline electrolytes for Edison cells are distinguished as "First-Fill Electrolyte," "Refill Electrolyte," and "Renewal Electrolyte." The first is a 21 per cent solution of potassium hydroxide in water with 50 grams of lithium hydroxide to the liter of solution. The second is also a 21 per cent solution of potassium hydroxide with x grams[36] of lithium hydroxide. The last is a 25 per cent solution of potassium hydroxide with 15 grams of lithium hydroxide per liter. The uses of these solutions are indicated, in a general way, by their names. The initial filling of the cells at the factory is done with the "First-Fill Electrolyte," with extra dry lithium hydroxide added according to the type of cell, to provide the correct amount per unit of the positive active material. The second kind of electrolyte is used mostly for export batteries, which after formation are shipped dry. The Refill electrolyte is also used to replace losses caused by spillage, or when it is necessary to replace the electrolyte because of impurities. The third or "Renewal" electrolyte is used when the previous solution has reached

[36] The amount of lithia is made approximately equal to the quantity found after formation in the original electrolyte.

the low limit of specific gravity. It is more concentrated than the first-fill, in order to compensate for the dilution caused by the old electrolyte held in the pores of the plates.

Before these electrolytes are placed in the cells, their normal specific gravities are as follows:

	Potassium Solutions, Sp. Gr. at 60° F
First-fill electrolyte	1.228 to 1.230
Refill electrolyte	1.212 to 1.216
Renewal electrolyte	1.248 to 1.250

The electrolyte for the nickel-iron batteries is shipped in steel containers. The smaller amounts, up to 22 pounds, are contained in cans similar to the containers for the cells. Larger amounts of 50 to 1100 pounds are shipped in steel drums.

Rebates are usually allowed for the return of the drums and carboys when they are empty.

When the electrolyte has decreased in the course of service to a specific gravity of 1.160, it should be renewed. In determining when this limiting value is reached, certain precautions must be taken. The electrolyte must be adjusted to the proper level and thoroughly mixed by giving the battery a full charge, and the sample taken must be free from gas bubbles at the time that the specific gravity is measured. Correction must be made for temperature if the measurements are made at other than the standard temperature of 60° F. Normally the electrolyte will require renewal two or three times during the useful life of the battery. A chemical analysis of the electrolyte or the water used for flushing the cells may sometimes be necessary.

Sufficient renewal electrolyte should be procured when the results of careful measurements indicate that renewal is necessary. When it is available, the battery should be discharged at the normal rate to zero voltage and then short-circuited in groups of not more than 5 cells for 2 hours. The old electrolyte is to be poured out while the battery is vigorously shaken to rinse the plates. It may be necessary to handle the cells separately, if they are large. Water should not be used for rinsing the plates. As each cell is emptied, the new electrolyte should be added immediately, and when all are filled the battery should be put on charge at the normal rate for 15 hours. During this operation observations of the temperature ought to be made at intervals, and if the battery should reach the limits set for normal operation (115° F) the charge should be interrupted until the battery has cooled.

Potassium hydroxide is variously called potassium hydrate, caustic potash, and potassa. It is a white solid substance which is very deliquescent and easily soluble in water. It dissolves the skin and many other organic substances. Potassium hydroxide, both in the solid state and in solution, absorbs carbon dioxide from the air. It is necessary to protect the solution in the battery, by gas valves in the vent plugs, from contamination by carbon dioxide. These valves are designed to prevent the ingress of gas but to permit the ready escape of gas generated within the cell. Potassium hydroxide may be prepared electrolytically from potassium chloride. Lithium hydroxide is obtained in solution by the action of lime on a solution of lithium carbonate. The solution is colorless and is very caustic. Crystals of lithium hydrate contain only 54 per cent lithium hydroxide.

Properties of Potassium Hydroxide Solutions

The data given in Table 32 are for pure solutions of potassium hydroxide without the addition of the lithium hydroxide. The amount of lithium hydroxide added is relatively small, and the electrolytes are, therefore, represented sufficiently well by the properties of the pure

TABLE 32. PROPERTIES OF POTASSIUM HYDROXIDE SOLUTIONS

| Per Cent | Density 15° C / 4° C | Resistivity | | Viscosity 18° C, centi-poises | Freezing Point, ° C | Specific Heat, cal | Grams of KOH per Ml of Solution |
		At 18° C ohm-cm	Temperature Coefficient				
1	1.008	....	......	1.08	− 1	0.98	0.0101
5	1.045	5.40	0.0186	1.17	− 3	.92	.0522
10	1.092	3.20	.0187	1.30	− 8	.87	.1092
15	1.140	2.34	.0190	1.48	−15	.83	.1710
20	1.188	2.00	.0196	1.72	−24	.80	.2376
25	1.239	1.86	.0206	2.05	−38	.77	.3097
30	1.290	1.84	.0220	2.50	−59	.74	.3870
35	1.344	1.96	.0240	....	....	.71	.4704
40	1.399	2.20	.0267	....	....	.69	.5596
45	1.456	2.56	.0298	....	....	....	.6552
50	1.514	....	......	....	....	....	.7570

solutions for which accurate information is available. The percentage composition is given for densities at 18° C, since most of the original material is given on this basis. To correct the observed results at any temperature to the standard temperature, 1 unit in the third decimal place is to be added for each 2 degrees Centigrade or 3 degrees Fahrenheit if the temperature is above the standard temperature, or subtracted if below it. These corrections are approximately the same as for the acid electrolyte used in the lead batteries.

Resistivity of Electrolyte

The resistivity of the alkaline electrolyte varies with the concentration, the temperature, and the amount of lithium hydroxide added to the solution. The electrical resistance depends on the resistivity, the length of the path, and the cross-sectional area through which the current flows, according to the equation $R = \rho(l/s)$, where R is the resistance, l the length, s the cross section, and ρ the resistivity. The unit of resistivity is the ohm-centimeter, which was explained in connection with the resistivity of sulfuric acid solutions on page 109. The resistivity of solutions of pure potassium hydroxide at 18°C (64.4°F) is given in Table 32. The resistivity of the solutions actually used is slightly higher than the figures given in the table, because of the addition of the lithium hydroxide. The percentage increase in the resistivity of a 21 per cent solution of potassium hydroxide containing 50 grams of lithium hydroxide per liter was determined by Turnock[37] to be 21 per cent. Smaller amounts produce effects shown in Table 33.

TABLE 33. PERCENTAGE INCREASE IN RESISTIVITY AND CAPACITY USING 21 PER CENT SOLUTION OF POTASSIUM HYDROXIDE WITH LITHIUM ADDED

Grams of LiOH, per Liter	Per Cent Increase in Resistivity	Per Cent Increase in Capacity
10	7.1	5.1
20	11.4	7.3
30	15.4	9.3
40	18.5	10.5
50	21.0	12.0

The addition of the lithium hydroxide increases the capacity of the cells, as stated on page 159, but it drives back the ionization of the potassium hydroxide according to the law of mass action and decreases the conductivity of the solution. By plotting the results given in Table 32, it will be seen that the shape of the resistivity-concentration curve for potassium electrolyte is somewhat similar to the resistivity curve for sulfuric acid. Minimum resistivity is attained at a concentration corresponding to a density of 1.270.

Freezing Points

The freezing points of potassium hydroxide solutions without lithium are given in Table 32.

[37] L. C. Turnock, Effect of lithium upon the capacity of the Edison storage battery, *Trans. Am. Electrochem. Soc.*, *32*, 405 (1917).

Concentration Limitations

As with lead batteries, there are rather definite limitations to the concentration of solutions that can be used successfully in the alkaline cells. Below 1.200 the resistivity begins to increase rapidly, which would impair the electric output. The cells tend to become sluggish in a weak solution and fail to give their rated capacity. For several reasons, therefore, the lower permissible limit of density has been set at 1.160 for Edison batteries. High concentrations, on the other hand, are also detrimental because of the increased solubility of the iron electrode which becomes noticeable at the higher temperatures. High concentrations, that is, above 1.270, result in increased resistivity. Aside from the matter of cost, therefore, the proper density of the potassium hydroxide electrolyte is the result of compromise and is fixed within rather narrow limits.

Effect of Impurities

Carbonates formed in the electrolyte as a result of the absorption of carbon dioxide from the air or introduced in the water used for flushing the cells are detrimental to the operating characteristics of the battery. Carbonates, when present, result in increasing the resistivity of the solution. The maximum permissible limit is usually stated to be 50 grams of carbonate per liter of electrolyte. The presence of carbonates has been assigned as one of the causes of sluggishness, but no quantitative data are available on this point.

Acid radicals are detrimental to the positive plates. It is quite obvious that the density or specific gravity of the alkaline solutions should not be measured with the same hydrometer that is used to measure acid solutions, unless it is washed entirely free from acid. Good practice requires a separate hydrometer for each kind of electrolyte.

Metallic impurities more positive than iron would deposit on the active material of the negative plate and tend to produce local action. In that event the iron would be oxidized and hydrogen would be liberated at the surface of the impurity. The detrimental effects of these impurities are mitigated in many cases by several factors. If the overvoltage for hydrogen discharge at the surface of the impurity is in excess of the potential for the discharge of hydrogen on the iron itself, local action cannot occur. Many metals form insoluble hydroxides in the alkaline electrolytes, and these will not be effective in producing local action.

Iron is very nearly insoluble in the electrolyte under normal operating conditions but may be present in the electrolyte if it is too concentrated or at too high a temperature. Iron in the electrolyte affects the positive plate resulting in a loss of capacity. At high temperatures hydrogen may be produced as a result of a reaction between iron and the electrolyte.

Modifications of the Electrolyte

The principal modification of the potassium hydroxide solutions used in alkaline batteries is the addition of lithium hydroxide, mentioned above, for use in Edison batteries. Vail[38] mentions efforts to make the electrolyte unspillable. A viscous mixture of sodium metasilicate and sodium hydroxide was used. This should not be confused with the formation of a gel sometimes employed in lead-acid batteries. In any kind of battery, whether of the lead-acid or alkaline type, the diffusion of the electrolyte is an important factor. It is hardly to be expected, therefore, that the batteries could operate as satisfactorily with a viscous or solid electrolyte as with the ordinary solutions.

Alkaline Electrolyte for Nickel-Cadmium Batteries

A solution of caustic potash in distilled water having a specific gravity of 1.210 is normal. This is without the addition of lithium hydroxide specified for use in Edison cells. During charge and discharge there is no appreciable change in density of the solution, but a gradual change occurs over a long period of time, and change to fresh electrolyte is indicated when the specific gravity has fallen to 1.190 as the lower limit for good operation. The change becomes mandatory at 1.170. To make the renewal, discharge the cells at the 7-hour rate to a voltage of 0.5 volt per cell and empty out the electrolyte, immediately refilling with "Renewal" electrolyte of 1.250 sp. gr., and charge at the 14-hour rate. "Refill" electrolyte of 1.225 sp. gr. is used to replace spillage or to fill cells in the dry condition for export. For the last, the cells should stand 24 hours after filling before charging at the 7-hour rate for a period of 14 hours.

Specific-gravity measurements should be referred to the normal state: $\frac{3}{4}$ inch above plates at 72° F. To make corrections, add 0.005 to the specific gravity reading for every $\frac{1}{2}$ inch above tops of plates and 0.001 for every 4° F above 72° F (subtract if below 72° F).

Opinions differ as to how much carbonate is permissible in the electrolyte, but this may depend in part on the conditions of service. Some place the limit at 30 grams per liter of solution while others may

[38] J. G. Vail, *Soluble Silicates in Industry,* p. 140, Chemical Catalog Co., New York, 1928.

permit as high as 90 grams per liter in batteries operated at currents not above the normal rate. Hauel[39] has found, as a result of single-electrode measurements, that the detrimental effects of carbonates are at the negative plates whereas the positives are unaffected. She concludes, therefore, that poorly conducting layers of $CdCO_3$ form over the active materials of the negative plates and produce the sluggishness characteristic of cells containing excess carbonates.

BIBLIOGRAPHY

Lunge, *Manufacture of Acids and Alkalis*, Gurney and Jackson, London, 1923.

Martin and Foucar, *Sulphuric Acid and Sulphur Products*, Appleton and Co., New York, 1916.

Sullivan, *Sulphuric Acid Handbook*, McGraw-Hill Book Co., New York, 1918.

De Wolf and Larison, *American Sulphuric Acid Practice*, McGraw-Hill Book Co., New York, 1921.

"Domke's Tables for Sulphuric Acid," *Wiss. Abh. Normal Eichungs Kom., 5,* Julius Springer, Berlin, 1904.

Partington, *The Alkali Industry*, Van Nostrand, New York, 1919.

"Testing of Hydrometers," *Natl. Bur. Standards, Circ.* 16.

"Standard Density and Volumetric Tables," *Natl. Bur. Standards, Circ.* 19.

"Standard Baumé Hydrometer Scale," *Natl. Bur. Standards, Circ.* 59.

Wells and Fogg, "Manufacture of Sulphuric Acid in the United States," *Bur. Mines, Bull.* 184.

Fairlie, *Sulphuric Acid Manufacture*, Reinhold Publishing Corp., 1936.

[39] Anna P. Hauel, The cadmium-nickel storage battery, *Trans. Electrochem. Soc., 76,* 435 (1939).

4

Theory of Reactions, Energy Transformation, and Voltage

ELEMENTARY THEORY OF ELECTRIC CELLS

The simplest form of cell may be made of a strip of copper and one of zinc immersed in water acidulated with sulfuric acid. If the zinc is sufficiently pure to be free from local action, there will be no visible effect until the zinc and copper are connected by a wire. The strips are, however, at different potentials with respect to each other, and when they are connected by a wire a current of electricity will flow in the wire. As this action progresses the strip of zinc will pass into solution and bubbles of gas will form at the copper electrode and collect on its surface. This gas is hydrogen which is formed from the electrolyte. The electric current flows from the copper strip through the wire to the zinc strip and from the zinc strip through the electrolyte to the copper.

As the action of this cell progresses, the sulfuric acid of the electrolyte is gradually replaced by zinc sulfate which is formed at the dissolving zinc electrode. This diminishes the voltage of the cell. A decrease is caused also by the collection of gas bubbles on the copper. Both effects produce "polarization." Such a cell is of little practical importance.

The Daniell cell, devised in 1836, is a modification of the foregoing and possesses important advantages since it eliminates the decrease in voltage caused by polarization. This cell consists of a strip of copper immersed in a solution of copper sulfate and a zinc strip in a solution of zinc sulfate, the two liquids being separated by a porous diaphragm to prevent mixing. As the action of this cell progresses the zinc electrode dissolves, forming more zinc sulfate, and the copper electrode is increased by the deposition of copper from the electrolyte. The Daniell cell will be referred to in the pages that follow, because it is useful in explaining parts of the theory.

Ions and Electrons

The current is transported through the electrolyte by particles of molecular size that carry electric charges. These are called "ions,"

166

a term that was originated by Faraday well over a hundred years ago. The ions are formed from the electrodes or from the substance that is dissolved in water to make the electrolyte. When zinc sulfate, for example, is dissolved in water, a spontaneous decomposition of a certain proportion of the zinc sulfate molecules takes places. The products are zinc ions and sulfate ions. This process is called "electrolytic dissociation." The zinc sulfate in the molecular state is electrically neutral, but as the molecules dissociate, positive and negative charges in equivalent amounts appear on the ions. In general, the metallic ions and hydrogen ions carry the positive charges; the non-metallic elements and radicals carry the negative charges. The sign of the charges has been ascertained by the direction in which the various ions move when between electrodes of opposite polarity, as in a simple electrolytic cell.

The electron is conceived to be a discrete and definite charge of negative electricity. One or more electrons surrounding a nucleus of equivalent positive charge are a part of each atom. The electrons are thought to be in shells at definite energy levels. Those in the outer shell play an important part in electrochemical processes. As long as an element has its normal number of electrons it is electrically neutral, but if it loses one or more electrons it becomes ionized, charged positively. Other elements may take on one or more electrons and become negatively charged ions. The alkali metals are characterized by a single electron in the outermost energy level. This is easily removed and ions formed. The reaction is monovalent. Zinc and cadmium, for example, have 2 electrons in the outer shell. Losing these, they become ions with 2 positive charges. The reaction is divalent. Quite differently, the halogens have 7 valence electrons in their outer shell. None of these are easily lost; rather than such a reaction they take on an electron, becoming ions with a negative charge. The charges carried by all ions are simple multiples of a unit charge.

We may write an equation for the ionization of such a substance as zinc sulfate, as follows:

$$ZnSO_4 \leftrightharpoons Zn^{++} + SO_4^{=}$$

The metallic ion, Zn^{++}, differs from the zinc atom both physically and chemically. Compared with an atom, it lacks 2 electrons; it is in solution; it moves under the influence of a potential gradient; and, finally, it has a marked tendency to attract and combine with oppositely charged ions. Probably the ions of opposite charges are combining and being formed continuously, but the proportion present at any time is determined by equilibrium conditions.

Ions of the same kind are identical in their properties. No difference is observable between the zinc ions formed from a zinc electrode and those formed by the dissociation of zinc sulfate. All the hydrogen ions in a storage battery behave in the same way, irrespective of whether they came from dissociated molecules of water or sulfuric acid. The same is true of the lead ions, which can be derived from the lead sulfate or from the negative plate.

The motion of an electric charge constitutes an electric current. The motion of an ion in the electrolyte, or of an electron in a wire, is a transfer of a definite amount of electricity from one point to another. The actual amount of electricity transported by a single electron or a monovalent ion is very small. Its value has been found to be

$$1.60186 \times 10^{-19} \text{ coulomb}$$

The number of ions required to transport 1 ampere flowing for 1 second (1 coulomb) is, therefore, enormously great. It is preferable to deal in magnitudes more easily comprehended. As a unit there has been chosen the aggregate charge carried by a number of ions equal in weight to the number of grams representing the atomic or molecular weight of the ion, which for convenience is designated as a "gram-ion." If the ion in question is monovalent, the total charge in round numbers is 96,500 coulombs; if divalent, the charge is twice this amount, and so on. This is further discussed under Faraday's law on page 173.

In a storage battery there are many kinds of ions present to carry the current. The sulfuric acid dissociates by two steps into hydrogen ions H^+ and sulfate ions $SO_4^=$, there being more of the hydrogen ions. There may be also ions of the composition HSO_4^- from the initial step in the dissociation process. The water present contributes hydrogen ions, H^+, and hydroxyl ions, OH^-. A few lead ions, including both the divalent and the tetravalent states, are present. These are Pb^{++} and Pb^{++++}. According to some theories, there are also lead dioxide ions, $PbO_2^=$. The lead and lead dioxide ions are relatively unimportant in carrying the current because there are few of them, but they are of great importance in fixing the potential differences between the plates and the electrolytes.

Potential Differences

When a strip of zinc is immersed in a solution of zinc sulfate, as in the Daniell cell, zinc ions carrying positive charges are thrown into the solution. The strip of zinc becomes negatively charged because of an excess of electrons that remain after the departure of the positively charged ions from its surface. This reaction cannot continue indefinitely, since the positive charges of the ions in solution repel the positively charged ions leaving the surface of the zinc. The latter are also held back by the attraction of the negatively charged zinc strip. The magnitude of the charges on the strip of zinc and of the surrounding electrolyte increases as more and more zinc ions are formed, until a state of equilibrium is eventually reached when the solution pressure of the metal is balanced by the electrostatic forces. This prevents the further formation of zinc ions. Figure 56 shows diagrammatically the double layer of charges. The ions that pass into the solution from the strip of zinc increase the osmotic pressure of the zinc ions that were formed from the zinc sulfate when it was dissolved in the water.

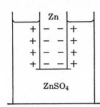

Fig. 56. Zinc in a solution of zinc sulfate becomes negatively charged.

The copper electrode of the Daniell cell is immersed in a solution of copper sulfate. The solution pressure of the copper is very small as compared with the osmotic pressure of the copper ions in the electrolyte. The result is a deposition of copper ions upon the surface of the strip of copper. When these ions are deposited their deficiency in electrons is imparted to the copper electrode. The electrode thereby acquires a positive charge, and the solution becomes negatively charged because of the excess of negative ions that remain. A double layer is again formed, as shown in Fig. 57. The equilibrium condition is reached when the electrostatic forces prevent the further deposition of copper ions on the electrode.

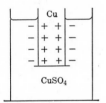

Fig. 57. Copper in a solution of copper sulfate becomes positively charged.

The result of practical importance that arises from the formation of the zinc ions at the zinc electrode and the deposition of copper ions at the copper electrode is the potential difference between the electrodes and the electrolyte. When the strip of zinc is immersed in a solution containing 1 gram-ionic weight of zinc ions, the electromotive force at the surface of contact between the solution and the zinc attains an equilibrium value of about 0.76 volt in a direction corresponding to the passage of current from the metal into the solution. The actual potential of the zinc with respect to the solution is negative. The copper, on the other hand, in a similar solution of copper ions, has a potential difference of 0.34 volt in a direction corresponding to the passage of the current from the solution to the metal. The copper is positive with respect to the solution. If a closed circuit between the zinc and the copper is formed, and at the same time the electrolytes are brought into contact through a porous diaphragm that will permit the passage of ions without allowing the two solutions to mix, a cell having a voltage equal to the algebraic difference of the potentials of the two electrodes with respect to the solution is formed. In the case of the Daniell cell, the combined voltage is $0.34 - (-0.76) = 1.10$ volts, neglecting the small potential difference that exists at the surface of contact between the solutions.

The polarity of the cell is defined by the polarity of the individual electrodes with respect to the electrolyte and, therefore, their potential relations to each other. In the Daniell cell the positive is the copper, although this is sometimes called the electronegative metal. In dealing with the storage battery, in which the current flows in opposite directions when on charge and on discharge, there is further chance for confusion. Fortunately, however, the brown dioxide plates are so universally called the positives, and the gray, metallic sponge-lead plates, the negatives, that this terminology may be considered fully standardized. When the storage cell discharges, the dioxide plate is the positive in the same sense as the copper electrode of the Daniell cell.

The Electrochemical Series

Lists of the elements arranged in the order of increasing potentials have been given by many authorities, but these differ somewhat both as to the order of the elements and the values assigned for their potentials. Single-electrode potentials are always measured in connection with some standard electrode. The hydrogen electrode is generally chosen as the standard, and its single potential is stated arbitrarily to be zero. Although the hydrogen electrode is the standard, its use is not always convenient or possible. Other electrodes that are used with storage batteries are described in connection with the measurement of plate capacities in Chapter 5.

A selected list of elements, with the approximate values of their potentials on the hydrogen scale, is given in Table 34. Metals following lead, such as copper, silver, and platinum, are the elements that deposit on the negative plates of storage batteries; but the elements that precede it will not be deposited under ordinary circumstances. The position of the various elements with reference to lead has a bearing on the effects they produce when present as impurities in the electrolyte. This is discussed in Chapter 3.

A more complete list of the elements, showing the probable order in the electrochemical series, is as follows: potassium, barium, strontium, calcium, sodium, magnesium, aluminum, manganese, zinc, iron, cadmium, cobalt, nickel, tin, lead, hydrogen, antimony, bismuth, arsenic,

TABLE 34.   ELECTRODE POTENTIALS

(Calculated for electrochemical reactions of elements immersed in solutions containing 1 gram-ion per liter.   The potentials are expressed as the potential of the electrode minus the potential of the solution.)

Element	Valence	Potential in Volts
Potassium	1	−2.92
Calcium	2	−2.87
Sodium	1	−2.71
Magnesium	2	−2.37
Manganese	2	−1.18
Zinc	2	−0.76
Iron	2	−0.44
Cadmium	2	−0.40
Nickel	2	−0.25
Tin	2	−0.14
Lead	2	−0.12
Iron	3	−0.04
Hydrogen	1	0.00
Antimony	3	+0.10
Copper	2	+0.34
Copper	1	+0.52
Silver	1	+0.80
Mercury	2	+0.80
Platinum	2	+0.98
Gold	3	+1.50

copper, silver, mercury, palladium, platinum, gold, silicon, carbon, boron, chlorine, sulfur, oxygen.

In using the list of elements, the order cannot be considered to hold true invariably.   The order depends somewhat on the nature, concentration, and temperature of the solutions.   Usually any metal will displace from solution the others that follow, and it will be displaced by those that precede.   Secondary reactions may interfere, however.

The Flow of Current

When the terminals of a cell are connected through an external circuit by a wire, the current that flows is proportional to the potential difference of the electrodes and inversely proportional to the resistance of the whole circuit, including the resistance of the battery as well as that of the external circuit.   This is in accordance with Ohm's law.

The Migration of Ions.   The flow of current is continuous throughout the entire circuit.   The current in the wire is believed to consist of a stream of electrons flowing from the anode, negatively charged with respect to the solution, to the cathode.   In the solution the current is carried by ions both positively and negatively charged, traveling in opposite directions toward their respective electrodes. The positively charged ions, called "cations," move to the cathode which in the Daniell cell is the copper. The negatively charged ions, called the "anions," migrate to the anode. It is obvious that at the surface of the electrodes, where one form of conduction meets the other, exchanges of electrons between the electrodes and the ions in the solution must take place if the electric current is to flow continuously. The reactions that involve both the current-carrying ions and electrons on the one hand and the chemical composition of the electrodes and electrolyte on the other are called electrochemical reactions.

Across any imaginary surface that may be drawn in the solution, the current is carried by the ions, each kind of ion carrying its proportionate share. The ions that are the most numerous and travel the fastest carry a greater amount than the others, but all take part. At the electrodes, however, the actual transfer of current from the solution to the electrode is usually through the medium of only one kind of ion.

When a connection is made by a wire, a path is provided for the electrons to flow to the positively charged copper. The motion of the electrons is opposite to the direction in which the current is said to flow. The departure of the electrons from the zinc disturbs the state of equilibrium between it and the electrolyte, and more zinc ions are thrown into the solution. Coincident with this, additional electrons are liberated in the zinc, and they follow on through the wire after the others as long as the circuit remains closed. Two electrons, each with one negative charge, are freed, with the formation of each zinc ion. This process is represented by the ionic equation

$$Zn - 2e = Zn^{++} \tag{1}$$

The electrons, reaching the copper terminal after their passage through the wire, neutralize an equivalent number of positive charges, thereby disturbing the equilibrium of this electrode and permitting the further discharge of copper ions. Two electrons neutralize two positive charges and permit the deposit of one ion of copper, which gives up its charges. This fact is represented by the equation

$$Cu^{++} + 2e = Cu \tag{2}$$

The zinc cations move toward the cathode, carrying their share of the current through the electrolyte. Probably most of them combine with the sulfate anions moving in the opposite direction to form zinc sulfate according to the equation

$$Zn^{++} + SO_4^{=} = ZnSO_4 \tag{3}$$

The relative motion of the electrons, anions, and cations is shown in Fig. 58.

In the solution around the cathode there are several different kinds of ions carrying positive charges, and therefore several different reactions are possible. The transfer of positive charges is accomplished under ordinary circumstances, however, by the discharge of only one kind of ion. The factors that determine the preferential reaction are the relative electrolytic potentials and the relative concentrations of the

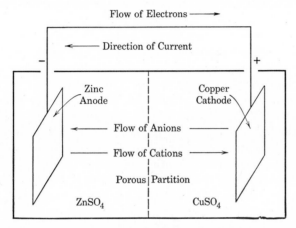

Fig. 58. Relative flow of ions, electrons, and the direction of the current in a Daniell cell.

ions. When each electron of the stream of electrons moving from the anode to the cathode through the wire reaches the cathode, it reduces the positive potential of the cathode by a small amount, and ions having the highest equilibrium potential, deposit. In the case of the Daniell cell these are the copper ions, Cu^{++}.

The unequal speeds of the ions produce concentration differences in the region of the electrodes, which have an important effect upon the performance of a battery. In a storage battery this effect is in addition to the actual consumption of the acid as it combines with the active materials of the plates when current is produced.

Faraday's Law of Electrolysis. In 1834 Faraday formulated his fundamental law of the relation between the amount of electrochemical action and the quantity of electricity passing through the cell. Many experiments since that time have proved the validity of this law, which is of the greatest importance in discussing the storage battery.

The first part of Faraday's law states that the chemical effects produced at the electrodes by the passage of an electric current are in direct proportion to the magnitude of the current and the time that it flows, that is, to the quantity of electricity. This law establishes a relationship between the output of a battery and the amount of active materials that it contains. It therefore denies the possibility of drawing current from the battery indefinitely (perpetual motion), or of charging it by merely changing the solution. Not all the materials within a battery may be used for the production of electric current, because of practical limitations, but interesting comparisons on the basis of Faraday's law may be made for the various types and kinds of batteries.

The second part of Faraday's law states that quantities of various substances, which are liberated or consumed by the action of a given quantity of electricity, are in all cases proportional to the equivalent weights. This part of the law enables the experimenter to determine with exactness the amount of lead, or lead dioxide, or sulfuric acid that takes part in the reaction when a specified amount of electricity is passed through the cell. The quantity of electricity that liberates 1 gram-equivalent weight is the same for all substances. It is equal[1] to 96,500 coulombs. This quantity has been named the "faraday," and it is the fundamental constant of electrochemistry. Lead has an atomic weight of 207.21 and a valence of 2, whence the gram-equivalent weight of lead that takes part in the reactions when 1 faraday of electricity is passed through the solution is $207.21/2 = 103.60$ grams.

THEORY OF THE LEAD-ACID CELLS

A number of theories have been proposed to account for the reactions taking place in the lead-acid batteries. The controversies that have arisen over these theories have been acrimonious, but the so-called "double-sulfate" theory, which will be described in detail, is now generally accepted.

The Double-Sulfate Theory

The *double-sulfate theory* was first proposed by Gladstone and Tribe in 1882.[2] Their theory met with violent opposition, but it has survived to the present day. They discovered that lead sulfate was

---

[1] G. W. Vinal and S. J. Bates, Comparison of the silver and iodine voltameters and the value of the faraday, *Sci. Papers, Natl. Bur. Standards*, 10, 425 (1914). The revised value, as of July 1, 1951, recommended for the Faraday constant by a subcommittee on Fundamental Constants of the Committee on Physical Chemisty of the National Research Council, is 96,493.1 absolute coulombs per gram equivalent. The difference between this and the old value is immaterial for the present discussion.

[2] J. H. Gladstone and A. Tribe, Chemistry of the Planté and Faure accumulators, *Nature, 25*, 221 and 461; *26*, 251, 342, and 602; *27*, 583 (1882–1883).

formed at both plates as part of the process of discharge. They also discovered that the electrolyte became more dilute during the discharge, but Frankland[3] is believed to have been the first to suggest utilizing the change in specific gravity to indicate the state of charge of the battery.

The double-sulfate theory is most conveniently stated by the equation for the reaction.

$$PbO_2 + Pb + 2H_2SO_4 \rightleftarrows 2PbSO_4 + 2H_2O \qquad (4)$$

From left to right this equation represents discharge, and from right to left, charge. The significance of the term "double sulfate" lies in the fact that lead sulfate is formed at both the positive and negative plates during the process of discharge.

Proof of the theory rests (1) on the identification of the materials, (2) on the quantities of these involved in the reactions, and (3) on the thermodynamic relation of the electrical energy produced to the chemical energy of the chemical reactions.

Gladstone and Tribe observed the formation of lead sulfate at the plates during discharge and its subsequent oxidation at positive plates and reduction at negative plates during charge. MacInnes[4] found that PbO_2 is the substance produced by anodic oxidation of lead, and that the oxide so formed has the same chemical composition and electrical potential as the active material of the positive plates. More recently Mazza[5] identified lead dioxide and lead sulfate in the battery by the use of X-rays with no indication of conflicting substances. He confirmed the double-sulfate theory. Barrett[6] also used X-rays and found lead sulfate on both positive and negative plates. He confirmed the double-sulfate theory.

Sinn[7] made allowance for secondary reactions and then found the consumption of acid at the positive plate to agree with requirements of the double-sulfate theory to within less than 1 per cent. Cohen and Overdijkink[8] by a new and novel method confirmed the double-

[3] E. Frankland, Contribution to the chemistry of storage batteries, *Proc. Roy. Soc. London, 35,* 67 (1883).

[4] D. A. MacInnes and E. B. Townsend, An electrovolumetric method for lead, *J. Ind. Eng. Chem., 14,* 421 (1922).

[5] L. Mazza, Products formed during functioning of the lead accumulator, *Atti reale accad. naz. Lincei, 4,* 215; *5,* 117 and 688 (1926–1927).

[6] C. S. Barrett, X-ray studies on lead-acid storage batteries, *Indus. Eng. Chem., 25,* 297 (1933).

[7] V. Sinn, Reaction at positive plate of lead accumulator, *Compt. rend., 206,* 1801 (1938).

[8] E. Cohen and G. W. Overdijkink, Piezodynamic proof of Gladstone and Tribe's theory of the mechanism of the lead storage battery, *Proc. Acad. Sci. Amsterdam, 42,* 834 (1939).

sulfate theory and measured the volume change due to a passage of
2 faradays. They determined the mole volume of the solids by X-rays.
The pressure coefficient of electromotive force at $0°$ C was stated to
be -3.07×10^{-6} volt per atmosphere.

Denina and Ferrero[9] weighed positive plates before and after discharge and measured the amount of acid used. They, like the others,
confirmed the double-sulfate theory at normal rates of discharge.

Beck and Wynne-Jones[10] reexamined the evidence and concluded
from their thermodynamic study that the double-sulfate theory cor-
rectly represents the reactions taking place in the lead-acid battery.

Many experiments have been made to determine the consumption
of H_2SO_4 per faraday. Some of these have shown 2 equivalents of
acid to be used per faraday, as required by the double-sulfate theory,
but others have indicated smaller amounts, with the result that the
validity of this theory has been questioned. One of the difficulties
in making experiments of this kind lies in determining the amount of
acid in the cell. The free electrolyte can be measured easily, but to
this must be added the electrolyte in the pores of the plates and in
the separators. The porosity of the plates changes during discharge
and thus makes corrections uncertain. If the plates are washed or
dried before the experiment, their condition is not normal.

A more accurate method of determining the amount of acid in the
cell and the quantity of it taking part in the reaction was described
by Vinal and Craig.[11] Their method is called the "method of mixtures." In principle it is based on the fact that, if to a solution of
known concentration, but unknown weight, a carefully measured portion of water or another solution differing in concentration is added,
and the concentration of the resulting mixture is determined, the exact
weight of both the original and the final solution can be calculated.
It is not necessary to dismantle the cell or to interfere with its operation. This method enabled the authors to measure not only the
equivalents of acid consumed but also the equivalents of water that
are formed.

The application of the method is illustrated by the following example. The mixture is made within the cell itself, care being taken
to equalize the acid throughout the cell. Evidence of complete equal-

[9] E. Denina and G. Ferrero, Gravimetric investigations of the lead storage
battery, Z. Elektrochem., 45, 314 (1939).

[10] W. H. Beck and W. F. K. Wynne-Jones, Behavior of the lead dioxide electrode, Trans. Faraday Soc., 50, 136 (1954).

[11] G. W. Vinal and D. N. Craig, Chemical reactions in the lead storage battery,
J. Research Natl. Bur. Standards, 14, 449 (1935).

ization of the acid is obtained when successive readings of electromotive force and density are constant. The concentrations in the following example are expressed as weight fractions.

Initial concentration of electrolyte	0.22783
Acid added, concentration 0.59820	
Weight added 117.684 grams	
Concentration of resulting mixture	0.33447

Let y equal weight of electrolyte after mixing

$$\frac{0.33447y - (0.59820 \times 117.684)}{y - 117.684} = 0.22783$$

Whence	$y = 408.73$
Deduct for sample removed	60.66
Net weight of electrolyte available	348.07 grams

A similar method was used to determine the weight of electrolyte at the end of the discharge.

The cell used in these experiments contained eleven small pasted plates. All the grids were pure lead, which reduced local action to a minimum. Each plate was provided with a long lug which extended through the cover, connections being made outside the cell in order to provide the best insulation. Beginning with the eighth experiment the cell was kept under a sealed bell jar to prevent evaporation of water which amounted otherwise to about 0.5 gram per day.

No preliminary assumptions regarding the correctness or applicability of any theory of chemical reactions in the battery are necessary. The number of equivalents of acid consumed or water formed as a result of the cell discharging a measured number of coulombs was measured directly. Let w and w' represent the weight of electrolyte at the beginning and at the end of discharge, respectively; let p and q be the weight fractions similarly; also, let c represent the coulombs of electricity discharged and F the value of the faraday. The equivalent weights of sulfuric acid and water are, respectively, 49.04 and 9.01 grams. All these quantities are known or can be measured. The number of equivalents of acid consumed per faraday is equal to

$$\frac{F}{c} \frac{(pw - qw')}{49.04} \tag{5a}$$

The number of equivalents of water formed per faraday is equal to

$$\frac{F}{c} \frac{(pw - qw') - (w - w')}{9.01} \tag{5b}$$

The experimental results as given by the authors in two tables have been combined into one, Table 35.

TABLE 35. EQUIVALENTS OF ACID CONSUMED AND WATER FORMED PER FARADAY DURING DISCHARGE OF THE LEAD STORAGE BATTERY

Experiment	Date	Discharge Data		Percentage of H_2SO_4 (by Weight)		Weight of Solution		Equivalents per Faraday	
		Time	Coulombs	Start	Finish	Start	Finish	H_2SO_4	H_2O
	1934	Hours				Grams	Grams		
1	Mar. 9	10.42	45,252	35.53	24.82	......	311.2	2.04	......
2	Mar. 20	7.81	34,992	36.60	29.17	358.77	330.60	1.96	2.05
3	Mar. 27	8.00	35,833	28.061	18.933	341.78	307.65	2.07	1.06
4	Apr. 6	8.83	39,116	30.885	21.684	358.71	324.73	2.03	1.75
5	Apr. 13	4.83	31,232	32.518	25.607	364.20	337.88	2.01	1.92
6	Oct. 5	10.60	42,088	28.21	17.89	355.33	317.20	2.03	1.36
7	Oct. 17	5.84	23,130	42.714	38.246	......	345.92	2.02	......
8a	Oct. 31	5.03	19,786	20.951	15.850	347.73	......	2.03	......
8b	Oct. 31	5.80	22,316	15.850	8.898	......	284.98*	2.01	......
9	Nov. 23	5.75	22,638	41.24	36.72	360.85	340.28	2.07	1.56
10	Nov. 30	19.87	44,173	33.948	23.584	351.22	315.14	2.00	2.14
11	Dec. 7	6.73	31,237	33.447	26.276	348.07	323.13	1.99	2.25
12	Dec. 27	13.42	38,152	31.141	22.298	359.09	326.87	2.01	1.89
						Mean of all		2.02	1.78
						Mean of last four		2.02	1.96
						Average deviation		±0.03	±0.19

* Between the two discharges 18.62 grams of electrolyte were removed for density determinations.

It is apparent that 2 equivalents of acid are consumed per faraday and that 2 equivalents of water are formed at the same time. These are precisely the requirements of the double-sulfate theory. It is apparent also that the results given in Table 35 are independent of the concentration of the electrolyte, the extent of the discharge, and the rate of the discharge. No evidence of the formation of basic sulfate was found, and no spontaneous change in the final products of the reaction was observed after the discharge was discontinued. The double-sulfate theory of Gladstone and Tribe is supported, therefore, to the exclusion of the other theories that have been proposed.

Fery[12] published many papers in support of a theory that the active material of the positive plates is an oxide higher than the dioxide, but there is little, if any, support for this theory now.

Thermodynamic transformations, which are discussed in later sections, afford further proof of the double-sulfate theory.

[12] Ch. Fery, Théorie chimique et fonctionement physique de l'accumulateur au plomb, Rev. gén. élec., 1, 10 (1917); 19, 296 (1926).

Reactions at the Positive and Negative Plates

Although the double-sulfate theory states that lead sulfate is formed at each plate during the discharge, and gives us an equation for calculating the performance of the cell, it leaves us in the dark about the actual processes that go on at the positive and negative plates. It does not tell how the substances are formed, or explain their relation to the transfer of electricity through the cell.

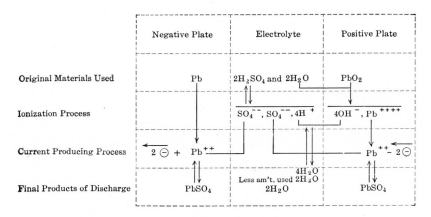

Fig. 59.   Discharge reactions.

Figure 59 is a diagram which shows in detail the probable processes that result in final products of discharge, according to the double-sulfate theory.   At the negative plate the process is relatively simple; the solution pressure of lead causes it to throw lead ions into solution (see p. 169).   These are in the divalent state, carrying 2 positive charges.   Coincident with the departure of each of the positive lead ions from the surface of the electrode, the electrode itself acquires 2 negative charges.   The lead ions react with sulfate ions, which have charges of equal magnitude but opposite sign, forming lead sulfate. This is so nearly insoluble in the electrolyte that it immediately precipitates out of the solution and is deposited on the electrode within space of molecular dimensions.

Reactions occurring at the positive plate are not so well understood. They may be explained, however, in several ways.   The tetravalent-ion theory, which seems the most probable, was proposed by LeBlanc[13]

[13] M. LeBlanc, *Lehrbuch der Elektrochemie,* 1st ed., W. Knapp, Halle, 1895. Tr. by W. R. Whitney, The Macmillan Co., New York, 1896.

many years ago. Lead dioxide, although insoluble in appreciable amounts in sulfuric acid under ordinary conditions, may pass into solution to a limited extent when the current flows. Small amounts of it, in combination with water, ionize into tetravalent lead ions and monovalent hydroxyl ions, according to the equation:

$$PbO_2 + 2H_2O \rightleftarrows Pb(OH)_4 \rightleftarrows Pb^{++++} + 4OH^- \qquad (6)$$

The current-producing process is associated with potential differences at the electrodes that depend, first, on the equilibrium potentials for the ionic reactions, and second, on the ionic concentrations. At the positive electrode the ionic reaction is

$$Pb^{++++} + 2e \rightarrow Pb^{++}$$

and for this the potential difference is about $+1.75$ volts.[14] Ionization of the electrolyte occurs substantially as follows:

$$H_2SO_4 \rightleftarrows H^+ + HSO_4^-$$

and

$$HSO_4 - \rightleftarrows H^+ + SO_4^=$$

As a last step in the process of discharge, the combination of lead and sulfate ions takes place, with the formation of lead sulfate, which, being practically insoluble in the electrolyte, deposits as a solid substance on the plates. This reaction is represented by equation (7):

$$Pb^{++} + SO_4^= \rightleftarrows PbSO_4 \qquad (7)$$

These reactions are shown diagramatically in Fig. 59. It should be noted that 4 molecules of water are really formed for each 2 molecules of sulfuric acid consumed, but, as 2 molecules of water are used up at the same time, the net change is the production of 2 molecules of water, which is in accordance with the equation for the double-sulfate theory (4).

The single potential of the positive plate is given by the equation:

$$E_p = 1.75 + \frac{RT}{2F} \log_e \frac{[Pb^{++++}]}{[Pb^{++}]} \qquad (8)$$

The concentrations of the ions inclosed in the brackets are small quantities for which no directly measured values are available. R in this

---

[14] S. Glasstone gives the value 1.75 volts for the normal plumbic-plumbous potential in his Physical chemistry of the oxides of lead, Part V, Electromotive behavior of lead dioxide, *Trans. Chem. Soc.*, *121*, 1478 (1922).

equation is the gas constant, T the absolute temperature, and F the faraday.

Similarly, the potential of the negative electrode is obtained. The reaction in this case is

$$Pb - 2e \rightarrow Pb^{++}$$

for which the equilibrium potential value (Table 34) is -0.12 and the single potential

$$E_n = -0.12 + \frac{RT}{2F} \log_e [Pb^{++}] \tag{9}$$

The electromotive force of the cell is the difference between these two single potentials and is given by the expression

$$E = E_p - E_n = 1.75 + 0.12 + \frac{RT}{2F} \log_e \frac{[Pb^{++++}]}{[Pb^{++}]^2} \tag{10}$$

Since $R = 8.32$, $T = 291°$, and $F = 96{,}500$, equation (10), after changing to common logarithms, becomes

$$E = 1.87 + \frac{0.058}{2} \log \frac{[Pb^{++++}]}{[Pb^{++}]^2} \tag{11}$$

A more practical approach to normal battery operation is found in the well-known equations for these oxidation-reduction couples.

$$PbO_2 + SO_4^= + 4H^+ + 2e^- = PbSO_4 + 2H_2O \quad E° = +1.685 \text{ volts} \tag{12}$$
$$Pb + SO_4^= = PbSO_4 + 2e^- \quad E° = +0.356 \tag{13}$$

$$\overline{PbO_2 + 2SO_4^= + 4H^+ + Pb = 2PbSO_4 + 2H_2O \quad E° = +2.041 \text{ volts} \tag{14}}$$

The potentials are in volts referred to the hydrogen-hydrogen ion couple as the zero for unit activities and temperature 25° C. Adding the equations gives the reaction for the complete cell and a voltage approximating the observed electromotive force of a commercial cell.

The charging process is shown diagrammatically in Fig. 60. Starting with the products of discharge, the lead sulfate at both electrodes passes into solution and ionizes. The water also is ionized as rapidly as equilibrium conditions permit. The divalent lead ions at the negative plate, which is now the cathode, take up 2 electrons, neutralizing their charges, and then are deposited as lead in the solid state. The divalent lead ions at the positive plate or anode are forced by the charging current to give up 2 electrons, which changes them to the tetravalent state. Each of these ions may then unite with 2 oxygen ions, through an intermediate step, to form lead dioxide, which is deposited upon the plate. The sulfate ions formed at each plate unite with 2 hydrogen ions, with an intermediate step, forming sulfuric acid at each plate.

Lead sulfate, which plays such an important role in the chemical reactions of the lead storage battery, is very sparingly soluble in the sulfuric acid electrolyte. The amount in solution under equilibrium conditions is so small that accurate measurements of it are difficult to make. This probably accounts for the meager data that have hitherto

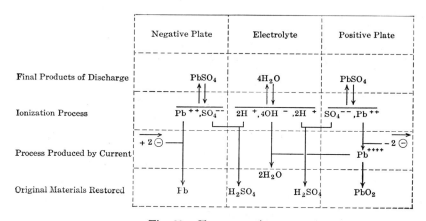

Fig. 60. Charge reactions.

been available. Diphenylthiocarbazone, commonly called dithizone, has been introduced as a sensitive reagent for lead. This organic reagent is sensitive to less than one-millionth of a gram of lead when used with a spectrophotometer or with a photronic cell and a suitable color screen. A dithizone solution in chloroform is normally green, but when in contact with an aqueous ammoniacal solution containing an excess of lead it becomes a bright cherry red. Craig and Vinal[15] have used this reagent to determine the solubility of lead sulfate in sulfuric acid solutions up to 50 per cent H_2SO_4. The solutions were stirred continuously for several days, after which the samples were drawn through fritted-glass filters and evaporated in quartz dishes. The lead sulfate residues, after evaporation, were taken up in nitric acid solutions and carried through a titration process with both aqueous and organic phases present. The results are contained in Table 36, which also includes a determination of the solubility of $PbSO_4$ in water, taken

[15] D. N. Craig and G. W. Vinal, Solubility of lead sulfate in solutions of sulfuric acid determined by dithizone with a photronic cell, *J. Research Natl. Bur. Standards*, *22*, 55 (1939).

TABLE 36. SOLUBILITY OF LEAD SULFATE IN
SOLUTIONS OF SULFURIC ACID

Concentration of H_2SO_4			Weight of $PbSO_4$ per Liter	
Per Cent	Molarity 25°	Specific Gravity $\frac{25°}{25°}$	At 25° C, mg	At 0° C, mg
0.0	0.00	1.000	45.0	28.0
0.5	0.05	1.003	4.60	2.06
1.0	0.10	1.006	4.91	2.10
5.0	0.52	1.033	6.15	2.46
10.0	1.08	1.067	6.68	2.86
15.0	1.68	1.102	6.28	2.63
20.0	2.32	1.140	5.18	2.21
25.0	3.00	1.179	3.76	1.76
30.0	3.72	1.219	2.75	1.27
35.0	4.48	1.260	2.02	0.84
40.0	5.30	1.303	1.52	0.53
45.0	6.16	1.348	1.23	....
50.0	7.10	1.395	1.08	....

from other sources. Lead sulfate is 10 times as soluble in water as in 0.5 per cent sulfuric acid. The effect of the acid in depressing the solubility is a good illustration of the common-ion effect, but the solubility increases again in stronger acid to a maximum at 10 per cent acid. At higher acid concentrations the solubility decreases again, probably as a result of changes in the equilibrium between the sulfate and bisulfate ions of the acid solution.

ENERGY TRANSFORMATIONS

A storage cell of any kind stores electrical energy by virtue of the chemical reactions taking place at the electrodes. Electrical energy is not stored as electrical energy, but as chemical energy. During the charging process the electrical energy is converted into chemical energy, and when the cell is discharged at a later time this chemical energy is reconverted into electrical energy. The law of the conservation of energy governs these transformations. The energy cannot be created from nothing, nor can it be annihilated; therefore, the electrical energy that the cell can supply bears a definite relation to the amount of chemical energy that it contains. We cannot determine the total amount of chemical energy of the cell, but it is possible to determine the change in chemical energy that occurs as the cell is charged or discharged and, by relating this to the electrical measurements, the electrochemical theory of the storage cell may be developed. This theory includes the theory of the chemical reactions and the theory of the energy transformations.

Chemical reactions in general are accompanied by the evolution or absorption of heat in varying amounts, depending on the nature of the reaction. The heat of reaction expresses the difference between the heat content of the reacting substances and that of the products of the reaction. If heat is liberated, the reaction is said to be "exothermic," but if heat is absorbed the reaction is "endothermic." For each reaction the amount of heat liberated or absorbed per gram-molecule depends on the initial and final states and may be called the heat of reaction.

The heat of reaction of the lead-acid storage cell as it would occur in a calorimeter depends on the heat of formation of each of the constituents. Basic data can be found in such books as the *International Critical Tables*, or in the more recently published book (1936), *Thermochemistry of Chemical Substances* by Bichowsky and Rossini. In calculating the heat of the reaction thermochemically, we shall choose as a starting point a dilute solution of sulfuric acid, having the proportions of 1 mole of acid to 200 moles of water. The heats of formation of the various constituents are written beneath the respective symbols in the equation

$$PbO_2 + 2(H_2SO_4 + 200H_2O) + Pb \rightleftarrows 2PbSO_4 + 402H_2O$$
$$\text{65,000} \qquad \text{2(211,500)} \qquad \text{0} \quad \text{2(218,500)} \quad \text{2(68,370)}$$

The heats of formation are values at 18° C. From these numerical values the heat of reaction, Q, is 85,740 calories. To convert this to the equivalent at 25° C, at which temperature most of the electrical measurements have been made, a correction factor is computed on the basis of the heat capacities of all the constituents. This amounts to −45.4 calories per degree for the whole reaction as expressed by the equation. The heat of the reaction then becomes 85,420 calories at 25° C.

The solution is the variable factor, and values for the heat of reaction for other concentrations will differ by changes of the heat of formation of H_2SO_4 and H_2O in the respective concentrations. Considered from a purely chemical point of view, the changes in the heat of formation of sulfuric acid solutions from one concentration to another vary by the amount of heat liberated as the solutions are diluted. Eventually a point is reached where no more heat is liberated and the solution is infinitely dilute and its volume infinitely great. The pressure is

assumed to be constant.   Values of the relative apparent molal heat
content at 25° C are given in Table 37.[16]   On the other hand, if we

TABLE 37.   RELATIVE APPARENT MOLAL HEAT CONTENT OF
SULFURIC ACID SOLUTIONS
(Values are in kilogram calories.)

Concentration of Solution		Relative Apparent Molal Heat Content	Relative Partial Molal Heat Content*	
n, Moles H_2O	Per Cent H_2SO_4	H_2SO_4*	H_2SO_4	H_2O
∞	0.00	0.00	0.000	0.000
25,600	0.02	1.54	2.234	0.000
6,400	0.08	2.90	3.725	0.000
1,600	0.34	4.04	4.811	−0.000₅
400	1.34	5.02	5.638	−0.001
200	2.65	5.41	5.842	−0.002
100	5.16	5.62	5.888	−0.003
50	9.82	5.78	6.065	−0.006
25	17.88	6.07	6.681	−0.024
15	26.63	6.55	7.896	−0.090
10	35.25	7.30	9.632	−0.233
5	52.13	9.45	13.282	−0.766
3	64.47	11.66	16.089	−1.477
1	84.48	16.72	21.451	−4.731
0	100.00	23.54	23.540	....

* The convention as to algebraic signs of these quantities requires some explanation.   In this book a positive sign has been used to indicate that energy
is liberated when the reaction, expressed by the thermochemical equation, is
from left to right.   In the paper[16] referred to, from which Table 37 is taken,
the opposite convention has been followed.   That is, a positive sign indicates
an increase in heat content of the solutions.   Values given in Table 37 should
have the signs reversed for the computations given here.   Since the symbol for
the acid appears on the left side of the equation (page 184), a negative value
would mean that the quantity should be added to the net value on the right
side.   The negative sign for the water, which appears in the table, becomes a
positive sign, and since the symbol for water is on the right side of the equation this is also added.   Both acid and water, therefore, contribute to increase
the heat of reaction for all concentrations above infinite dilution.

start with a given solution and add a small quantity of the solute, the
heat change per mole is the partial or differential heat effect.   The
relative partial molal heat contents are also given in Table 37.   At
infinite dilution these are zero and the heat of reaction in the cell is pre-

[16] D. N. Craig and G. W. Vinal, Thermodynamic properties of sulfuric acid
solutions and their relation to the electromotive force and heat of reaction of the
lead storage battery, *J. Research Natl. Bur. Standards*, *24*, 475 (1940).

cisely the same as the thermochemical or calorimetric heat of reaction. To calculate the heat of reaction for a state of infinite dilution of the sulfuric acid, we must deduct from the value above the relative apparent molal heat content for the solution $H_2SO_4 \cdot 200H_2O$. As shown in Table 37, this is 5.41 kilocalories or 5410 calories per mole. Two moles are involved, whence $85,420 - 10,820 = 74,600$ calories as the heat of reaction at infinite dilution. This applies to either the thermometric or the battery reaction and is a convenient value from which to calculate the heat of reaction of a battery at any other concentration of the solution.

From a practical standpoint we wish to know the heat of reaction for the battery at finite and useful concentrations of the electrolyte. The discharge reaction involves the consumption of acid represented by a term on the left of the equation and the formation of water on the right side of the equation. With due regard for algebraic signs, the heat of reaction then equals the heat of reaction at infinite dilution, 74,600 calories plus twice the sum of the relative partial molal heat contents of the acid and the water. The factor 2 comes from the use of 2 moles each of acid and water for each mole of lead dioxide.

Table 38, which is also based on the work of Craig and Vinal,[16] gives in parallel columns the heat of reaction calculated from thermochemical data and from the best available electrical data. The two are entirely independent sources of information, but the agreement of the results is remarkable throughout the range of acid concentrations from 3 to 40 per cent. In the battery the calories, representing the heat of reaction, are one form of energy. If only heat energy were involved, this would mean that the battery would become very hot, but this is obviously not the case. The cell may even cool during discharge. These apparently contradictory facts are reconciled by the knowledge that the change in energy of the reacting materials, which has been calculated as calories, is transformed into electrical energy when the battery discharges. It does not appear as heat.

The method of computing the values from electrochemical data will appear after a brief discussion of the thermodynamic relationships between electromotive force, temperature, and heat of reaction.

ELECTROMOTIVE FORCE

Relation of Electromotive Force to Temperature

The Gibbs-Helmholtz Equation. The maximum useful work that the cell can deliver is the quantity of electricity (current $\times$ time) expressed in coulombs, multiplied by the electromotive force. This is sometimes called the free or available energy, to distinguish it from the

TABLE 38.   COMPARISON OF THE HEATS OF REACTION OF THE LEAD STORAGE
BATTERY CALCULATED FROM THERMOCHEMICAL AND ELECTROCHEMICAL DATA

Concentration of Solutions			Heat of Reaction at 25° C, Thermochemical, kg cal	Heat of Reaction at 25° C, Electrochemical (Electrode Measurements), kg cal
n Moles of H_2O to 1 Mole of H_2SO_4	Molality	Per Cent H_2SO_4		
400	0.1388	1.34	85.88	85.65
200	0.2775	2.65	86.29	86.24
100	0.5550	5.16	86.38	86.23
80	0.6937	6.37	86.43	86.43
60	0.9251	8.32	86.61	86.62
50	1.110	9.82	86.74	86.75
40	1.388	11.98	87.02	87.01
30	1.850	15.36	87.52	87.58
25	2.220	17.88	88.01	88.13
20	2.775	21.40	88.88	89.12
15	3.700	26.63	90.57	90.88
12	4.626	31.20	92.59	92.68
10	5.550	35.25	94.33	94.35
8	6.937	40.49	96.88	96.85

Average difference ± 0.11 per cent

total energy as represented by the heat of the reaction. The available energy will be designated by W.

By the first law of thermodynamics, the change in the total energy of cell Q, considered as a system, is equal to the sum of the external work W, done by the system, and the heat evolved, q. This is expressed by the equation

$$Q = W + q \qquad (15)$$

The heat of the reaction, Q, is measured in calories. The calorie is also called the gram calorie. This unit of heat energy may be converted into its electrical equivalent of energy, by the following relation: 1 calorie (20° C) = 4.1840 volt-coulombs.

The available energy of the cell, W, is naturally expressed in volt-coulombs. The quantity q will be positive or negative according as W is less or more than Q. In the case of the lead storage battery, W exceeds Q, and therefore q must be subtracted from W. The significance of this is that the cell has a tendency to cool on discharge, or must absorb heat from the surroundings if its temperature is kept constant.

188 Storage Batteries

It is a well-known fact that the electromotive force of any kind of voltaic cell varies more or less with the temperature. The change in electromotive force per degree change in temperature for a lead-acid storage cell is so small that for practical purposes it may be neglected, but it is of importance from a theoretical point of view. Before attempting to determine the actual magnitude of the temperature coefficient, we shall derive,[17] from theoretical considerations, the celebrated Gibbs-Helmholtz equation, which shows the relation of the energy transformations, electromotive force, and temperature. It will then be possible to make comparisons of the observed quantities with those calculated from thermodynamic reasoning.

Let it be assumed that a storage cell of the lead-acid type has an electromotive force of E when the absolute temperature is T, and that the cell is in equilibrium with its surroundings. It is also assumed that a quantity of electricity, designated by e, may be passed through the cell either as a charge or discharge, and that the cell may be kept at a uniform temperature (isothermal process) or completely insulated to prevent the transfer of heat (adiabatic process), as we may desire.

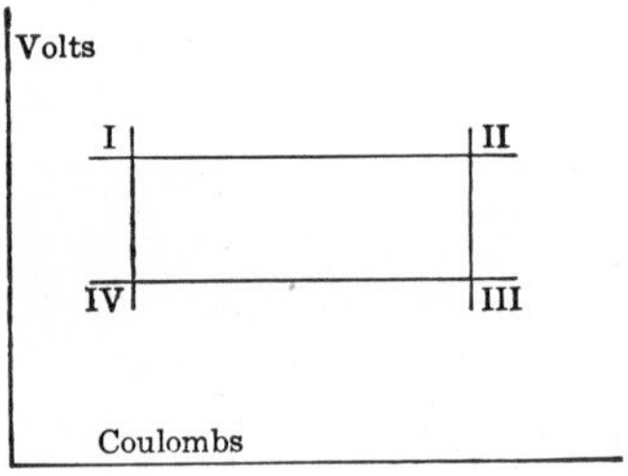

The following cyclic process is to be carried out:

I. Let the initial state of the cell be described by the electromotive force E and the temperature T, at the point marked I on the diagram (Fig. 61).

Fig. 61. Cyclic process.

Then let a quantity of electricity, e, pass through the cell in the discharging direction while the temperature is held constant at T by furnishing a small amount of heat as required by the term q in the equation above. The current must be small during this process, so that irreversible heat effects due to the ohmic resistance of the cell may be neglected. The work, in volt-coulombs, done by the cell during this part of the cycle is $W_1 = Ee$.

II. When the quantity of electricity, e, has passed through the cell, let the isothermal process be changed momentarily to adiabatic, during which time the temperature of the cell will fall by an infinitesimal amount δT, becoming $T - \delta T$. The electromotive force becomes $E - \left(\dfrac{\partial E}{\partial T}\right)\delta T$, where $\left(\dfrac{\partial E}{\partial T}\right)$ is the temperature coefficient when no current is flowing.

[17] E. Buckingham, *Theory of Thermodynamics,* The Macmillan Co., New York, 1900. A discussion directly applicable to this part of the theory is given on p. 139, "Electromotive force of a reversible galvanic cell."

III. Let the quantity of electricity, e, pass through the cell in the charging direction while the temperature is held at $T - \partial T$. The amount of work done on the cell during this process is

$$W_3 = \left[E - \left(\frac{\partial E}{\partial T} \right) \delta T \right] e.$$

IV. When the quantity of electricity, e, has passed through the cell, let the isothermal process be changed to adiabatic while the temperature of the cell increases by the amount δT, which restores the cell to its original temperature. The total external work done by the cell during the cycle is

$$W_1 - W_3 = Ee - \left[E - \left(\frac{\partial E}{\partial T} \right) \delta T \right] e = e \left(\frac{\partial E}{\partial T} \right) \delta T = \delta q$$

The adiabatic processes are infinitesimal in length and are negligible.

Considering, for a moment, the cell as a purely chemical apparatus, we shall suppose chemical reactions to take place in amount and direction equivalent to the discharge of 1 coulomb. The heat of this reaction is to be measured in volt-coulombs and designated by Q'. For a discharge equivalent to e coulombs, the heat of the reaction will be $Q'e$. The electrical energy of the corresponding reactions at temperature T is equal to Ee. The difference in these two quantities gives at once the amount of heat supplied by the constant temperature reservoir at temperature T during the first step of the cyclic process. This is sometimes called the latent heat of the cell q. It is equal to the difference between the electrical energy and the heat of the reaction:

$$q = Ee - Q'e$$

According to Carnot's cycle, the ratio of the amount of heat converted into useful energy (which in this case is electrical) to the amount of heat received by the system when held at the upper temperature during the cyclic process is related to the absolute temperature according to the equation

$$\frac{\delta q}{q} = \frac{\delta T}{T}$$

Substitution of the values for δq and q may now be made in this equation

$$\frac{e \left(\frac{\partial E}{\partial T} \right) \delta T}{Ee - Q'e} = \frac{\partial T}{T}$$

or

$$T \left(\frac{\partial E}{\partial T} \right) = E - Q'$$

the relation of Q' to Q when measured in calories for n faradays is

$$\frac{96,500nQ'}{4.1840} = Q$$

and the final equation becomes

$$\frac{Q}{23,070n} = E - T\left(\frac{\partial E}{\partial T}\right) \tag{16}$$

This is the celebrated Gibbs-Helmholtz equation applied to a reversible electric cell. It shows that the term q in equation (15) is proportional to the absolute temperature and the temperature coefficient of the electromotive force. This equation correlates heat of reaction, electromotive force, and temperature coefficient. It was used to compute the electrochemical heat of reaction shown in the last column of Table 38.

Variation of the Electromotive Force with Acid Concentration

The electromotive force, or open-circuit voltage, as it is often called, varies with the concentration of the sulfuric acid contained in the cell. The reasons for this are apparent from the discussion of the heat content of the constituents of the cell. The effect of changing the acid concentration on the potentials of the plates may be seen by considering the equations on page 181. The product of the concentration of the hydrogen and hydroxyl ions is equal to a constant, K.

$$[H^+][OH^-] = K$$

If the percentage of sulfuric acid in the electrolyte is changed, an increase in the specific gravity means an increase in the concentration of the hydrogen ions and a corresponding decrease of hydroxyl ions. Applying this to equation (6) on page 180, an increase in acidity causes the reaction to proceed to the right, and equilibrium conditions require an increase in the concentration of the tetravalent lead ions in proportion as the hydroxyl ions decrease. Equation (8) for the potential of the positive plate shows that the potential increases as the concentration of the tetravalent lead ions increases. Coincident with the increase in concentration of the hydrogen ions there is an increase in concentration of the sulfate ions. Equilibrium conditions for equation (7) on page 180 show that this condition must be accompanied by a decrease in the concentration of the divalent lead ions. This also results in increasing the potential of the positive plate, as equation (8) shows. Turning now to the potential of the negative plate, by equation

(9) it is apparent that a decrease in the concentration of divalent lead ions, accompanying an increase in concentration of the electrolyte, results in a smaller value for the potential of the negative plate; but, as this is to be substracted from the potential of the positive plate to obtain the voltage of the cell, the effect is in the same direction as for the positive plates.   Experiments by Dolezalek[18] have shown that, for an increase in acid strength from 6 to 16 per cent, the increase in potential of the positive plates is 0.06 volt, and the decrease in potential of the negative plates for the same change in concentration is only 0.004 volt.   The change in voltage of the cell is, therefore, the sum of these values, or 0.064 volt.   It is at once apparent that the positive plates are much more sensitive to the changes in concentration of acid than the negatives.   Considering the cell as a whole, as in equation (10), it is found that the voltage increases as the concentration of the tetravalent lead ions increases and as the square of the concentration of the divalent lead ions decreases.   The observed facts are in agreement with the theory, but reliable data are lacking as to the magnitude of the concentrations of the ions upon which a computation of the quantitative effects can be based.

Many measurements of the relation of acid concentration to electromotive force have been made, but it is difficult to compare the results, because the earlier experimenters have employed various means of expressing the acid concentration and the electrical measurements are subject to uncertainty.   An effort was made to calculate the various determinations reported in the literature to a uniform basis, and these were tabulated in earlier editions of this book.   Newer and more reliable data are now available on batteries and on electrodes from which the electromotive force of the batteries can be calculated.   A critical and perhaps somewhat arbitrary selection of the earlier values has been made.   Table 39 contains a group of determinations on batteries and another group of measurements on electrodes for corresponding acid concentrations.   The mean values of the former are in close agreement with those of the latter group.   In the last column of the table the mean of all measurements included in the table is given as representing the best data on the relation of electromotive force to acid concentration.   The value of such data is twofold.   From the theoretical standpoint it is necessary for calculations based on thermodynamic theory, and from the practical standpoint it is of use in fixing the conditions of floating batteries.

[18] F. Dolezalek, *Theory of the Lead Accumulator,* p. 80, tr. by von Ende, John Wiley & Sons, New York, 1904 (out of print).

Storage Batteries

TABLE 39. ELECTROMOTIVE FORCE OF THE LEAD STORAGE BATTERY AS A FUNCTION OF THE ACID CONCENTRATION
(All values in volts.)

Sp. Gr. 25°C / 25°C	Per Cent H₂SO₄	Determinations on Storage Batteries					Calculated from Electrode Measurements				Mean of All
		Dolezalek Calculated to 25°C	Thibaut Calculated to 25°C	Vinal and Altrup at 25°C	Craig and Vinal at 25°C	Mean	Harned and Hamer at 25°C	Vosburgh and Craig at 25°C	Craig and Vinal at 25°C	Mean	
1.020	3.05	.....	(1.833)	.....	1.855	1.855	1.855	1.856	.....	1.855	1.855
1.030	4.55	.....	(1.851)	.....	1.876	1.877	1.878	1.878	1.882	1.879	1.878
1.040	6.04	.....	(1.867)	1.890	1.892	1.892	1.895	1.893	1.896	1.895	1.893
1.050	7.44	1.906	(1.879)	1.903	1.905	1.905	1.908	1.909	1.908	1.908	1.907
1.100	14.72	1.963	(1.941)	1.956	1.962	1.960	1.960	1.962	1.961	1.961	1.961
1.150	21.38	2.012	(1.944)	2.000	2.005	2.005	2.005	2.006	2.007	2.006	2.006
1.200	27.68	2.054	2.043	2.045	2.050	2.048	2.050	2.049	2.051	2.050	2.049
1.250	33.80	2.098	2.094	2.091	2.098	2.095	2.097	.....	2.094	2.095	2.095
1.280	37.40	2.125	2.126	2.119	2.128	2.125	2.127	.....	2.121	2.124	2.125
1.300	39.70	2.143	2.148	2.138	2.148	2.144	2.147	.....	2.140	2.144	2.144

Notes on Table 39

1. All emf values have been read from curves for the percentages given in column 2, for which equivalent specific gravities at 25° C are given in col. 1.

2. Dolezalek's values in col. 3 have been calculated to 25° C and corrected for the difference between his value for the Weston cell and the value in use now.

3. Values in parenthesis, col. 4, were omitted in taking the mean values.

4. Determinations in col. 6 were made on cells containing pure lead grids; other cells are believed to have contained lead-antimony grids.

5. Electrode measurements are based on combining values of the separate electrode potentials, experimentally determined.

6. References: Dolezalek, *Theory of the Lead Accumulator*, p. 55, 1904; Thibaut, *Z. Elektrochem., 19*, 881 (1913); Vinal and Altrup, this book, 1st ed., p. 166, 1924; Harned and Hamer, *J. Am. Chem. Soc., 57*, 33 (1935); Vosburgh and Craig, *J. Am. Chem. Soc., 51*, 2009 (1929); Craig and Vinal, *J. Research Natl. Bur. Standards, 24*, 475 (1940).

Temperature Coefficient of Electromotive Force

By the temperature coefficient is meant the differential change in electromotive force, or open-circuit voltage, of the cell with change in temperature of the cell. This should not be confused with the change in capacity of the cell that is caused by change of temperature. The coefficient is conveniently expressed as the change in voltage per degree change in temperature. It is positive within the working range, that is, a rise in temperature is accompanied by a rise in voltage. The coefficient is constant or very nearly so, which means that the same values for the change in voltage per degree change in temperature may be expected for a given concentration, regardless of the actual temperature of the cell.

A critical review of the temperature coefficients for different acid concentrations, Table 40, has been made as in the case of the electromotive force data. The temperature coefficients are more difficult to determine accurately, but it is important that they should be known for theoretical reasons as well as for more practical matters of battery operation. Three sets of determinations on batteries are given in the table. The mean values agree reasonably well with the mean of three sets of determinations on electrodes. As the most probable values, the mean values of all are given in the last column. The temperature coefficients for acid concentrations below 4.3 per cent are negative, but this is below the operating range of storage batteries. For all concentrations above 4.3 per cent the coefficient is positive and reaches a maximum value at approximately 21 per cent (1.150 sp. gr.).

Having the necessary data on electromotive force and temperature coefficient, the heat of the reaction for different concentrations of electrolyte may be calculated from electrochemical measurements, using the Gibbs-Helmholtz equation (16) given on page 189. These values for the heat of the reaction are contained in Table 38 where they are compared with the results of thermochemical calculations.

Reversible and Irreversible Heat Effects. If a storage cell of the lead-acid type be placed in a calorimeter which will permit the making of accurate measurements of heat evolved or absorbed during charge or discharge, it will be found that the cell absorbs a small amount of heat from its surroundings during discharge and gives out a similar amount during charge. This means that on discharge the cell delivers more useful work than is represented by the heat of the chemical reactions taking place in it, and that correspondingly more energy is required to charge it. Stated in another way, it means that the cell on discharge draws heat energy from its surroundings and converts it into

TABLE 40. CHANGE IN ELECTROMOTIVE FORCE OF THE LEAD STORAGE BATTERY WITH TEMPERATURE

(Values of $\frac{\Delta E}{\Delta T}$ are in millivolts per degree Centigrade.)

Sp. Gr. 25°C/25°C	Per Cent H₂SO₄	Determinations on Storage Batteries				Calculated from Electrode Measurements				Mean of All
		Dolezalek 0 to 24°C	Thibaut 15 to 50°C	Craig and Vinal 0 to 40°C	Mean	Harned and Hamer 0 to 40°C	Craig and Vinal 0 to 40°C	Vosburgh and Craig 20 to 40°C	Mean	
1.020	3.05	......	−0.06	......	−0.06	......	......	−0.06	−0.06	−0.06
1.030	4.55	......	+0.02	......	+0.02	......	+0.03	+0.01	+0.02	+0.02
1.040	6.04	−0.04	+0.08	+0.10	+0.05	......	+0.08	+0.06	+0.07	+0.06
1.050	7.44	+0.06	+0.13	+0.15	+0.11	+0.08	+0.10	+0.10	+0.09	+0.10
1.100	14.72	+0.34	+0.26	+0.29	+0.30	+0.19	+0.23	+0.21	+0.21	+0.25
1.150	21.38	+0.36	+0.29	+0.33	+0.33	+0.22	+0.26	+0.24	+0.24	+0.28
1.200	27.68	+0.31	+0.28	+0.30	+0.30	+0.22	+0.25	+0.22	+0.23	+0.26
1.250	33.80	+0.24	+0.25	+0.22	+0.24	+0.20	+0.22	......	+0.21	+0.23
1.280	37.40	+0.19	+0.22	+0.19	+0.20	+0.20	+0.19	......	+0.20	+0.20
1.300	39.70	+0.16	+0.20	+0.18	+0.18	+0.17	+0.16	......	+0.17	+0.17

Notes on Table 40

1. Values in col. 5 are for cells with pure lead grids. Range of temperatures is indicated approximately.
2. The temperature coefficient is zero for an acid concentration of about 4.3 per cent H₂SO₄.
3. For references see Table 39.

useful work. Several names have been proposed for the heat liberated or absorbed during the charge or discharge of the cell, but none of them has come into common use. One of these is "reversible heat," which distinguishes this heat from the irreversible heat that is generated in the cell during its operation owing to the ohmic resistance. The latter is proportional to the square of the current. The reversible heat effect is directly proportional to the current and time, because it is dependent on the extent of the electrochemical action. Another name which has been proposed is the "latent heat" of the cell, suggested by the analogy of the heat energy liberated or absorbed when water freezes or ice melts. A third proposal has been to call it the "Helmholtz heat," to distinguish it from that due to the I^2R loss within the cell, for which the name "Joule heat" has been sometimes used. This reversible heat effect has been designated by q.

All the determinations of the temperature coefficient mentioned above were made directly on the voltage changes accompanying a change in temperature. There is another, but indirect, method that can be used, since it was shown above that the reversible heat absorbed during discharge and liberated during charge is dependent on the temperature coefficient multiplied by the absolute temperature. When 2 faradays pass through the cell,

$$q = (23070 \times 2) \; T \; dE/dT$$

Streintz[19] made such a measurement, using an ice calorimeter to determine the heat liberated. He could separate the reversible from the irreversible heat effect, since the former is positive on charge and negative on discharge, whereas the latter is positive in both cases. The irreversible heat effect depends on the square of the current and the resistance. Streintz therefore obtained the following equations from which to calculate the value of q for unit time:

Total heat effect, discharge $h = I^2R - qI$

Total heat effect, charge $h' = I^2R + qI$

whence

$$q = \frac{1}{2I} \; (h' - h)$$

By making I the same on charge and discharge and keeping it a small quantity, so that there would be a negligible difference in the resistance

[19] F. Streintz, Beitrag zur Theorie des Secundärelementes, *Wied. Annalen, 49*, 565 (1893).

of the cell, he obtained two values of the coefficient at the specific
gravity 1.155, which were 0.000345 and 0.000326.

THEORY OF THE NICKEL-IRON CELLS

It is generally agreed that the reactions taking place in the nickel-
iron cells consist of the transfer of oxygen from one plate to another.
When the cells discharge, oxygen is taken from the nickel oxide or
positive plate and added to the iron or negative plate.   During charge
the reverse action takes place.   The electrolyte as a whole does not
appear to change in composition or density; but there are, nevertheless,
changes in the electrolyte within the pores of the plates.

Equations for the Reaction

Although the general character of the reactions is known, the exact
nature of the reacting substances and the chemical formulas for them
are in some doubt.   The process is not as well understood as that in
the lead-acid cells.

Much of the available theoretical material in the past was based on
the researches of Zedner, Foerster, Herold, and Schoop, but other experimenters more recently have contributed to it.   A great variety of
equations to represent the reactions has been proposed.   Some of these
are for certain phases of the operation of the cells, but none of them
covers the complete reactions in a single equation as is possible with
the lead-acid cells.

The active material of the positive plate is an oxide of nickel or a
combination of such oxides.   Two oxides of nickel are generally recognized: NiO and NiO_2.   It has been supposed that the oxides of nickel
form solid solutions. Ni_2O_3 and Ni_3O_4, however, are sometimes regarded as mixtures of NiO_2 and NiO.   Besides these oxides, there are
two hydroxides, nickelous hydroxide, $Ni(OH)_2$, and nickelic hydroxide,
$Ni(OH)_3$.   LeBlanc and Sachse[20] found that nickel oxide, NiO, can
take on astonishingly large amounts of oxygen in an active state without a change in the lattice.

Although some authors have stated definitely that Ni_2O_3 is a proven
compound, such is not the view of the Edison Company.   From a final
product of charge which contains at least a portion of NiO_2, a stable
state is reached at a ratio of oxygen to nickel that approximates
Ni_4O_7.   The transition is gradual, and this suggests a solid solution or
mixture of NiO with a higher oxide, probably the peroxide, NiO_2.   To
say this, however, involves the assumption that all the higher oxide

[20] M. LeBlanc and H. Sachse, Versuche zur Darstellung von rein Nickel-
monoxyde; Über schwarzes Nickeloxyd. Z. Elektrochem., 32, 58 and 204 (1926).

formed is not in solid solution with NiO and that some excess of NiO_2 which is less stable is therefore formed.

When the positive plate is charged the proportion of oxygen to nickel is increased. Foerster[21] has found that nickel peroxide, NiO_2, is the primary product of the electrolytic oxidation process and determines the charging potential. The nickel peroxide combines, as soon as it is formed, with unchanged nickelous oxide to give an intermediate product that approximates Ni_2O_3. Towards the end of the charge, however, the available supply of nickelous oxide becomes greatly decreased and nickel peroxide begins to accumulate in the plate. The nickel peroxide is not stable, and it passes over to a lower state of oxidation, while oxygen is given off and the potential of the plate decreases by more than a tenth of a volt. The peroxide is believed to be present in a solid solution in the active material, since the potential of the plate falls gradually during a period of several days. The open-circuit voltage of the nickel-iron cell when freshly charged is about 1.48 volts and after standing for some time about 1.35 volts.

Foerster writes the equation for the complete cell, neglecting the initial decomposition of NiO_2,

$$Fe + Ni_2O_3, 1.2H_2O + 1.8H_2O \rightleftarrows Fe(OH)_2 + 2Ni(OH)_2$$

Foerster has shown that the density of the solution is slightly higher after discharge than before. The difference is so small, however, that it may easily escape detection by the ordinary syringe-hydrometer measurements.

Kammerhoff[22] and others have quoted the equation for the Edison cell as follows:

$$2Ni(OH)_3 + Fe \rightleftarrows 2Ni(OH)_2 + Fe(OH)_2$$

It is probable that both the iron and nickel oxides in the alkaline solution are hydrated, and this equation is referred to by Allmand as the usual equation. Foerster, in referring to this equation, states that the nickel-iron cell is not independent, theoretically, of the electrolyte. Practically, however, the cell is independent of the electrolyte, and this equation has much to recommend it.

Roeber[23] proposed the formula

$$NiO_2 + Fe \rightleftarrows NiO + FeO$$

[21] F. Foerster, Die Vorgänge im Eisennickel-superoxydsammler, *Z. Elektrochem.*, **14**, 285 (1908); *Elektrochemie wässeriger Lösungen*, 3d ed., p. 270, Johann Ambrosius Barth, Leipzig, 1922.

[22] M. Kammerhoff, *Der Edisonakkumulator*, p. 13, Julius Springer, Berlin, 1910.

[23] E. F. Roeber, On the theoretical concentration changes in the new Edison battery, *Elec. World*, **37**, 1105 (1901).

which is delightfully simple. On the basis of this formula he developed theoretical discussion of the changes that occur in the electrolyte at the two electrodes during charge and discharge.

Crennell and Lea[24] have concluded, as a result of their critical study of the theories of alkaline cells, that the main part of the discharge is provided by a reduction of the hydrated sesquioxide to a lower state of oxidation and the simultaneous formation of ferrous oxide at the negative plates. Since the products of the discharge are hydrated to a different degree than in the charged state, small changes in the concentration of the electrolyte do occur. They give as the equation for the complete cell reaction:

$$Ni_2O_3, 6H_2O + Fe + 3H_2O \rightleftarrows 2NiO, 9H_2O + FeO$$

The discharged material of the negative plate may be given, however, as ferrous hydroxide.

Turning to some of the more recent studies of the operation of nickel-iron batteries, Briner and Yalda[25] gave as the equation for charge and discharge reactions the same as that quoted above from Kammerhoff's book. On discharge 69.6 calories are disengaged, and this, together with Thompson and Richardson's temperature coefficient for the electromotive force, enables them to apply the Gibbs-Helmholtz formula. As they gave it,

$$E = \frac{Q}{23.03n} + \frac{T \, dE}{dT}$$

in which $Q = 69.6$, $n = 2$, $T = 291°$ K, and $dE/dT = 0.00024$. From this they calculate $E = 1.58$ volts, but experiment indicates the value to be 1.48. The agreement is, therefore, only approximate.

Experiments by Glemser and Einerhand[26] led to the conclusion that the primary product at the positive plate as a result of charge is $NiO_2 \cdot xH_2O$. This then reacts to form β-$NiOOH$ by decomposition and reaction with $Ni(OH)_2$. On discharge the $NiO_2 \cdot xH_2O$ causes a rapid initial fall in potential to that of β-$NiOOH$. On slow discharge they noticed a potential step corresponding to $Ni_3O_2(OH)_4$.

[24] J. T. Crennell and F. M. Lea, *Alkaline Accumulators,* p. 94, Longmans, Green and Co., London, 1928.

[25] E. Briner and A. Yalda, Sur le fonctionnement aux basses températures de l'accumulateur au plomb et l'accumulateur au nickel-fer, *Helv. Chim. Acta, 25,* 416 (1942). M. deK. Thompson and H. K. Richardson, On the Edison storage battery, *Trans. Am. Electrochem. Soc., 7,* 114 (1905).

[26] O. Glemser and J. Einerhand, Die chemischen Vorgänge an der Nickelhydroxydanode des Edison Akkumulators, *Z. Elektrochem., 54,* 302 (1950).

These authors give as the equation for nickel-iron batteries

$$Fe + 2NiOOH + 2H_2O \rightleftarrows Fe(OH)_2 + 2Ni(OH)_2$$

The spontaneous reaction has certain interesting consequences, as shown by a study of the performance of cells of the nickel-iron type. The decomposition of NiO_2 is accompanied by the liberation of oxygen, which causes the popping of the caps for a considerable time after charging has ceased. The discharge of the cell soon after the completion of the charge will show a slightly greater average voltage and capacity than a similar discharge after one or more days. This is an important point to recognize in testing nickel-iron batteries, since differences in procedure will produce variations in voltage, watt efficiencies, and capacities. There is abundant experimental evidence to show that little gas escapes from the cells if they are put on discharge soon after the termination of the charge, as compared with the amount of gas that escapes if the cells stand idle. This indicates that the oxygen that escapes spontaneously can be utilized. It also explains why the "noon-hour boost" for motive-power cells in heavy service is so effective in making them "lively," aside from the mere number of ampere-hours that are added to their capacity. Lastly an explanation is offered for the well-known fact that the rate of loss capacity of the nickel-iron cell is considerably greater during the first two days after charging than subsequently. It is often stated that a partially discharged nickel-iron battery will lose less charge proportionately, when standing idle, than one fully charged.

Briner and Yalda (*loc. cit.*) experimented with cooling the positive and negative plates separately, and found the effect on the emf greater on positives than on negatives. When both plates were cooled together the emf at the end of charge reached 2.04 volts at $-40°$ C but decreased exponentially with time after the cessation of charging to normal values. No ozone was detected at these low temperatures, as is sometimes observed with lead-acid batteries.

It is possible to construct a diagram, showing the probable reactions at the plates, as has been done for the lead-acid batteries. Foerster gives such a diagram, and a slightly different diagram, based on Crennell and Lea's equation, was published in an earlier edition of this book (2d ed., p. 179). Both of these assumed the principal reaction to involve the oxide Ni_2O_3. As the more recent researches of the Edison Storage Battery Division of Thomas A. Edison have indicated that a definite compound of this composition is doubtful and that the active material of the positive plate stabilizes at a ratio of oxygen to nickel approximating Ni_4O_7, which is probably a solid solution or mixture of NiO_2 and NiO, the following diagrams for charge and discharge reactions are based on NiO_2, the characteristic product of charging reactions. It must be noted, however, that the diagrams (Fig. 62), prepared and published by the Edison Division, probably do not represent the complete reactions, nor do they indicate the hydrated state of the active materials which probably exists. The highest electromotive force, immediately after charging, is characteristic of the peroxide, but this falls eventually over an extended time to a more

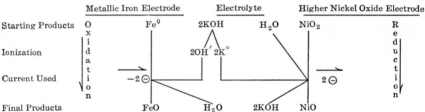

Fig. 62. Charge and discharge reactions, Edison cells.

stable value of about 1.35 volts per cell, the cell being on open circuit. This change depends on the potential of both positive and negative plates, but the greatest change is at the positives, and it is accompanied by evolution of oxygen. Such diagrams as those given in Fig. 62 suggest the mechanism of the reactions, but they cannot be regarded as the only or necessarily correct explanation of what occurs at the individual plates.

Certain irreversible effects, resulting in the liberation of gas during the charging period, occur in the nickel-iron batteries as in the lead storage battery. Gassing of the negative plate begins soon after charging commences. Hydrogen is liberated and plays a most important role in the operation of the battery, since it makes the iron active in spite of its tendency to become passive in an alkaline solution. A liberal proportion of the hydrogen is probably absorbed by the iron.

The necessity for the liberation of hydrogen at the negative plate explains the saying, familiar among battery service men, that if the battery does not gas it is not taking the charge. The rate of liberation of the gas is dependent on the amount of the current flowing, and therefore the rate of charging can affect the subsequent output of the battery. During the latter part of the charging period, oxygen is also liberated, the amounts changing slowly to 2 parts of hydrogen to 1 of oxygen as the end of the charge is reached. This means, of course, that at the end the charging current is merely electrolyzing the water present in the cell.

Schoop[27] made an interesting series of experiments on this type of cell, with electrodes suspended from the arm of a balance so that changes in weight could be measured as the cells were charged and discharged. He found changes in the volume of the active material to occur during charge and discharge. His experiments showed that the active material of the positive plate decreases in volume during discharge and increases during charge. Because of this change in volume the active material of positive plates in the Edison batteries is enclosed in tubes reinforced by iron rings. Changes in volume at the negative plate are smaller under normal operating conditions and therefore of less importance.

Secondary Discharge Reactions

The capacity of the alkaline cells is ordinarily limited by the positive plates. When the knee of the discharge curve has been reached the practical capacity of the battery has been exhausted. Beyond this there is some evidence of second-stage reactions which occur at lower voltages.

Positive plates containing graphite show a second-stage reaction occurring at a potential lower by about 0.5 volt than that corresponding to the first stage. This effect has been ascribed to oxygen absorbed by the graphite as it has not been observed in positive plates which do not contain graphite.

If a second-stage reaction occurs in the negative plates, it is not important, because the capacity of alkaline batteries is normally limited by the positive electrode. The presence of ferric iron in the active material of the negative plate is detrimental, the plates losing capacity. Edison found that the presence of mercury was very beneficial in stabilizing the capacity of these plates and in preventing reactions of the second stage.

[27] M. U. Schoop, A contribution to the theory of accumulators with constant electrolyte, *Electrochem. Ind., 2,* 272 and 310 (1904).

Relation of Voltage to Concentration

Since the electrolyte, considered as a whole, does not change, it might be supposed that the electromotive force of the nickel-iron cell would be independent of the concentration. Such is very nearly true. The researchers of Foerster have shown that the reactions are not entirely independent of the electrolyte, however, and he has found changes of a few millivolts in the voltage of the cells when the specific gravity of the electrolyte is varied through limits wider than are ever found in practice. The effect of changing concentration on the voltage is therefore negligible.

THEORY OF THE NICKEL-CADMIUM CELLS

Nickel-cadmium cells have many points of similarity with nickel-iron cells. The active materials of the positive plates, the electrolyte, and some features of construction are essentially the same. The chief difference lies in the negative plates, which contain cadmium or a mixture of cadmium and iron as the active materials. Jungner[28] patented a cell having cadmium in the negative plate about the same time Edison patented his nickel-iron cell.

The fundamental chemical reactions in a nickel-cadmium cell are somewhat uncertain, especially in regard to the hydrated state of the active materials and the part that iron[29] takes in the reactions at the negative plates. The net result of the reaction, however, is the transfer of oxygen from the active material of one plate to that of the other, without measurable change in the electrolyte as a whole. The reactions are usually written as follows:

$$\text{Cd} + \text{Ni}_2\text{O}_3 \rightleftarrows \text{CdO} + 2\text{NiO} \quad \text{(main reaction)}$$

$$\text{Cd} + \text{NiO}_2 \rightleftarrows \text{CdO} + \text{NiO} \quad \text{(secondary reaction)}$$

The principal part of the current is supplied by the main reaction, but the secondary reaction may supply part, if discharge is made within a few hours after completing the previous charge.

There has been a tendency to explain reactions in alkaline storage batteries in terms of the hydroxides. Glemser's equation[30] for the nickel-iron cell is an example. For the nickel-cadmium cell, the

---

[28] British patent 7768 (1900); German patents 110,210, 113,726, and 114,905 (1899).

[29] Anna P. Hauel, The cadmium-nickel storage battery, *Trans. Electrochem. Soc., 76,* 435 (1939).

[30] Glemser, *loc. cit.*

manufacturers of Nicad batteries give the following equation:

$$2Ni(OH)_3 \cdot 5H_2O + 2KOH + 2H_2O + Cd \rightleftarrows$$
$$2Ni(OH)_2 \cdot 6H_2O + 2KOH + 2H_2O + CdO$$

When more thermochemical data on the system of nickel oxides are available, a thermodynamic proof of the reaction may become possible.

BIBLIOGRAPHY

Gladstone and Tribe, *Chemistry of the Secondary Batteries of Planté and Faure,* Macmillan and Co., London, 1883.

Cooper, *Primary Batteries,* Electrician Printing and Publishing Co., London, 1916.

Clark, *Determination of Hydrogen Ions,* Williams and Wilkins Co., Baltimore, 1928.

Allmand and Ellingham, *Applied Electrochemistry,* Longmans, Green and Co., New York, 1924.

Lewis and Randall, *Thermodynamics,* McGraw-Hill Book Co., New York, 1923.

Dolezalek, *Theory of the Lead Accumulator,* tr. by von Ende, John Wiley & Sons, New York, 1904 (out of print).

Foerster, *Elektrochemie Wässeriger Lösungen,* 3d ed., Johann Ambrosius Barth, Leipzig, 1922.

Wade, *Secondary Batteries,* Electrician Printing and Publishing Co., London, 1902.

Crennell and Lea, *Alkaline Accumulators,* Longmans, Green and Co., London, 1928.

5

Capacity

METHODS OF RATING STORAGE-BATTERY CAPACITY

Ampere-Hour and Watt-Hour Capacity

The capacity of a storage battery may be expressed either as the ampere-hour capacity or as the watt-hour capacity. The ampere-hour capacity is a measure of the electrochemical reactions taking place within the cell in accordance with Faraday's law. In this sense the term capacity means the quantity of electricity that the battery is able to deliver. The watt-hour capacity, on the other hand, is a measure of the energy or ability to do work. The watt-hour capacity is obtained by multiplying the ampere-hour capacity by the average value of the voltage during the discharge period. In stating the capacity of any battery, it is necessary to specify the rate at which the battery is discharged, the temperature, and the final or cut-off voltage. It will be shown in this chapter that these three factors affect the capacity to a marked degree, apart from the effect of other factors which involve the design or condition of the battery.

The ampere-hour capacity of storage batteries is more often stated than the watt-hour capacity. It is simpler to measure; but a more probable reason for its almost universal use is the fact that in nearly all applications of storage batteries the current requirement is an important and controlling factor. The torque exerted by a motor, the heating of a lamp filament, or the intensity of the field within a solenoid are proportional to the current. In such instances the current becomes a primary consideration, and it is necessary to know the ampere-hour capacity. If the ampere-hour capacity and the number of cells in any battery are known, an approximation of the watt-hour capacity, sufficient for ordinary purposes, can be made immediately, since the nominal voltage of the lead-acid type of battery is 2 volts per cell and of the nickel-iron type 1.2 volts per cell.

Time and Current Ratings

Batteries may be rated for ampere-hour capacity in several ways. That most commonly adopted is the so-called time rating, by which the capacity of the battery is stated to be a certain number of ampere-hours that can be delivered within a specified time. For example, stationary batteries are rated at 8 hours, and starting and lighting batteries are rated at 20 hours. Current ratings consist in specifying the capacity that may be obtained at some particular current. The time rating is preferred to the current rating because it makes the capacites of different sizes of batteries comparable, whereas a current rating irrespective of the size of the battery imposes a less severe tax on the larger sizes and their capacity appears more than proportionately greater.

Fluctuating and Intermittent Ratings

The ampere-hour capacity of the battery is most easily ascertained when the current is constant during the period of discharge. In this case the ampere-hours are equal to the product of the current in amperes by the time in hours. If the current fluctuates with time, it is necessary to determine the integral

$$C = \int_0^t I \, dt$$

This can be obtained most conveniently by plotting the current as a function of the time and integrating the curve. The capacity C is that obtained during a discharge lasting for a time t. In general t will be limited by the falling voltage of the cell, but it may be chosen as a purely arbitrary quantity.

Choice of Final Voltages

The entire theoretical capacity of a battery cannot be obtained for several reasons. The electrolyte does not diffuse into the pores of the plates with sufficient rapidity when the pores are partially clogged with the lead sulfate; the resistance of the active material and the electrolyte increases as the discharge progresses; and, finally, it is not practical to discharge the battery to zero voltage.

As the battery discharges, the voltage at the terminals falls gradually from its open-circuit value, or slightly below, until the end of the discharge is approached, when it begins to fall much more rapidly. This point indicates that the exhaustion of the cell is near. If a curve is plotted, showing the voltage throughout the period of the discharge, the rapid fall begins at what is commonly known as the knee of the curve. The discharge may be continued slightly beyond this point, but only a small percentage of the total capacity can be obtained after the knee of the curve is passed. The amount of the actual capacity that remains after the knee of the curve is reached depends on the rate at which the battery is being discharged, the percentage being greater the higher the current. Standard practice has fixed some of the final voltages for discharges at various rates. As an illustration of the variation of voltage with the rate of the discharge, the figures in Table 41 for cells of the motive-power type are given for various multiples of the normal 6-hour rate of discharge.

TABLE 41. FINAL VOLTAGES FOR CELLS OF THE LEAD-ACID TYPE

(The table applies primarily to motive-power cells.)

Multiples of Normal Ampere Rate at 6 Hours	Volts per Cell
$\frac{1}{2}$	1.78
1	1.76
2	1.72
3	1.68
5	1.59
10	1.38

TABLE 42. FINAL VOLTAGES FOR EDISON CELLS

(The voltages are for various types and sizes of cells discharging at multiples of the normal rate, which for types A, B, C, and D is 5 hours.)

Multiples of Normal Ampere Rate	Volts per Cell*
$\frac{1}{2}$	1.05
1	1.00
2	0.91
3	0.82
4	0.73
5	0.64

* The final voltage for Edison cells is often specified as 1.0 volt per cell, irrespective of the rate; see Fig. 71.

It is not economical to discharge the battery beyond the proper final voltage.

FACTORS DETERMINING CAPACITY

The principal factors that affect the capacity of storage cells are as follows: the amount of material within the cell, the thickness of the plates, the rate of the discharge, the temperature, the quantity and concentration of the electrolyte, the porosity of the plates, the design of the plates, and the previous history of the plates. These factors will be taken up, one by one, in the succeeding pages.

The Amount of Material within the Cell

Application of Faraday's Law.[1] According to Faraday's law, 96,500 coulombs transform 1 equivalent of lead. Since the atomic weight of lead is 207.2 and the valence 2, the equivalent weight of lead is 103.6 grams. From this we may readily calculate the number of grams of lead, on the negative plate, that are transformed into lead sulfate during the passage of 1 ampere-hour. As 96,500 coulombs are equivalent to 26.80 ampere-hours, we have 3.866 grams of lead corresponding to 1 ampere-hour. Similarly, 4.463 grams of the dioxide, or active material of the positive plate, take part in the reaction per ampere-hour. Two molecules of sulfuric acid in the electrolyte are transformed for each molecule of lead or lead dioxide, in accordance with the equation

$$PbO_2 + Pb + 2H_2SO_4 = 2PbSO_4 + 2H_2O$$

The molecular weight of sulfuric acid is 98.076, and, since the valence is 2, there are 98.076 grams of acid reacting for each equivalent of the lead. This amounts to 3.660 grams per ampere-hour. The net change in the weight of the electrolyte during charge or discharge differs from 3.660 grams per ampere-hour because of the formation of 2 molecules of water during discharge for each 2 molecules of the acid that take part in the reaction. When the battery is on charge the reverse takes place; that is, 2 molecules of water disappear for each 2 molecules of acid that are formed. The net change on discharge is, therefore,

$$- 2H_2SO_4 + 2H_2O \rightarrow - 2SO_3$$

This equation shows that the actual change in weight is 160.12 grams for every 2 gram-molecules taking part in the reaction. This loss in weight of the electrolyte is balanced by a corresponding gain in weight of the plates.

The mass of the electrolyte reacting at the positive plates during discharge is slightly greater than at the negative plates. This may be seen from the formula for the chemical reactions taking place at the positive and the negative plates, as follows:

At + $\quad\quad$ $PbO_2 + H_2 + H_2SO_4 = PbSO_4 + 2H_2O$

At − $\quad\quad\quad\quad$ $Pb + SO_4 = PbSO_4$

The actual consumption of sulfuric acid at the two plates differs more

---

[1] See p. 173 for discussion of Faraday's law and the newly determined value of the Faraday constant. The round value is used here, as the difference from the new value is immaterial for the present purpose.

than the formula indicates, because it is necessary to take into account also the concentration changes caused by the migration of the hydrogen (H^+) and the sulfate ($SO_4^=$) ions.

During discharge sulfate ions are consumed at both the positive and the negative plates. Sulfate ions migrate toward the negative plates and away from the positive plates. Hydrogen ions, on the other hand, migrate away from the negative plate toward the positive plate, where part of them are consumed in the formation of water. The net change taking place is therefore a greater loss of acid at the positive plate than at the negative. If this is considered together with the amount of water formed at the positive plate, it is found that the positive requires about 1.6 times the amount of acid that the negative plate requires during the discharge period. This fact is a reason for placing the corrugated side of the separators next to the positive plate in order to allow more space for acid. It is sometimes noticed, when separators are renewed in old batteries, that the capacity of the battery is considerably increased. This is undoubtedly due in part to the additional acid space provided next to the positive plates by new separators, since the ribs of the old wood separators are generally worn down almost to the web of the separator by the time that they require renewal.

Limitations to the Use of the Materials. The foregoing figures showing the relation of lead, lead dioxide, and sulfuric acid consumed per ampere-hour are based entirely upon theoretical considerations. The practical amounts of the active materials required are considerably greater than those calculated from theory. There are several reasons for the limited use of the materials in actual service. The lead sulfate that is formed during the process of discharge is a non-conductor and increases the resistance of the active material of the plates. When the active material contains 50 per cent of sulfate, the resistance has risen to a very high value. Another reason is the stoppage of the pores of the plates by the lead sulfate, which hinders the diffusion of the electrolyte. A third reason is the increasing resistance of the electrolyte itself. It was shown in Chapter 3 that the minimum resistivity of the electrolyte occurs at a specific gravity of approximately 1.225. As the specific gravity decreases below this point the resistivity increases, slowly at first, and then more rapidly as the concentration falls below 1.100. A fourth reason, which involves the design of the plate itself, is the limited contact between the active material of the plate and its support.

The ratio of the amount of active material taking part in the reactions to the total amount of active material present in the plates is called the coefficient of use. This coefficient differs markedly with the thickness and porosity of the plate, the rate of discharge, and the temperature. A cell of good quality will have normally a coefficient of 0.25 or more. Lower values are to be attributed to deficient amount of electrolyte, lack of porosity of the plates, or improperly designed separators.

The reactions of discharge penetrate into the plate only a portion of the way. For this reason, cells that contain thin plates are found to have greater capacity than cells of similar size containing thicker plates. This is particularly true at high discharge rates. Thin plates do not have a greater capacity *per plate* than thick plates. As a matter of fact, the capacity per plate is somewhat less, but, owing to the larger number of plates that can be used in a cell of given size, the capacity of the cell as a whole may be considerably greater. For example, when two cells of equal size containing plates which are $\frac{14}{64}$ inch in thickness and $\frac{9}{64}$ inch in thickness are compared, it is found that the capacity of the thinner plate is 91 per cent of the capacity of the thicker plate, although the thinner plate has only 64 per cent of the thickness of the other. In a given jar, therefore, 56 per cent more of the thinner plates can be installed, assuming a corresponding reduction in the thickness of the separators. The relative capacity of the two cells, neglecting the effect of separators, can be calculated by multiplying 156 per cent by 91 per cent. This gives 142 per cent as the capacity for the cell containing thin plates, as compared with the capacity of the same cell with the thicker plates taken as 100 per cent. The increase in capacity is, therefore, 42 per cent, which agrees very closely with laboratory measurements on such cells.

Maximum Output Obtainable. The maximum output obtainable from storage cells can be computed. It will be necessary to assume, for this estimate, thin plates of high porosity discharging at a low rate over a long period of time. The grid of a pasted plate amounts to 35 to 50 per cent of the weight of the finished plate. It was found on page 207 that 3.866 grams of lead and 4.463 grams of lead dioxide are required per ampere-hour. This is a total of 8.329 grams of active material as the theoretical requirement for each ampere-hour. We shall assume that the coefficient of use, as defined above, is 50 per cent. This is rather more than can ordinarily be obtained. Dividing 8.329 grams by 0.50 gives 16.65 grams of active material as the best possible equivalent per ampere-hour. Assuming that the grid is only 35 per cent of the finished weight of the plate, the active materials will constitute 65 per cent, and, dividing 16.65 by 0.65, we obtain 25.62 grams per ampere-hour as the weight of the plates. It is necessary now to ascertain the relation between the weight of the plates and the weight of the complete cell. This varies considerably among the cells made by different manufacturers, but in Table 43 are given the weight of each of the different parts of the cell in grams and the per cent of the total weight.

TABLE 43. RELATIVE WEIGHTS IN GRAMS AND PERCENTAGES OF VARIOUS PARTS OF THE COMPLETE CELL

(Figures are for cells of the same size containing 15, 17, and 21 plates. The data for the 15- and 17-plate cells are the mean of three different makes; and for the 21-plate cells the mean of two different makes.)

Plates		Total Weight	Posi- tives	Nega- tives	Straps	Elec- trolyte	Wood Separa- tors	Rubber Separa- tors	Jar	Cover, Vent, and Sealing Com- pound
Thick	15	21,237	7660	6657	982	4021	420	153	1150	194
	Per Cent	36.1	31.3	4.6	18.9	2.0	0.7	5.4	0.9
Medium	17	21,528	7738	6203	1133	4561	403	269	1281	320
	Per Cent	35.9	28.8	5.3	21.4	1.9	1.2	5.9	1.5
Thin	21	20,913	7077	5970	1240	4710	354	221	1187	154
	Per Cent	33.9	28.5	5.9	22.5	1.7	1.0	5.7	0.7

In the table it is shown that the positive and negative plates constitute from 62 to 67 per cent of the total weight of the cell. Resuming the example, the weight of the cell per ampere-hour is therefore 25.62 divided by 0.624, or 41.1 grams per ampere-hour, as the result computed for the 21-plate cells. This corresponds to a maximum value of 24.2 ampere-hours per kilogram or 11.0 ampere-hours per pound of the cell. If we may assume that the average voltage during the discharge of the cell is 1.95 volts, the energy capacity becomes 47.2 watt-hours per kilogram or 21.4 watt-hours per pound.

Having calculated the maximum output obtainable, we shall now compare it with the actual output of several types of storage cells. Batteries containing soft, medium, hard, and very hard plates have been chosen for this table. Column 2 shows the weight of the positive active material, since it is the positive plate which limits the capacity of the cell.

The effect of hardness of the plates on the utilization of the active materials is better illustrated by these batteries of 30 years ago than by batteries of the present day, which differ much less in hardness of plates. The batteries containing medium plates, Table 44, correspond to 13-plate batteries of 96 ampere-hours, discharging at the 20-hour rate.

TABLE 44. THEORETICAL AND ACTUAL CAPACITY OF STORAGE BATTERIES

(Batteries [7-plate cells] discharged to 1.70 volts per cell at 5-hour rate and 1.50 volts at 20-minute rate, assuming the grid to be 50 per cent of the weight of the positive plates.)

Kind of Plate	Weight of Positive Active Material, g	Complete Theo-retical Capacity, amp-hr	Actual Capacity 5-Hour Rate, amp-hr	Coeffi-cient of Use, %	Actual Capacity 20-Minute Rate, amp-hr	Coeffi-cient of Use, %
Soft	504	113	49	43	28	25
Medium	552	123	40	33	23	19
Hard	528	118	35	30	17	14
Very hard	603	135	27	20	12	9

The output of storage cells of the motive-power type per unit of weight and space is given for the 5-hour rate of discharge in Table 45. The last figure given for the watt-hour capacity per cubic foot of the 21-plate cells indicates about 2 hp per cubic foot of the space occupied by the battery.

TABLE 45. OUTPUT OF STORAGE CELLS OF THE MOTIVE-POWER TYPE PER UNIT OF WEIGHT AND SPACE, 5-HOUR RATE OF DISCHARGE

(The cells were all nominally of the same external dimensions. The ampere-hours and watt-hours per unit of weight are based on the weight of a single cell. The watt-hours per unit of volume are calculated from observations on a battery of 12 cells. The plates were 21.9 cm by 14.6 cm [$8\frac{5}{8}$ by $5\frac{3}{4}$ inches].)

Number of Plates per Cell	Thickness of Positive Plates		Ampere-Hours per		Watt-Hours per			
	Cm	In.	Kg	Lb	Kg	Lb	Dm³	Ft³
15	0.60	15/64	10.9	4.9	21.2	9.6	45	1263
17	0.52	13/64	11.7	5.3	22.6	10.3	48	1350
21	0.44	11/64	13.1	5.9	25.4	11.5	51	1440

Thickness of the Active Material

The capacity of a storage cell increases with the thickness of the active material of the plates at moderate rates of discharge, assuming that the plates have sufficient porosity for the electrolyte to reach the inner recesses. The effect of thickness cannot be considered apart from the rate of the discharge, because the faster the discharge the more nearly is the total output of the cell confined to the layers of active material that are in immediate contact with the free electrolyte. At

Storage Batteries

excessively high rates of discharge, the output of the cell becomes practically a surface phenomenon. This is because there is insufficient time for the electrolyte to diffuse into the pores of the plates, and the

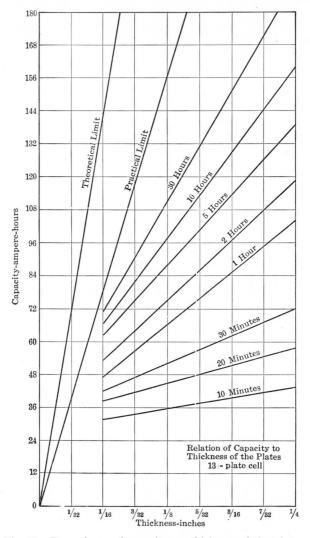

Fig. 63. Dependence of capacity on thickness of the plates.

sulfate forming at the surface clogs the pores. At low rates of discharge, on the other hand, almost any depth of the active material may become effective. For example, Fig. 63 shows the relation of capacity to the plate thickness for a certain make of starting and lighting battery containing plates varying from $\frac{1}{16}$ inch to $\frac{1}{4}$ inch in thickness. These results are calculated for a battery containing 13 plates to the cell. The very great difference between discharges made at high rates and at low rates for plates of different thickness may be illustrated as follows: At the 10-minute rate of discharge, the $\frac{1}{4}$-inch plate gives only 38 per cent more capacity than the plate $\frac{1}{16}$ inch in thickness, but at the 30-hour rate it gives 170 per cent more.

The negative plates are more sensitive to changes in thickness than the positive plates. This means that, if the capacities of the positive and negative plates are equal at any high rate of discharge, the negative will exceed the capacity of the positive at any lower rate of discharge.

Area of the Plates

For any specified amount of material and thickness of the plate, the area of the plates is necessarily determined. The area is, therefore, not an independent factor in determining the capacity, but, since it is most easily measured, it is desirable to discuss the output in its relation to the surface of the plates. The area of the plate requires careful definition, since it may be taken to mean the area as calculated from dimensions of width and height, or the entire developed surface in the Planté plates, which is 7 to 10 times the area calculated from the dimensions. The area of pasted plates is calculated from the width and height and doubled to allow for both sides of the plates.

The output obtained from a given area of plate varies with the type of plate and the rate of the discharge. For Planté plates an output of 1 ampere-hour for each 100 to 125 square centimeters (15 to 20 square inches) of developed surface is about the average. The output for pasted plates of the motive-power type is given in Table 46, show-

TABLE 46. OUTPUT PER UNIT OF PLATE AREA FOR MOTIVE-POWER TYPE CELLS OF NOMINALLY SAME SIZE

Plates per Cell	Area Positive Group, sq cm	Capacity 5-Hour Rate, amp-hr	Ampere-Hours per	
			Sq Cm	Sq In.
15	4480	228	0.051	0.329
17	5120	244	0.048	0.309
21	6400	260	0.041	0.264

ing that the capacity per unit of area is somewhat less for the thin plates than for the thicker plates. In the aggregate, however, the capacity for the thin-plate cells is greater, because the increased surface of the plates more than offsets the decreased capacity per unit of area.

The values given in the table are for motive-power cells discharging at the 5-hour rate.

In order to study the relation of plate area to the rate of discharge, we refer to Fig. 63 in the preceding section, which is calculated for a cell of 13 plates. We shall take as an example a cell containing 13 plates, $\frac{3}{16}$ inch in thickness, and compare with it another cell of similar size having double the number of plates, which are $\frac{3}{32}$ inch in thickness. It is obvious that the volume of the plates in each is the same, but that the area of the $\frac{3}{32}$-inch plates is double that of the $\frac{3}{16}$-inch plates. At the 30-hour rate, the cell containing the $\frac{3}{32}$-inch plates has an aggregate capacity of 182 ampere-hours. The cell containing the $\frac{3}{16}$-inch plates has a capacity of 152 ampere-hours. Halving the thickness of the plates, and thereby doubling the area, has increased the capacity 20 per cent at this low rate of discharge. At the 10-minute rate of discharge, the cell containing $\frac{3}{32}$-inch plates has a capacity of 67 ampere-hours and the other cell with $\frac{3}{16}$-inch plates has a capacity of 40 ampere-hours. At this high rate of discharge, halving the thickness and doubling the area has increased the capacity of the cell 68 per cent.

Efforts to get greater area of plates are made particularly for airplane batteries, for which plates as thin as 0.050 inch are often used. Thin plates possess two advantages for high-rate service. The first is increased capacity for unit of space and weight. The second advantage is decreased resistance, which permits the cell to deliver very large currents with a small waste of energy within the cell itself.

Rate of Discharge

It is a familiar fact that storage cells will not give as great capacities when discharging at high rates as when discharging at lower rates. The causes of the decreased capacity at the high rates are the following: the sulfation on the surface of the plates, which closes the pores; the limited time available for diffusion of the electrolyte; and the loss of voltage because of internal resistance of the cells.

Variation of Capacity with Time and Rate. Figure 64 shows the discharge characteristics of one type of storage cell designed for starting and lighting service. The capacity is expressed as ampere-hours per positive plate, as a function of the current discharged per positive plate. That is to say, if a cell contains 6 positive plates, the ampere-hour capacity indicated for any given rate or time of discharge is to be multiplied by 6. The capacities are shown for all rates of discharge, from the 5-minute rate to the 30-hour rate, but it should be noted that the scale of the abscissa is arbitrarily changed at 10 amperes per positive plate, in order to make it possible to show the complete characteristics of cells of this type within the limits of a single illustration. The diagonal lines intersecting the curve of capacities give the time of discharge.

Successive Discharges. The capacity of storage cells is seen from Fig. 64 to increase as the rate of discharge becomes lower. The physical significance of this effect lies chiefly in the diffusion phenomena.

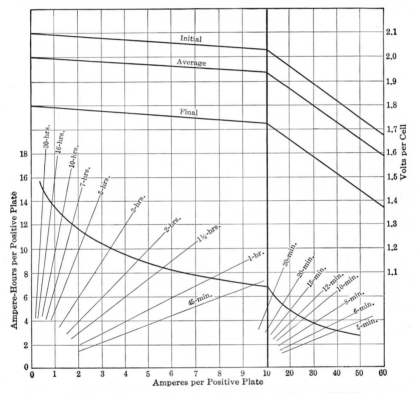

Fig. 64. Discharge characteristics of a plate of the starting and lighting type.

After a battery has been discharged to its final voltage at any given rate, it may be further discharged at any lower rate. That is to say, the total capacity of the battery is equal to the summation of terms in accordance with the following equation:

$$C = \Sigma I_1 t_1 + I_2 t_2 + I_3 t_3 \cdots I_n t_n$$

where I_1, I_2, I_3, etc., are successively lower values of the current, and the t's represent the time during which the battery was discharging at the rate indicated by I with corresponding subscripts. Deception has been practiced upon some automobile owners in order to sell special electrolytes for which extraordinary claims are made, the deception consisting in the failure to state that the same result may be obtained with the ordinary electrolyte as with the special product whose merits are exploited. A storage cell for demonstration purposes is discharged to thê point where it fails to crank the engine, and then is allowed to rest for a period of a few minutes. During this time the more concentrated electrolyte in the cell is diffusing into the pores of the plates, and it is possible to discharge the battery again. This may be repeated a number of times, and so it may appear, in the absence of exact measurements, that the battery has acquired an unlimited capacity because of the exploited material it contains.

Equations to Relate Current and Time. A number of attempts have been made to develop equations which would relate the current to the time of discharge, in order that the capacity of batteries might be computed for any rate or time of discharge. The most widely used of these equations is Peukert's equation,[2]

$$I^n t = C \qquad (1)$$

In this equation, n and C are constants which may be evaluated by tests made on any cell or battery at two different rates of discharge. It will be assumed that the different rates are I_1 and I_2, the times of these discharges corresponding to t_1 and t_2. The values for n and C may therefore be calculated as follows:

$$\left. \begin{array}{l} I_1{}^n t_1 = C \\ I_2{}^n t_2 = C \end{array} \right\} \qquad (2)$$

$$\left. \begin{array}{l} n \log I_1 = \log C - \log t_1 \\ n \log I_2 = \log C - \log t_2 \end{array} \right\} \qquad (3)$$

$$n = \frac{\log t_2 - \log t_1}{\log I_1 - \log I_2} \qquad (4)$$

The numerical value for n being found for any particular type, the other constant C may be determined by solving either of the equations (2).

Peukert's equation was based on and originally applied to batteries made in European factories. It will be interesting, therefore, to apply Peukert's equation to American-made batteries.

Peukert's equation, being logarithmic, can be plotted as a straight line on log-log paper. Similarly the observed relation of current and time of a battery's discharge should plot as a straight line if the rela-

---

[2] W. Peukert, Über die Abhängigkeit der Kapacität von der Entladestromstärke bei Bleiakkumulatoren, *Elektrotech. Z., 18,* **287** (1897).

tionship is strictly a logarithmic function. Many observers have reported, however, that the experimental results deviate slightly from calculated values based on Peukert's formula. These deviations are attributed to ohmic resistance of the cells and to a fixed or an incorrect choice of cut-off voltages. Table 47 is based on observations made

TABLE 47. RELATIVE CAPACITY OF LEAD-ACID STORAGE BATTERIES AT VARIOUS RATES OF DISCHARGE AND TEMPERATURES

(Percentages based on the capacity at 5-hour rate and 80° F taken as 100 per cent.)

Rate of discharge	10 hr	5 hr	3 hr	2 hr	1 hr	40 min	30 min	20 min	10 min	5 min
Cut-off voltage	1.77	1.75	1.73	1.71	1.64	1.60	1.54	1.46	1.26	0.95

Temperature		Relative Capacity in Per Cent for Rates of Discharge and Cut-Off Voltages Specified above								
°C	°F									
27	80	120	100	88	78	65	58	53	46	38	30
10	50	95	78	68	60	49	44	40	35	29	23
0	32	79	65	56	50	41	37	33	29	23	18
−10	14	64	52	45	40	32	28	26	22	17	13
−20	−4	50	40	34	30	24	21	18	15	11	6
−30	−22	36	28	23	20	15	13	11	8	4	..
−40	−40	23	17	13	10	6	4	3	1	..	..
−50	−58	11	5	2	0	..	..	..	..	..	..

on a group of small cells having 9 plates to the cell and wood separators. Temperatures were measured by thermocouples immersed in the electrolyte, and the cut-off voltages were determined by the knee of the discharge curves at each rate. The values for n and C in Peukert's equation having been determined from data in Table 47, observed and calculated values are compared in Table 48. The accuracy with which Peukert's formula represents the discharge of these cells is remarkable.

Example 1. Compute the values of the constants n and C in Peukert's equation from test results at the 1- and 10-hour rates as given in Table 47.

$$I_1 = 65 \text{ amperes} \qquad\qquad t_1 = 1 \text{ hour}$$
$$I_2 = 12 \text{ amperes} \qquad\qquad t_2 = 10 \text{ hours}$$

$$\begin{array}{ll} \log 65 = 1.813 & \log 10 = 1.000 \\ \underline{\log 12 = 1.079} & \underline{\log\ 1 = 0.000} \\ \qquad\quad 0.734 & \qquad\quad 1.000 \end{array}$$

$$n = 1.000/0.734 = 1.36$$

$$C = (65)^{1.36} = 291.4$$

Example 2. Using the values of n and C obtained in the preceding example, compute the length of time, t, of discharge at 39 amperes, and compare the results obtained with the experimental data given in Table 48.

$$t = 291.4/(39)^{1.36} = 2.00 \text{ hours} \quad \text{(table shows } 2.0\text{)}$$

TABLE 48. COMPARISON OF CALCULATED TIME OF DISCHARGE, USING PEUKERT'S FORMULA, WITH OBSERVED VALUES

(The constants calculated for Peukert's formula are: $n = 1.36$; $C = 291.4$.)

Current, amp	Time of Discharge, hr	
	By Formula	Observed Values, Table 47
12	9.94	10.0
20	4.96	5.0
29.3	2.95	3.0
39	2.00	2.0
65	1.00	1.0
87	0.67	0.67
106	0.51	0.50
138	0.36	0.33
228	0.18	0.17
360	0.097	0.083

Other equations[3] which have been published, relating the capacity of storage cells to the current at which they are discharged, are by the following: Schroeder; Liebenow; Rabl; Davtyan. These equations in the experience of the author are not as representative of American-made cells as the equation of Peukert, and two of them are much more difficult to use.

Temperature

The temperature plays an important part in determining the capacity that can be delivered by a storage cell under any specified conditions of rate and final voltage. Because batteries are commonly used when atmospheric temperatures are low, interest centers on improving their output under such conditions. Much has been accomplished by the use of "organics," usually wood extracts, as expanders in the negative plates. Chemical reactions do not take place readily at

[3] L. Schroeder, Berechnung von Akkumulatoren für Elektricitätswerke, *Elektrotech. Z.*, *12*, 585 (1891). C. Liebenow, Über die Berechnung der Kapazität eines Bleiakkumulators bei variabeler Stromstärke, *Z. Elektrochem.*, *4*, 63 (1897). M. Rabl, Berechnung der Kapazität von Bleiakkumulatoren bei Teilweiserentladung bis zu beliebigen Spannungsgrenzen Anderung de Kapazitätskurven mit der Temperatur, *Z. Elektrochem.*, *42*, 114 (1936). O. K. Davtyan, Capacity of Acid Storage Batteries, *Bull. acad. sci. U.R.S.S., Classe sci. tech.*, 737 (1946).

extremely low temperatures, and much below 0° F it becomes impossible to charge the batteries.

Table **47** gave the comparative capacities of batteries at various rates of discharge and temperatures. These data may be considered conservative. Some of the published data indicate higher percentages

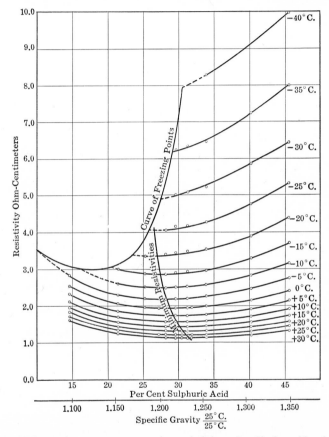

Fig. 65. Effect of temperature on the resistivity of sulfuric acid solutions.

than those given in this table, and there is a rather wide difference between the highest and lowest values. Part of this may be attributed to the various makes and types of batteries. As it is not possible to furnish data that will be universally applicable, a composite table, **49**, based on four independent researches, is provided. This illustrates the principle and gives approximately correct data for the more common types of cells.

Figure 65 shows graphically the effect of temperature on resistivity of the electrolyte. The concentration is limited to about 1.280 sp. gr. when the battery is charged, if the negative plates are to function properly. When the battery is discharged the specific gravity cannot be allowed to fall below the limits set by the freezing-point curve. The resistivity increases very rapidly at temperatures below 0° C (32° F) as shown by the curves.

TABLE 49. RELATIVE CAPACITY OF LEAD-ACID STORAGE BATTERIES AT NORMAL AND LOW TEMPERATURES

Temperatures		Percentage of Capacity at 80° F, 20-Hr Rate	
°C	°F	20-Hr Rate	20-Min Rate
27	80	100	46
15	60	90	39
4	40	77	31
−7	20	63	24
−18	0	49	16
−29	−20	35	9
−40	−40	21	1
−51	−60	9	..

Low temperatures increase the viscosity of the electrolyte, impairing its circulation in the pores of the plates. The increase in viscosity becomes very rapid at temperatures below 0° C (32° F) as shown in Fig. 66. At −50° C (−58° F) the viscosity is nearly 30 times as great as at ordinary temperatures. This illustration shows rather better than the preceding figure the limitations imposed by the two arms of the freezing-point curve. The range of concentrations within which the electrolyte remains a liquid becomes increasingly narrow as temperature is decreased. The extreme low temperatures shown in these figures are not below possible atmospheric temperatures.

Temperature Coefficient of Capacity. Many experiments have been made to determine the temperature coefficient of capacity of storage cells. One per cent per degree Centigrade (0.56 per cent per degree Fahrenheit) for cells containing Planté plates is a value that has been given in several previous books. Liebenow[4] states that he found the capacity to be a linear function of the temperature, and the increase for 1° rise in temperature to be 1 per cent of the capacity at 15° C. Data given in Table 49 for pasted plates indicate an approximately similar value, 0.64 per cent per °F (1.2% per °C).

[4] C. Liebenow, *Dependence of Capacity on Current Strength*, Inaug. Dissert., Göttingen, p. 6, 1905.

Variation of the Coefficient with the Rate. Since temperature affects the rate of diffusion and the resistivity of the electrolyte, it may readily be expected that the temperature coefficient of capacity will vary with the rate of discharge. Using the data in Table 49, at the 20-hour rate the capacity may be taken as 100 ampere-hours at 80° F and 49 ampere-hours at 0° F. The available capacity is, therefore,

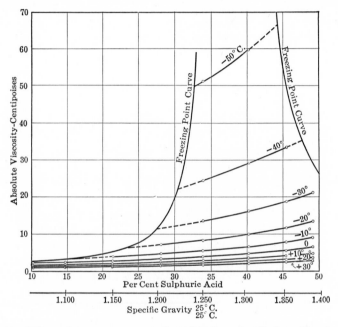

Fig. 66. Effect of temperature on the viscosity of sulfuric acid solutions, and the limitations imposed by freezing points.

49 per cent of the capacity at 80° F. At the 20-minute rate, the capacities at 80 and 0° F are 46 and 16 ampere-hours, respectively, and the available capacity at the lower temperature is 35 per cent of that at the higher temperature. Stated in another way, the percentage of available capacity at lower temperatures decreases as the rate of discharge increases.

Concentration of the Electrolyte

Change in Capacity with Concentration. The concentration of the acid in the pores of the plates is a vital factor in determining the voltage and capacity of a cell. Unless a sufficient amount of sulfuric acid can be maintained in the pores during discharge the voltage at the terminals of the cell will decrease rapidly and the cell will become exhausted. The concentration affects the capacity, first, because it determines the potential of the plates; second, because it affects the resistance of the electrolyte to the passage of the electric current; third, because it affects the viscosity of the electrolyte and thereby the rate of diffusion; fourth, because differences in the concentration of the electrolyte in the pores of the plates and outside also affect the rate of diffusion.

Liebenow's Experiment. Liebenow[5] carried out a celebrated experiment which has often been quoted to show the influence of concentration on the capacity of the plates. He welded a negative plate of the pasted type to form a window in the side of a lead vessel which he then placed in a larger vessel. By varying the height of the electrolyte in the two vessels, it was possible to exert a hydrostatic pressure to force the electrolyte through the plate, which served as a window of the inner compartment, in either direction, or, if the electrolyte were at the same level in the two vessels, to eliminate the hydrostatic pressure. Liebenow suspended a positive plate inside of the inner vessel and determined the capacity of this arrangement when the electrolyte had the same level in the two vessels. The capacity was found to be about 14 ampere-hours. After this the level of the electrolyte in the inner compartment was maintained higher than in the outer vessel, so that the acid was continually passing through the plate. He again determined the capacity and found it to be approximately 42 ampere-hours, or 3 times as great. This experiment shows that, if the concentration of the electrolyte in the pores of the plate can be maintained during the discharge period, the cell will have a much greater capacity. In other words, the capacity of the cell ordinarily is limited by the supply of electrolyte. Only under extreme conditions, when the discharge takes place at very low rates and the coefficient of utilization of the active material is very high, can the capacity be said to be limited by the amount of active material available.

A somewhat similar experiment was made more recently by Hauel,[6] who used a negative plate as a microporous diaphragm, forcing electrolyte through it to observe changes that occur during formation. The rate of flow varied with changes in the structure of the active material.

Maximum Capacity. Experiments to determine the concentration of electrolyte for which the capacity of any given storage cell is a maxi-

---

[5] C. Liebenow, Über die Berechnung der Kapazität eines Bleiakkumulators bei variabeler Stromstärke, *Z. Elektrochem.*, *4*, 61 (1897).

[6] Anna P. Hauel, Microporous lead plates in storage batteries, *Trans. Electrochem. Soc.*, *78*, 231 (1940).

mum have been made by numerous observers.   Their conclusions have
varied from concentrations of 1.100 to 1.270 sp. gr.   The wide diver-
gence of their results is probably accounted for by the varying conditions of their experiments.

The capacity of the cells increases as the concentration of electrolyte
increases, within the range now employed.   This is particularly true at
high rates of discharge when the capacity of the cells is limited by the

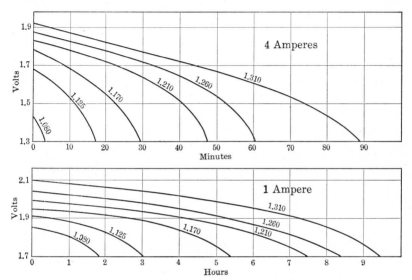

Fig. 67.   Effect of concentration of the electrolyte on the capacity of cells
discharging at 1 ampere and at 4 amperes.

positive plates.   Figure 67 shows a series of experiments made on small
cells of the couple type having pasted plates and containing electrolyte
of specific gravities ranging from 1.080 to 1.310 at the beginning of
the discharge.   The figure shows the characteristic curves for discharges at 4 amperes and 1 ampere.   The plates were 3 inches wide by
4 inches high.   This figure shows that, when the rate of discharge was
decreased from 4 amperes to 1 ampere, the time of discharge for the
cell containing the highest specific gravity was increased 6.4 times, but
the cell containing the lowest specific gravity gave 35 times the length
of discharge at the lower current rate.

**Comparison of Effects of Concentration and Temperature on Posi-
tive and Negative Plates.**   Although a high concentration of the
electrolyte is favorable for the positive plates, it may be detrimental to
the negative plates.   The capacity of the negative plates in an electrolyte of 1.315 sp. gr. is less than in an electrolyte of 1.140 sp. gr., particu-

larly at high rates of discharge and low temperatures. The negative plates may thus become the limiting factor. Cases are on record showing that it has been necessary to reduce the concentration of electrolyte in airplane batteries, for example, in order that they might meet the capacity requirements specified for them. The reduced concentration

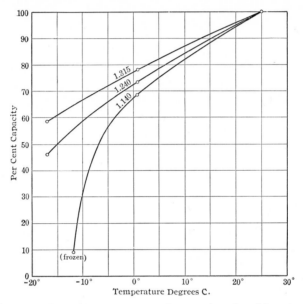

Fig. 68. Effect of temperature on the capacity of positive plates.

may reduce slightly the capacity at normal rates and temperatures when the positives limit the capacity, but this is more than offset by the gain in capacity under more severe conditions that result in limitations imposed by the negatives.

In order to study the effects of temperature and concentration on the positives and negatives separately, Vinal and Snyder[7] made use of small cells in which the capacity could be limited arbitrarily by the plates of either polarity. Figures 68 and 69 show the results for positive and negative plates discharging at a rate that approximated the 5-hour rate at normal temperatures. The concentrations stand in reverse order in the two figures. Furthermore, the temperature coefficient of the negative plates as measured by the slope of the curves is greater than that of the positives. It must not be concluded, however, that a specific gravity as low as 1.140 is always the most favorable

[7] G. W. Vinal and C. L. Snyder, Effect of temperature and other factors on the performance of storage batteries, *Trans. Am. Electrochem. Soc.*, **53**, 233 (1928).

for the negative plates. The same cells discharging at a higher rate (2 amperes) gave the best results when 1.240 sp. gr. acid was used. The possibility of freezing when the specific gravity is too low should not be overlooked. In any event the choice of the strength of the solution involves a compromise, depending on the conditions of service.

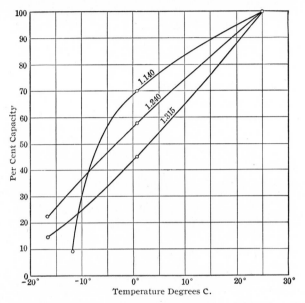

Fig. 69. Effect of temperature on the capacity of negative plates.

Porosity of the Plates

The importance of porosity of the plates in facilitating the access of electrolyte to the active material has been treated in connection with other factors previously discussed. The porosity of the finished plate varies with the expansion of the material from which it was made and with its state of charge. The aggregate porosity of the plate is a matter of 50 per cent of the entire volume of the plate, but the individual pores are probably little more than molecular in size.

Although the term porosity is generally understood to indicate the possession of absorption qualities similar to those of a sponge, it is desirable to establish a definition for it. The porosity is equal to 1 minus the ratio of the apparent density of the active material to the real density. From this definition the percentage of porosity of the plates may be obtained in any case by multiplying by the factor 100.

Variation of Porosity with State of Charge. The porosity of the plate varies with the state of charge. Lead sulfate, which is formed as a product of the discharge of the cells, is less dense than either lead or dioxide of lead. It therefore occupies more space than the active materials, yet the apparent volume of the plate does not change. The reason for this is that the expansion is taken care of by the pores of the plates.

Measurement of Porosity. The relative porosity of storage-battery plates can be determined by the amount of water that can be absorbed by the pores when they are initially in the dry condition. Such a measurement, however, is seldom accurate, and it is preferable to determine the actual porosity by calculation as follows:

Example. Determine the porosity of a positive plate of an airplane battery. The three dimensions are as follows:

Length	12.46	cm
Width	14.28	cm
Thickness	0.1275 cm	

(In order to obtain a sufficiently accurate determination of the thickness, it is necessary to use a micrometer caliper and to take the mean of a number of readings at different parts of the plate.)

The volume of this plate is 22.632 cc. To this must be added the volume of the lug, 1.03 cc. The total volume of the plate is 23.662 cc. The volume of the grid equals the weight of the grid divided by its density. Its density is determined by Table 3, given in Chapter 2, after a chemical analysis has been made to find the percentage of antimony it contains. In this example the volume of the grid equals 81.215 divided by 10.85 = 7.485 cc. Deducting the volume of the grid from the volume of the plate, we obtain the space available for the active material, 23.662 − 7.485 = 16.177 cc. The weight of the fully charged plate washed and dried was 143.3 grams. Deducting the weight of the grid, 81.22, the weight of the active material is 62.08 grams. The actual volume of the active material is equal to the weight divided by the density = 62.08 divided by 8.8 = 7.06 cc. The pores of the plate are equal to the space available for the active material minus the actual volume of the material = 16.18 − 7.06 = 9.12 cc. The porosity equals the ratio of the pore space to the total space available for active material = 56.5 per cent. *Answer.*

A determination of the porosity of the same plate by the absorption method, the water being drained off for about 1 minute, gave the porosity as 50.5 per cent. The porosity as determined by water is consistently less than the calculated value because of the rapid evaporation of the water on the surface of the plates and the capillary action of the pores. Determining the porosity of the negative plates requires a special procedure, because the negative plates in the battery oxidize spontaneously when exposed to the air, accompanied by a marked rise in temperature. Fully charged negative plates have been successfully dried without oxidation in atmospheres of inert or reducing gases and in vacuum.

Samsonov[8] and his coworkers made porosity determinations based on true and apparent specific gravities of the active materials. They found benzene and toluene best suited for determinations of the volume of the active mass.

Diffusion Phenomena. The diffusion which takes place during both the charging and discharging periods is an important factor in the operation of the cell. According to the kinetic theory of diffusion, the molecules and ions in the electrolyte are in continual motion. The proof of this theory is found by examination of colloidal solutions in which the Brownian movement of the colloidal particles can be clearly seen. The motion of the particles to and fro is resisted by the viscosity of the solution.

Fick's law[9] states that the rate of diffusion varies as the difference in the concentration and as the area of the pores, and inversely as the length of the path to be traveled. We may, therefore, write the equation for the amount of acid diffusing in a given time as follows:

$$Q = D\Delta C\,(a/l)$$

Q = quantity of acid diffusing in a given time; D = diffusion coefficient; ΔC = difference in concentration between the electrolyte within the pores and that outside; a = aggregate cross section of the pores; and l = distance through which the acid must diffuse.

Diffusion is a slow process, and the diffusion coefficient is a constant unique for a particular substance. It probably varies slightly with the concentration. The electromotive force affords a sensitive and exact indication of the time when diffusion is complete. Twenty hours or more may be necessary for the emf to become constant to within 1 part in 20,000, as was found by Vinal and Craig.[10]

Other phenomena related to diffusion deserve mention, although they affect the operation of the cell to a very small extent. Electrical endosmosis undoubtedly takes place through the porous separators during both charge and discharge. It does not occur within the porous sections of the plates, however, because the active material of the plates is an electrical conductor, and therefore the potential differences are small in spite of the difference of concentration of the electrolyte in the pores of the plates and outside. There is an electromotive force at the

[8] P. D. Samsonov et al., Structure of the pores and the efficiency factor in the utilization of the active material of the plates, *J. Appl. Chem., U.S.S.R., 14,* 317 (1941). In French, p. 334.

[9] Adolf Fick, Über Diffusion, *Pogg. Annalen, 94,* 59 (1855).

[10] G. W. Vinal and D. N. Craig, Chemical reactions in the lead storage battery, *J. Research Natl. Bur. Standards, 14,* 449 (1935).

liquid junction between the two concentrations. This is caused by the unequal speed of the ions. The hydrogen ions move the most rapidly, carrying with them positive charges. Since the electrolyte in the pores of both positive and negative plates becomes depleted on discharging, the liquid-junction potentials are in opposite directions at the two plates and nearly equal in magnitude. Hence their combined effect is negligible.

Previous History of the Plates

Variations of Capacity during Life. When the plates in a storage battery are new, there is an increase in the capacity during the first few cycles of charge and discharge, up to a maximum value beyond which the capacity becomes relatively constant and then gradually falls off. Thin plates reach their maximum capacity after a smaller number of cycles than thick plates, but beyond this point their decrease in capacity is more rapid. After the capacity of the plates has fallen to 80 per cent of their initial capacity, they are considered worthless. During the later cycles in the life history of the plates, the decrease in capacity of the negative plates is caused primarily by a decrease in the porosity of the material or shrinkage of the plate itself. The loss of capacity of the positive plates is caused by corrosion of the grids and loosening of the active material. In addition to these effects, however, there is a third effect arising from the wearing of the separators. As the separators become thinner and thinner, particularly the ribbed portions, the space available for the electrolyte next to the positive plate becomes inadequate, and the capacity of the cell may be limited by insufficient acid for the action of the positive plate.

Effect of Previous Discharges. The capacity of the plates is affected by the discharges immediately preceding. For this reason it is desirable to run a few preliminary cycles at the same rate of discharge as is to be used in a formal test. This hysteresis effect was observed by Jumau[11] and has been confirmed at the National Bureau of Standards. Jumau found that the capacity delivered at any given rate depends on the previous discharge. The capacity is lower if the discharge has been preceded by a discharge at a higher rate, and it is higher if preceded by a discharge at a lower rate. Table 50 gives measurements on thin-plate airplane batteries illustrating this effect. There is also a sluggishness which is caused by inactivity for any considerable period of time. This may be overcome by charging and discharging the cells for a few cycles.

[11] L. Jumau, Sur la rôle de la diffusion dans le fonctionnement des plaques positives de l'accumulateur au plomb, *L'éclair. élect.*, *16*, 413 (1898).

TABLE 50. HYSTERESIS EFFECT

(Average measurements on four thin-plate batteries showing the variation of capacity with changes in the rate of discharge.)

Ampere-Hour
Capacities

Rate of Discharge	Batteries 1 and 2	Batteries 3 and 4
5-Hour rate	31.5	65.9
20-minute rate	17.8	41.6
Subsequent discharges at this rate	15.0	31.7
5-minute rate	6.9	14.7
Subsequent discharges at this rate	6.1	13.0
20-minute rate	13.7	32.2
Subsequent discharges at this rate	....	....
5-hour rate	30.8	63.9
Subsequent discharges at the 5-hour rate	31.6	64.6

CAPACITY OF EDISON BATTERIES

Rating

Edison batteries of types A, B, C, D, M, N, and F are normally rated at the 5-hour rate of discharge. The capacities for standard types and sizes of plates have been given in Table 10 in Chapter 2.

Amount of Material within the Cell

As with lead-acid batteries, the capacity of the Edison batteries is dependent on the amount of active material within the individual cells. The amount of nickelous hydroxide that is used in filling the tubes is about 4 times the amount theoretically required. Assuming the oxide of nickel, NiO_2, to be the highest reaction product and the nickel oxide, NiO, to be the lowest, the reaction may be expressed by the equation

$$NiO + O = NiO_2$$

Since the electrochemical equivalent of oxygen is 0.2984 gram per ampere-hour, it may readily be calculated that 1.3932 grams of NiO are required for the reaction. The NiO would require 124.12 per cent of its weight as the nickel hydrate, and, therefore, the amount of the nickel hydrate theoretically required is 1.729 grams per ampere-hour. Actually the amount necessary is much greater than this.

Representing the reaction at the negative plate by the equation

$$Fe + O = FeO$$

the theoretical amount of iron required per ampere-hour is calculated to be 1.042 grams. The actual amount used is about 5.5 times this.

The electrolyte of the nickel-iron type of battery does not change materially in chemical composition or density (considered as a whole) during charge and discharge. As will be shown later, the discharge of these batteries is terminated because of the exhaustion of the available active material of the positive plate, and not because of lack of electrolyte in the pores, as is true of the lead-acid type of battery. The capacity of the cell is independent of the amount of electrolyte it contains, provided the plates are covered.

Effect of the Rate of Discharge

The relation of capacity to the rate of discharge of Edison cells is quite different from that of the lead-acid batteries. Nearly full capacity of an Edison battery may be obtained, irrespective of the rate of the discharge, but the voltage will be lower if the current is higher than the normal rate. This effect is shown by the curves of Fig. 70 which apply to a cell of the A type. It will be observed that the capacity delivered at 180 amperes is approximately the same as the

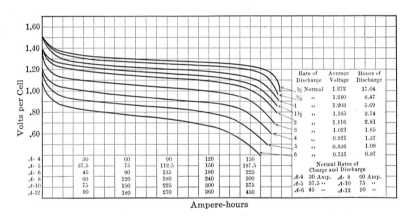

Fig. 70. Discharge curves of Edison type A cells at various rates subsequent to normal charge.

capacity at 15 amperes, but the voltage at the terminals of the cell is very much lower. The active material of the positive plate, which limits the discharge of an Edison cell, contracts during discharge. The pores of the plate are thus expanding during the discharging process and allow free access of the electrolye to the material. Although Edison batteries can deliver nearly full ampere-hour capacity even at high rates of discharge, there are practical limitations which make this effect of less importance than would otherwise appear. In operating the motor on a truck, for example, it is necessary that the voltage at the terminals of the battery be maintained above a certain minimum value if the truck is to maintain its speed.

If arbitrary final voltages are assigned, it will be found that the capacity of the Edison batteries to those voltages decreases as the rate of discharge increases, in much the same way as was found with the

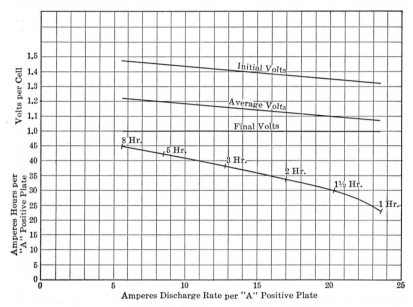

Fig. 71. Discharge characteristics of Edison nickel-iron-alkaline cells, type A, in fully active condition at rates ranging from the 8-hour to the 1-hour rate, all to a final voltage of 1.00 volt per cell, temperature 80 to 85° F.

lead batteries. (See Fig. 71.) The watt-hours delivered at high rates of discharge decrease as the rate of discharge increases because of the fall in the average voltage.

Effect of Temperature

The capacity of Edison batteries is decreased as the temperature is decreased. The relation does not, however, follow a nearly linear law as with lead batteries. The Edison batteries have a critical temperature which varies with the rate of the discharge. Below this critical temperature the output is small; above it, practically the full capacity of the battery may be obtained. The diminished capacity of the cells at low temperatures appears to be due to the temporary passivity of the iron electrode. Comparatively little information is available on

232   Storage Batteries

this subject, but a paper by Holland[12] gives data for cells of the type
A6 discharging at currents ranging from 15 to 75 amperes. The normal
rate of discharge for this size of cell is 45 amperes. At the normal
rate the critical temperature of the electrolyte is in the neighborhood
of 2° C (36° F). For rates of discharge in excess of this, the critical
temperature is higher, and for smaller rates it is lower.

This critical temperature does not mean that Edison batteries cannot be used at atmospheric temperatures below the critical temperature. The ohmic resistance during discharge produces sufficient heat
to warm the cell appreciably and keep it above the critical temperature
under ordinary conditions of operation. The cell will be operative
although used at ambient temperatures much below the critical value,
provided the temperature of the electrolyte within the cell does not fall
below the critical value.

Ventilation of battery compartments is desirable and in many installations necessary for control of the temperature of the battery as well
as to provide for the escape of gases. Edison cells are spaced in their
trays so that air can rise between the cells if openings are provided
in both top and bottom of the compartment. The amount of ventilation necessary will vary with the size of the battery, its average rate
of discharge, and the ambient temperature. Truck and tractor batteries that are used principally indoors and batteries of mine locomotives should be provided with maximum ventilation. They operate
in nearly even temperatures the year round. Street trucks and electric
locomotives that are used outdoors regardless of low atmospheric
temperatures should be provided with means to keep the electrolyte
temperature within a satisfactory range. The ventilation must be
sufficient to prevent excessive temperatures when the outside temperatures are high. It may be necessary to vary the amount of ventilation
from one time of the year to another. As a result of tests on Edison
batteries, Allen[13] recommended providing a ventilation area, in both
top and bottom of the battery compartment, of 3 square inches for
each A4 equivalent of battery capacity. As an example, if 21 cells of
Edison A6 are used, the ventilation area in top and bottom would be
$\frac{6}{4} \times 21 \times 3 = 94.5$ square inches.

Higher than the normal operating limits of temperature may result
in permanent loss of capacity if soluble iron compounds are formed.
Beside the loss of material at the negative plate, the iron in solution

[12] W. E. Holland, Effect of low temperature on the alkaline storage battery,
Central Station, 11, 135 (1911).

[13] E. W. Allen, *Ventilation of Battery Compartments in Motive-Power Service,*
Edison Storage Battery Division, Thomas A. Edison, 1935.

may react with the nickel oxides of the positive plates and impair their capacity.

Concentration of the Electrolyte

The concentration of the electrolyte in the Edison cell is of less importance. The concentration does not change during the charging and discharging of the cell but falls gradually over a long period of time. This decrease undoubtedly affects the rate of diffusion and also increases the resistivity. It is therefore advisable to renew the electrolyte when it has decreased to certain values, indicated in Chapter 3 (p. 160).

Previous History of the Plates

The previous condition of the plates affects the capacity of Edison batteries in much the same way as with lead batteries. Sluggishness occurs when the battery is allowed to stand idle for a considerable period of time, or if the battery is operated for a long period of time at rates considerably below normal. This sluggishness is caused by the slow oxidation of the iron in the pockets of the negative plates, which may result in temporary passivity. The loss in capacity resulting from this effect may require several cycles of charge and discharge to overcome. A battery of the Edison type which has stood idle for 2 months should receive the following treatment in order to restore it to its former capacity. Discharge it at normal rate through a suitable resistance to zero voltage. Then short-circuit the cells in groups of not more than 5 cells for at least 2 hours. Charge again at the normal rate for not less than 15 hours and discharge at the normal rate again to 1 volt per cell. Follow this by a charge at the normal rate for not less than the usual time, after which the battery is ready for service. Normally the capacity of the battery will increase during the first 125 cycles to about 115 per cent of rated capacity and then decrease slowly until the electrolyte is changed, when the capacity should immediately be increased. The useful life is ended when capacity falls below 80 per cent of the rated capacity.

If Edison cells are to be idle for some time, it is advisable that they be discharged to zero voltage and short-circuited in groups of not more than 5 cells before being placed in storage.

PLATE CAPACITIES

In addition to determining the capacity of storage cells, it is sometimes necessary to determine the capacity of the positive and negative plates separately, as a means of locating faults in the operation of the cells. The capacity of the cells of both the acid and alkaline types is normally limited by the capacity of the positive plates. When the capacity or one or more cells in a battery is found to be below normal, and the cause cannot be attributed to faulty insulation, the trouble may often be found in the condition of the active materials. One or the other of the plate groups may be only partially charged, although the battery as a whole is supposed to be fully charged. The nickel-iron batteries occasionally show inequalities in the plate capacities which may be due to periods of inactivity or to the materials.

The capacity of the positive and negative plates may be determined separately by the use of a constant auxiliary electrode immersed in the electrolyte of the cell. It is necessary that this electrode should maintain a constant potential with respect to the solution, so that the changes taking place in the positive and negative plates may be measured by the potential differences against this electrode.

Auxiliary Electrodes

A number of different electrodes have been proposed, but the one most generally used for the lead-acid batteries has been the cadmium electrode. The hydrogen and the mercurous sulfate electrodes are more accurate for laboratory purposes, but they are not well adapted to ordinary field testing.

The Hydrogen Electrode. This gas electrode consists of hydrogen gas absorbed in a layer of platinum black, which may be deposited on one of the noble metals, such as gold or platinum. When the electrode is immersed in a solution containing hydrogen ions, a difference of potential exists between the solution and the electrode. This difference depends on the concentration of the hydrogen ions in the solution. To maintain this potential difference constant at a standard value, the electrode is partly dipped in a solution having a known concentration of the hydrogen ions, such as a normal or a tenth-normal solution, and a stream of purified hydrogen is passed over its upper surface continuously. We cannot measure directly the difference of potential that exists between the electrode and the solution, but we can easily measure the difference in potential between the hydrogen electrode and another electrode with which it is in electrolytic contact. As the basis for an arbitrary scale, the potential of the electrode in a solution that is normal with respect to the hydrogen ions is usually assumed to be zero. With such an electrode as a standard, it is possible to measure the progressive changes that take place in the potential of the plates of an acid storage battery. For details of the hydrogen electrode and its use, reference should be made to Clark's *Determination of Hydrogen Ions*, 1928.

This electrode is adapted to use with the lead-acid batteries, because the solution used with it may be sulfuric acid which cannot contaminate the battery electrolyte in any way. The instrument is delicate and requires a potentiometer for measuring the voltage and a supply of hydrogen gas. It is, therefore, not readily portable.

The Mercurous Sulfate Electrode. This electrode is also suitable for use with the lead-acid storage batteries. The cell consists of pure mercury covered with a layer of pure mercurous sulfate and a solution of 1.250 sp. gr. sulfuric acid. This electrode, like the hydrogen electrode, is intended for laboratory use only. The constancy of this electrode may be judged from the fact that it is similar to the positive limb of a standard cell.

The Calomel Electrode. In appearance and structure, this electrode is very similar to the mercurous sulfate electrode, but as the solution it contains is potassium chloride it is adapted for use with the nickel-iron storage batteries. It consists of a layer of pure mercury covered with mercurous chloride, commonly called "calomel," and the solution of potassium chloride of a definite concentration. In using either of these electrodes, the end of the glass tube of the electrode is dipped into the electrolyte of the battery.

The Cadmium Electrode. Cadmium is a metal resembling zinc. It is usually obtained in sticks about $\frac{5}{16}$ inch in diameter, which is a convenient shape for the electrode. Pieces from one to several inches in length are used. The electrodes are prepared for service by keeping them immersed in sulfuric acid solution of about the strength used in the batteries. It is necessary that new electrodes should be in the acid for several days before being used to measure plate potentials. The sulfuric acid corrodes the surface of the electrode, and an equilibrium state is eventually reached.

The cadmium must be insulated so that it cannot come in contact with the plates of the cell, but at the same time the electrolyte of the cell must have free access to the cadmium. A perforated rubber separator from a storage cell is suitable for making the necessary cover for the electrode. A flexible wire for connection is attached to the cadmium. Readings are taken when the current of the battery is flowing, usually on discharge, but sometimes on charge also. The readings are meaningless if the cell is on open circuit.

The cadmium electrode is reproducible to about 0.02 volt if precautions are taken. It will remain constant during several hours to within 0.01 volt but may vary from day to day by 0.02 volt. Amalgamation of the electrode, which is often recommended, does not improve its reliability.[14]   If the electrode surface becomes dry, it must be soaked in the acid for at least $\frac{1}{2}$ hour before being used again.   The electrode should be immersed in the electrolyte so that as much as possible of the cadmium surface may be wet, to avoid polarization.

The greatest error in the use of the cadmium electrode, as found by Holler and Braham, is due to polarization.   The ordinary voltmeter of low voltage range has from 100 to 300 ohms resistance.   When such an instrument is used it permits sufficient current to flow to polarize the cadmium electrode.   The error thus introduced may amount to 0.1 volt, which would lead to false conclusions about the condition of the negative plates.   The polarization of the electrode is proportional to the potential being measured.   It is, therefore, many times greater in measuring the potential of the positive plates than the negative plates, because the potential difference of the electrode and the positive plates is about 2 volts, while the difference between the electrode and the negative plates is only one-tenth as great.   The error may be avoided entirely by measuring the voltage on a potentiometer or a high-resistance voltmeter.   Such voltmeters may be obtained with a resistance of 3000 ohms and a special scale of 0.25—0—2.65 volts.   In the absence of such means for measuring the voltage, the best procedure is to measure the potential of the negative plates, and from this and the cell voltage calculate the potential of the positive plates.   This is easily done, since the potential of the positive plates is the algebraic difference of the cell voltage and the potential of the negative plates.   Metallic impurities in the electrolyte, if they are electropositive to cadmium, may lead to false indications of the cadmium electrode.   Copper, which is often found in small amounts in storage-battery electrolyte, is an example of this class of impurities. The copper will deposit on the cadmium when it is put into the electrolyte and alter its potential.   Reliable readings of the plate potentials can only be obtained with care, and it is essential that the conditions of the experiment be understood to avoid drawing false conclusions in some cases.   Zinc is sometimes used in place of cadmium.   It is not recommended for lead-acid batteries but may be used in the alkaline type.

Plate Potentials during Discharge and Charge

Lead-Acid Cells.   The voltage of the cell is dependent on the potentials of the positive and negative plates with respect to the solution.

[14] H. D. Holler and J. M. Braham, The cadmium electrode for storage-battery testing, *Natl. Bur. Standards Technol. Paper 146* (1919).

Since these potentials may vary independently, there are a number of possible combinations of plate potentials that will give any particular cell voltage. For example, if a group of three cells is measured with a cadmium electrode when each cell has reached its cut-off voltage of 1.8 volts, the following values may be found:

Cell 1, Positive plate 2.10 ⎱ Cell voltage   1.80
        Negative plate 0.30 ⎰

Cell 2, Positive plate 2.00 ⎱ Cell voltage   1.80
        Negative plate 0.20 ⎰

Cell 3, Positive plate 1.95 ⎱ Cell voltage   1.80
        Negative plate 0.15 ⎰

These results require interpretation in the light of the cell's rated capacity and normal performance. The values of the potentials vary with the rate of the discharge, the concentration of the electrolyte, and to a less extent with other factors also. It must be assumed that the voltage drop within the cell, which is equal to the product of the current and the resistance, IR, is negligible. If I is not too large this will be so, since R is small.

Considering the example given above, the plate potentials of Cell 1 show that the negative plate has reached the limit of its capacity before the positive plate is fully discharged, or, in other words, the discharge has been terminated by the negative plate, which is not normal.

In considering the values for Cell 3, it is necessary to know whether the cell gave its rated capacity before reaching the final voltage 1.80. If it did not, the potential readings indicate the positive plates to be deficient in capacity. Although the plate potentials near the end of the discharge are the most important, it is desirable to take the readings throughout the discharge at frequent intervals and from them to plot curves. The values for Cell 2 are about normal for a cell in good condition.

Figure 72 shows in a graphical way the relation of plate potentials of a cell of the lead-acid type. The curves show that the capacity of the negatives exceeds that of the positives. They also show that very satisfactory agreement may be obtained between measurements of the plate potentials obtained by the hydrogen electrode, the mercurous sulfate electrode, and the cadmium electrode. In order to superpose the determinations of these three electrodes, it has been necessary to displace the zero reference line for each by an amount equal to the difference in potential of the several electrodes. When this is done, the observed values fall on the same curves with a high degree of accuracy.

When lead cells are charged, the relations of the plate potentials are somewhat different. In particular, the potential of the negative plate should reverse toward the end of charge, the lead sponge becoming negative with respect to the cadmium by about 0.20 volt. When

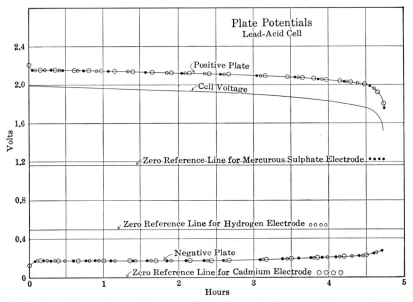

Fig. 72. Comparison of three auxiliary electrodes for measuring potentials of the plates.

this occurs the voltage of the cell becomes greater than the potential of the positive plate. At the end of charge, the plate potentials are about as follows:

$$\left.\begin{array}{ll}\text{Positive plate} & 2.50 \\ \text{Negative plate} & -0.20\end{array}\right\} \text{Cell voltage}\quad 2.70$$

In old cells it is sometimes difficult to make the potential of the negative plate reverse because of antimony, which accounts for the lower charging voltages of old cells.

Edison Cells. The capacities of the positive and negative plates of the nickel-iron batteries may be determined by the use of auxiliary electrodes in much the same manner as has been described for the lead-acid batteries. Suitable electrodes are the calomel electrode, for use in the laboratory, or a section of a positive tube taken from an alkaline battery. Connection to this electrode should be made by the same kind of metal as used for the support of the active material, or else the tube should be made long enough to project out of the liquid. The electrode must be insulated so that it cannot touch the plates. For this purpose a perforated rubber sheet is suitable.

Since it was shown, in the section on the theory of the alkaline batteries, that the positives suffer a spontaneous decomposition of the nickel peroxide to a lower state of oxidation after the charge is completed, it is necessary to discharge partially the intermediate electrode (tube) after charging it, in order to bring it to a stable state. In addition to this precaution, it is also desirable to season the electrode by allowing it to stand in the battery electrolyte for a day or two before use.

The decrease in voltage of the Edison cell during discharge is caused almost entirely by the fall of potential of the positive plate if the cell is in normal condition. The potential of the iron plate falls slightly at the beginning of discharge to a value which remains practically constant throughout the remainder of the discharge. If the active material of the negative plate is sluggish, however, the negative plate may limit the capacity of the battery, as can easily be determined by single potential measurements.

6

Operation

METHODS OF CHARGING[1]

Direct current alone can be used for charging. If alternating current only is available, it must be converted into direct current. This can be done by means of: (1) synchronous converter, (2) motor-generator, or (3) rectifier.

There are two basic systems of charging in general use: (1) constant-current, and (2) constant-potential or constant-voltage. The latter method is usually modified slightly by the addition of a fixed resistance of small value in series with the battery in order to limit the starting or inrush current and to improve the finishing rate. Such a system is called a semi-, or modified, constant-potential system.

The positive terminal of the charging source is to be connected to the positive terminal of the battery so that the charging current flows through the battery in the direction opposite to that of the discharge current.

Constant Current

In the constant-current system, as the name implies, the current is held constant by means of a rheostat in series with the battery or by controlling the voltage of the source.

The current is maintained at the normal rate by adjusting the rheostat, cutting out or decreasing the resistance as the charge progresses. This increases the voltage impressed on the battery. The value of the current flowing through the battery is dependent upon the difference between its voltage and that impressed on it.

[1] J. L. Woodbridge, Storage-battery charging, *Trans. Am. Inst. Elec. Engrs., 54,* 516 (1935); R. A. Harvey, The automatic control of lead-acid-battery charging equipment, *Proc. Inst. Elec. Engrs. London, 96,* 607 (1949); F. Dacos, *Rev. universelle mines, 2,* 3 (1946), Abstr., *Eng. Digest, 3,* 291 (1946); J. D. Huntsberger, Storage-battery charging, *Plant Eng., 3,* 43 (August 1949); *Recommended Specifications for Automatic Battery-Charging Equipment for Industrial Truck Motive-Power Service,* Electric Industrial Truck Association, Long Island City, N. Y., 1949.

Let the impressed voltage $= E$; the current flowing at any chosen instant $= I$; the counterelectromotive force of battery $= E_c$; the resistance of the battery $= R$. Then

$$E = E_c + IR$$

Whence

$$I = (E - E_c)/R$$

Therefore, when the voltage of the battery and the charging system are the same, no current will flow; when the voltage of the battery is lower than that of the charging system, current will flow into the battery and charge it; and when the voltage of the battery is higher than that of the charging system, current will flow out of the battery and discharge it. As the voltage of the battery increases gradually with the progress of the charge, it is apparent that the voltage impressed across its terminals must be increased in order to maintain a constant value for the charging current. For batteries of the lead-acid type, the specified current is maintained until all cells are gassing freely and then is reduced to a much lower value designated as the finishing rate, at which the charge is continued to the end. The value of the finishing rate for motive-power batteries is approximately 40 per cent of the 8-hour starting rate. The ampere-hours required to produce free gassing of a lead battery at the normal starting rate of charge will be approximately 90 per cent of the ampere-hours previously discharged. Gassing may be expected earlier if the previous discharge was incomplete. When free gassing occurs the current should be cut down to the finishing rate and the charge continued until the battery begins to gas freely again.

Figure 73 is a diagram of connections for a typical constant-current charging circuit.

In order to obtain most efficient results with batteries of the lead-acid type, the voltage of the charging circuit should be approximately 2.5 volts per cell[2] at normal temperatures with current flowing at one-half the value of the finishing rate. When the voltage of the circuit exceeds this value, the maximum resistance of the rheostat must be sufficient to permit of a reduction in voltage to this value. It is desirable that the ampere capacity be sufficient to permit a current value of four or five times the normal, to permit of boosting (see p. 253), provided that the wiring of the charging circuit can safely carry that current. In a new installation, this condition can be provided for

[2] Batteries with organic expanders in the negative plates and batteries free from antimony or its effects will normally have higher final charging voltages by 0.2 to 0.3 volt per cell.

readily. In modifying an existing installation, the current capacity of the rheostat need not exceed the safe current-carrying capacity of the circuit.

The value in ohms of the resistance equals

$$\frac{E - (B \times C)}{D}$$

in which E = voltage of system; B = number of cells in battery; C = volts per cell, a constant, 2.5 for all types and sizes of lead-acid cells; D = 50 per cent of finishing rate in amperes for lead batteries.

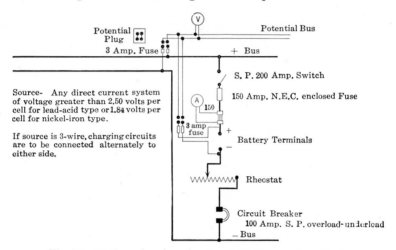

Fig. 73. Battery-charging circuit, constant-current method.
A, ammeter; V, voltmeter.

Example. Assume a 17-plate truck battery of 24 cells having a normal charging rate of 45 amperes and a finishing rate of 18 amperes; charging circuit 110 volts.

Resistance in ohms equals

$$\frac{110 - (24 \times 2.5)}{9} = 5.6$$

The current capacity equals

$$5 \times 45 = 225 \text{ amperes.}$$

Such a rheostat, unless properly designed, would be large and expensive. The excessive capacity may be reduced by designing the variable portion so that at each step the current-carrying capacity is just within the safe limit. A simpler arrangement, which would provide for the boosting current and also permit sufficiently good regulation for charging under ordinary conditions, could be made as follows:

Ten fixed resistances, each connected as required between the bus and the battery by a single-pole switch, are arranged so that the first three switches would connect in 6 ohms each; the next three, 3 ohms each; and the seventh to the tenth switches, 1.5 ohms each. This rheostat would provide, under the conditions given in the example, a maximum current of 210 amperes, and a minimum current of 8.3 amperes. The maximum kilowatt capacity when all resistances are in parallel would be 10.6 kilowatt, and the minimum capacity 0.4 kilowatt for the smallest charging current.

Some leeway is allowable in the constancy of the current. A constant-current charge may for practical purposes be defined as one in which the starting amperes do not exced 120 per cent of normal and the finishing rate is not less than 80 per cent of normal, provided that the average throughout charge is equal to the normal rate specified by the manufacturer.

Edison batteries are charged at the full normal rate, without reduction to a finishing rate. The maximum resistance is required, therefore, at the beginning of charge when the counterelectromotive force of the battery is a minimum, about 1.55 volts per cell. In using the above equation, C is taken as 1.55 and D as the normal charging rate.

Final Charging Voltage

The voltage of a storage battery rises during the charging period, reaching a maximum value when charge is complete. What this value will be depends on (1) the charging current, (2) the temperature, (3) the internal resistance of the cell, (4) the presence or absence of certain impurities in the electrolyte, (5) composition of the grid alloy, and (6) the presence of organic expanders in the negative plates.

Constancy of maximum voltage is a better criterion, therefore, than any particular value when one wishes to determine that charging is complete. There are other means of judging when charge is complete, such as: (1) the specific gravity of the acid electrolyte, (2) the extent and uniformity of the gassing, (3) plate potentials, and (4) measured input in ampere-hours. Figure 74 shows the variation of final voltage of the acid cells with temperature and rate of charge. These curves apply to a particular type of cell, but they illustrate the principles involved and show the necessity of temperature compensation if voltage relays are employed to terminate the charge of a battery.

There is a critical temperature for charging lead-acid batteries at about 120° F, where the charging current becomes unstable and may sometimes rise out of control to the detriment of both battery and charging equipment.

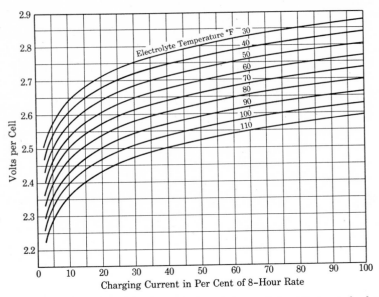

Fig. 74. Final charging voltages at various rates and temperatures, lead-acid
batteries. The interval between curves is 0.0035 volt per °F.

Ampere-Hour Law

Because the lead-acid storage battery in a normally discharged state
can absorb electrical energy very rapidly without overheating or excessive gassing, it is possible to begin charging at a high rate of current,
usually much in excess of the so-called normal or starting rate of charge.
In fact, the term "starting rate" for lead-acid batteries is rapidly dis-
appearing from use, but finishing rates are a matter of importance.
The practical limitations to the rate at which a battery can be charged
are (1) temperature rise to excessive values and (2) excessive gassing.
These are matters of judgment, and a more definite answer is needed
if we are to learn what is the shortest time in which a lead-acid battery
can be charged. Woodbridge[3] says: "As a result of numerous tests,
it has been found that if the charging rate in amperes is kept below a
value equal to the number of ampere-hours then out of the battery the
conditions as to gassing and temperature will be met." That is, if
200 ampere-hours have been discharged, the charging rate may begin
at anything less than 200 amperes, but obviously this must be progressively reduced so that the charging current in amperes is always less
than the number of ampere-hours the battery lacks to complete 100
per cent charge. This is known as the ampere-hour law.

[3] *Loc. cit.*

The older method of constant-current charging in two steps, beginning with a "starting rate" and ending with a "finishing rate" complied substantially with requirements of the ampere-hour law but failed to make use of possibilities for most rapid charging. If charging is done at successively lower rates according to the ampere-hour law, the process is sometimes referred to as "step method of charging." Practical limitations will dictate how many steps there shall be, but theoretically for an infinite number of steps the charging current follows an exponential law expressed by the equation $I = Ae^{-t}$, where I is the current in amperes, t the time in hours and A the number of ampere-hours out of the battery when charging begins, that is, when $t = 0$. Figure 75 from Woodbridge's paper shows that charging a battery in accordance with

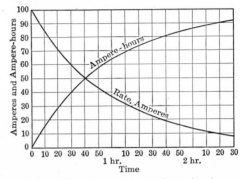

Fig. 75. Charging lead-acid batteries in accordance with the ampere-hour law.

this law restores 90 per cent of the ampere-hours previously withdrawn in 2 hours and 20 minutes. Practical tests reported by him show that the minimum time for a complete charge, including 15 per cent overcharge, is about 4 hours.

Because automotive batteries which are discharged can be recharged at high rates initially, many so-called "fast chargers" have appeared. These naturally have a strong appeal to the impatient public. When intelligently used with due regard for the limitations of temperature and hard gassing they may serve a desirable purpose, but it should be remembered that the charge put into the battery in 20 minutes or $\frac{1}{2}$ hour is not a complete charge. According to Fig. 75, in $\frac{1}{2}$ hour the most that can be given the battery is about 50 per cent of a full charge; the excess input is wasted in gassing. However, this amount is sufficient to enable the battery to perform its functions, and it should more gradually acquire the remainder of its full charge as the car is operated. Some types of fast chargers are described on page 253.

Another method of controlling step charging is to reduce the current at intervals when the voltage across the cell equals a predetermined value as 2.35 volts.

For batteries of the nickel-iron type, charging at constant current is carried on at the normal rate for the full period of time required, normally 7 hours, or until the voltage becomes constant, at a value usually between 1.8 and 1.9 volts per cell, the exact value depending on the temperature. As the specific gravity of the electrolyte of batteries of this type remains constant during charge and discharge, the only indications of complete charge are the condition of voltage as stated above and the length of time taken for the charge. If the battery shows a temperature exceeding 46° C (115° F), it should be cut off immediately and allowed to cool.

Constant-Potential

In the constant-potential or constant-voltage method, the voltage is maintained at a constant fixed value per cell. The value of the initial, or starting, current of a completely discharged battery when first put on charge is much in excess of the so-called normal rate. During the charge, as the voltage of the battery gradually rises, the current falls off to a value much below that of the normal rate, and at the end of the charge it is below that of the finishing rate of the constant-current system. The average value of the current is about equal to that of the normal rate.

For batteries of the lead-acid type, the maximum voltage for unmodified constant-potential charging should not exceed 2.35 volts per cell, and the minimum should not be less than 2.25 volts per cell.

With the average voltage thus established at approximately 2.3 volts per cell, a battery in any state of discharge may be put on charge, and it will automatically receive the proper charge without reaching the free gassing point or excessive temperature. Caution is necessary, however, as slight variations in the line voltage produce large variations in the charging current. The "modified" constant-potential method of charging, described below, is a safer method.

Modified Constant-Potential

The very large charging current at the beginning of an unmodified constant-potential charge makes it necessary to limit the initial or starting current, and to accomplish this a fixed resistance of small value is placed in series with the battery. This is then known as a semi-constant-potential or a modified constant-potential system.

In practice, when using the modified constant-potential method, the voltage at the bus may be kept constant at any value from 2.5 volts to 3.0 volts per cell, although 2.63 volts is the value commonly used in the United States for an 8-hour charge of lead-acid cells. In charging nickel-iron and nickel-cadmium batteries, the minimum bus voltage required is 1.85 volts per cell but may be as high as 2.30 volts.

Figure 76 is a diagram of connections for such a system with a three-wire supply. For a two-wire supply the connections are the same as shown on this diagram for circuit B, or between the two outside wires. Applying the modified constant-potential method to charging lead-acid batteries, typical curves showing the relative changes in voltage, current, temperature, and specific gravity of the electrolyte are obtained, and these are illustrated in Fig. 77. In this example the bus

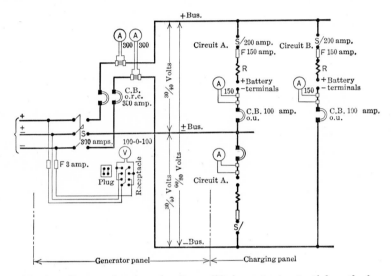

Fig. 76. Battery-charging circuit, modified constant-potential method.
A, ammeter; *V*, voltmeter.

voltage was 2.63 volts per cell, and the modifying resistance was 0.0091 ohm. The charge was completed in 8 hours. At the beginning of the charge the battery counterelectromotive force was at its lowest, and consequently the initial charging current was at its maximum. This decreased as the battery voltage gradually increased. After $2\frac{1}{2}$ hours the acid diffusing from the pores of the plates began to show its effect by increasing the specific-gravity measurements. This delay of several hours (and sometimes more) is important to notice, because to the unskilled this is often erroneously interpreted to mean that the battery is "not taking a charge." Gassing began at about $4\frac{1}{2}$ hours, accompanied by a more rapid rise in the battery's countervoltage and a corresponding fall in the charging current. The point 2.37 volts per cell is significant as being on the steep portion of the voltage curve and a suitable point for the operation of relays controlling the charge. After 7 hours the battery has received ampere-hours equal to its previous discharge, and the remainder of the time provides a small excess at the finishing rate. This is desirable and necessary, as it is a safeguard against overdischarge on following cycles.

In practice, automatic control is usually employed to terminate the charge. This may be the combination of a sensitive voltage relay and timing device or an ampere-hour meter. The voltage relay starts

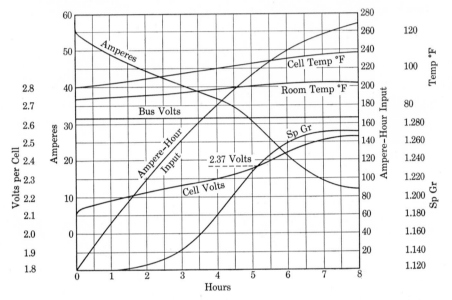

Fig. 77. Charging a motive-power cell (lead-acid type) by the modified constant-potential method. The bus voltage was 2.63 volts per cell, fixed resistance 0.0091 ohm.

a timing mechanism when the battery voltage has reached a predetermined value. The timer runs for a specified time, usually 2 to 4 hours, and then opens a switch that terminates the charge.

Conditions under which the modified constant-potential system may be employed vary, and it is necessary, therefore, to determine (1) the bus voltage available and (2) the proper modifying resistance according to the number of cells in the battery and their ampere-hour capacity. It is customary to base calculations on a battery of 100 ampere-hour capacity at the 6-hour rate. Then the needed values can be determined from Table 51, as follows: The indicated bus voltage per cell is multiplied by the number of cells in series in the battery. The normal modifying resistance (and its maximum required value) is multiplied by the number of cells in the battery, and this product is

TABLE 51. MODIFIED CONSTANT-POTENTIAL CHARGING OF
LEAD-ACID BATTERIES

(Based on 100 ampere-hour capacity at the 6-hour rate; Specifications of
The Electric Industrial Truck Association.)

Hours Available for Recharge	Bus Volts per Cell	Resistance Values per Cell		Ampere Rates per 100 Amp-Hr		Control Adjustment	
		Normal Ohms	Maximum Ohms	Start of Charge	Resistor Capacity	Voltage-Time Relay	Amp-Hr Meter % Overcharge
						Hours	
7.0	2.60	0.016	0.027	27.5	32.5	2.5–3.0	10
7.5	2.61	0.018	0.029	25.5	30.0	2.5–3.0	10
8.0	2.63	0.022	0.031	22.5	26.0	2.5–3.0	10
8.5	2.65	0.026	0.035	20.0	23.0	2.0–2.5	10
9.0	2.67	0.030	0.039	18.5	21.0	2.0–2.5	10
9.5	2.69	0.034	0.043	17.0	19.5	2.0–2.5	10
10.0	2.72	0.040	0.049	15.5	17.5	2.0–2.5	10
12.0	2.84	0.064	0.073	12.0	13.5	2.0–2.5	10
14.0	3.00	0.096	0.105	10.0	11.0	2.0–2.5	10
16.0	3.27	0.150	0.160	8.5	9.0	2.0–2.5	10

divided by the number of hundreds of ampere-hours of battery rated capacity.

For example, a 15-cell, 238-ampere-hour battery is to be charged in 8 hours at a bus voltage of 2.63 volts per cell. The required bus voltage is $2.63 \times 15 = 39.45$ volts, and the modifying resistance is $(0.022 \times 15)/2.38 = 0.138$ ohm with a maximum adjustment to 0.195 ohm. The ampere rate at the start of charge is given as the current per 100 ampere-hours capacity. In this example the initial amperes will be $22.5 \times 2.38 = 53.6$.

Values of initial current are subject to changes in the bus voltage for which a tolerance of ±3 per cent is usually specified. The initial current will be higher than indicated when the bus voltage is above normal, and to allow for this the rating of panel wiring and equipment should be based on current corresponding to a bus voltage of 103 per cent of normal. This is covered in the specifications referred to on page 240.

The ampere-hour meter is provided with a zero contact, a resetting device, and a fixed red hand set at the discharge limit. It should be adjusted to run slower on charge than on discharge. The meter can be connected in the battery circuit during discharge as well as during charge. If the ampere-hour meter is on the charging panel, the extent of the previous discharge must be determined by specific gravity readings on lead batteries or by the charge test fork for nickel-iron or cadmium cells.

For lead batteries, the higher the bus voltage the greater will be the loss in the resistance, with a decrease in efficiency and an increase in the length of time required for the charge. On the other hand, if the bus voltage is but little higher than that of the battery, the charging current will become more or less unstable and subject to considerable variation with changes in temperature of the battery and slight changes in the bus voltage. If the supply voltage is in excess of a value for which the charge can be completed in a specified time, counter cells may be used to reduce the voltage at the terminals of the battery to the proper value.

As the bus voltage and resistance are increased for the same number of cells in the battery, the charging current becomes more nearly constant throughout the charge, thus approaching the constant-current method.

For batteries of the nickel-iron type, the impressed voltage should not be less than 1.70 volts per cell. As with the lead-acid type, the voltage at the charging panel must be slightly higher than this value. In practice this can be 1.84 volts or more, with best conditions usually about 2.00 volts per cell. When batteries of this type are to be charged from a modified constant-potential system, the resistance R is found in Table 52.

Automatic Two-Rate Charging

A more modern method of decreasing the time necessary for completing the charge for lead cells when the bus or rectifier voltage is too high to permit this to be accomplished in the time available is to resort to "two-step charging." The first step, using a low fixed resistance, permits high charging rates during the early part of the charge when the battery can absorb energy very rapidly. A higher resistance for the second step reduces the charging current to safe rates during the latter part of the charge. The change from first to second step is accomplished automatically by a voltage relay or an ampere-hour meter with auxiliary contact.

The two-rate system is also well adapted to charging batteries when rectifiers are the source of current. When the bus or rectifier voltage is 2.70 volts per cell or higher and it is desired to complete charge in 8 hours, the two-rate system should be used.

When an ampere-hour meter is used, the position of the auxiliary contact is the determining factor in choosing the proper value of the external fixed resistance, because the current must be so chosen that it

TABLE 52. MODIFIED CONSTANT-POTENTIAL CHARGING OF
EDISON BATTERIES

(The fixed resistance to be placed in series with the battery is the product of
the number of cells in the battery and the ohms per cell that are required
for the type and the available bus voltage. Based on specifications of the
Electric Industrial Truck Association.)

Bus volts	1.84	1.90	2.00	2.10	2.20	2.30
Initial rate, % of normal	165	155	140	128	124	120
Final rate, % of normal	65	70	78	84	86	88

Type of Cell	Charge, Normal Rate, amp	Fixed Resistances, Ohms per Cell, for Modified Constant-Potential Charge				
B1	3.75	0.04000	0.05600	0.08270	0.10933	0.13600	0.16267
B2	7.50	.02000	.02800	.04133	.05467	.06800	.08133
B4	15.00	.01000	.01400	.02067	.02733	.03400	.04067
B6	22.50	.00667	.00933	.01378	.01822	.02267	.02711
A4	30.00	.00500	.00700	.01033	.01367	.01700	.02033
A5	37.50	.00400	.00560	.00827	.01093	.01360	.01627
A6, C4	45.00	.00333	.00467	.00689	.00911	.01133	.01356
A7	52.50	.00286	.00400	.00590	.00781	.00971	.01162
C5	56.25	.00267	.00373	.00551	.00729	.00907	.01084
A8	60.00	.00250	.00350	.00517	.00683	.00850	.01017
C6	67.50	.00222	.00311	.00459	.00607	.00756	.00904
A10	75.00	.00200	.00280	.00413	.00547	.00680	.00813
C7	78.75	.00190	.00267	.00394	.00521	.00648	.00775
A12, C8, D6	90.00	.00167	.00233	.00344	.00456	.00567	.00678
A14	105.00	.00143	.00200	.00295	.00390	.00486	.00581
C10	112.50	.00133	.00187	.00276	.00364	.00453	.00542
A16, D8	120.00	.00125	.00175	.00258	.00342	.00425	.00508
C12	135.00	.00111	.00156	.00230	.00304	.00378	.00452
A20, D10	150.00	.00100	.00140	.00206	.00273	.00340	.00406
A24, D12	180.00	0.00083	0.00116	0.00172	0.00228	0.00283	0.00339

will not exceed both the finishing rate and the requirements of the
ampere-hour law at any time before the change to the lower rate.
The low rate of charge, on the other hand, is determined by the finishing rate for the particular battery, regardless of the ampere-hour meter.
The auxiliary contact on the meter will be at a certain number of
ampere-hours on the scale or at a specified time if a time-relay is used.

The factors involved in two-rate charging are: (1) the available bus
voltage; (2) the time available for the recharge; (3) the number of

cells and 6-hour capacity of the battery; (4) the method of controlling the change in rate. At the beginning of the charge part of the resistance in series with the battery is shunted by a closed switch. Initial rate of charge is high but tapers off slightly. At a predetermined point the voltage relay or ampere-hour meter opens this switch, and the added resistance decreases the charging current until the final cut-off is reached. This system does mechanically what the modified constant-potential method does automatically as a result of the rising counterelectromotive force of the battery, but two-rate charging does what the other cannot do, it charges a battery in a relatively short time when the bus voltage is too high.

The proper values of high- and low-rate resistors, their current-carrying capacity, the expected charging currents, and the settings for time or ampere-hour control have all been worked out and tabulated for a variety of bus voltages and available time for charging. These tables are very large, and it is unnecessary to reproduce them here as they are readily available in the *Recommended Specifications for Automatic Battery-Charging Equipment for Industrial Truck Motive-Power Service* (page 240) and in publications of several of the battery manufacturers.

Equalizing Charge

As its name implies, an equalizing charge serves to correct any inequalities among cells of a battery that may develop in service. It is essentially a prolonged charge at the finishing rate or less. The frequency of giving equalizing charges depends on service conditions. For floated batteries once a month is usually often enough, and this is easily done by raising the bus voltage by several tenths of a volt per cell for a specified time. The longer the time the less increase in voltage is needed. For manually cycled batteries, one or more of the regular charges should be continued into an equalizing charge. A battery in daily use for operating a truck or tractor should receive an equalizing charge weekly. The time to terminate an equalizing charge is indicated when the hydrometer and voltage readings have been constant for several hours, usually 3, provided the charging rate has been held constant during that period.

Equalizing charges tend to nullify the effects of deep cycling on negative plates, and some manufacturers recommend that the best means of maintaining either negative or positive plates in a really healthy condition is to submit them to occasional deep discharges with a full equalizing recharge.

Once every 3 or 4 months, the voltage and gravity readings of each cell should be recorded. These will serve as an indication of trouble within the cells, such as sulfation or leakage, if there is a progressive change of the gravity readings.

The gravity of the individual cells should be adjusted to the proper value when the battery is first put in service or when it has been necessary to add electrolyte to any cell to replace electrolyte that has been spilled or otherwise lost. This should be done at the end of the equalizing charge. If the gravity of the cell is too high, a portion of the electrolyte may be withdrawn with the hydrometer syringe and replaced by distilled or approved water. Similarly, the gravity may be increased by replacing the portion drawn off by electrolyte of 50 points higher specific gravity. Before the adjustment is considered complete, the equalizing charge should be continued in order to mix the electrolyte of the cell. The final value is shown by several consecutive constant readings at 15-minute intervals. The gravity should be adjusted to within 5 points, corrected for temperature, of the proper value.

Equalizing charges are not required for Edison batteries if they are given the proper amount of excess charge over previous discharge on each cycle. When the specific gravity of the electrolyte for these has fallen to the limits set in Chapter 3 the electrolyte should be renewed.

Boosting

Under certain conditions, the ampere-hour capacity of the battery may be insufficient for the day's work. In that event, charging at a high rate of current for a short time may be resorted to. Such a charge is known as "boosting" and is usually given the battery during the noon hour. Figure 78 shows the value of current, with time, to be used for boosting batteries of the lead-acid type.

The so-called fast chargers provide what is in effect a boosting charge. For automotive batteries they generally provide a current of 100 amperes initially, tapering to about 80 amperes. Some of these devices are time-controlled, and others are temperature- and voltage-controlled. The temperature is very important in this high-rate charging, and the limit for the electrolyte is 125° F. For controlling charges when thermostatic control is not otherwise provided, small monitors are available. This device contains a thermostat set for 125° F which is immersed in the battery's electrolyte (center cell) and receives power from the cell itself to sound a loud buzzer and light a signal light when this temperature is reached. All vents of the cells should be in place and open, and the voltage across the battery terminals should not exceed 8.5 volts when 100 amperes are flowing into the battery. Batteries which have a high internal resistance owing to sulfation, low temperature, or other cause will exceed 8.5 volts unless the charging current is cut to a much lower figure. These should be given a slow charge for a long time. Cold batteries can be charged by as large a current as will not cause the voltage to exceed 8.5 volts.

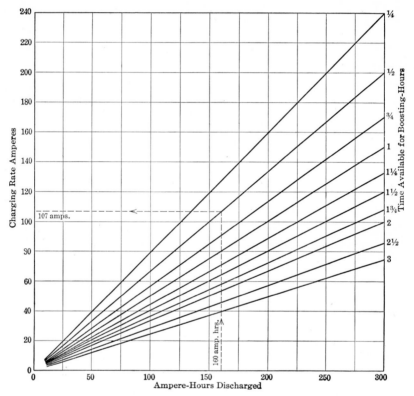

Fig. 78. Maximum rates for boosting lead-acid cells of any capacity.

Example: If 160 ampere-hours have been discharged and ½ hour is available for boosting, follow the dash lines in the direction of the arrows. The proper value of the current to use is found to be **107** amperes.

Edison batteries of the nickel-iron type may be boosted, provided the temperature of the inside cells does not exceed 46° C (115° F). The rates of charge recommended by the manufacturer are:

5 minutes at 5 times normal rate.
15 minutes at 4 times normal rate.
30 minutes at 3 times normal rate.
60 minutes at 2 times normal rate.

Frothing at the filler opening is an indication that the boosting has been carried too far.

Trickle Charge

A trickle charge, as the name implies, is a continuous charge at a low rate sufficient to compensate for the internal losses of the battery and suitable to maintain the battery in a fully charged condition. Low-rate charges are satisfactory for lead batteries provided the total amount of charge received by the battery is sufficient to cover the local action. Several applications have been made of trickle-charging. One of them is to put lead batteries in so-called "wet" storage. At regular intervals, the filling plugs should be removed and water added to the cells if necessary. At the beginning of the charge the specific gravity of each cell should be read and recorded, and this operation should be repeated at intervals. The charging current may then be reduced to the lowest limit that will keep the specific gravity constant.

The term trickle charge is also applied to very low rates of charge which are sufficient not only to compensate for the internal losses of the battery but to restore intermittent discharges of small amount.

Trickle-charging may also be applied to batteries of the Edison type. The recommended rate, depending on the nature of the application, may vary from 0.125 to 0.25 ampere per positive plate of the A type, and proportionately more or less for larger or smaller plates. The actual value of the trickle-charge rate employed in any particular case depends on the discharge rates to which the battery is subject. Higher trickle-charge rates are employed when discharges are at the normal 5-hour rate or above, and the lower limit specified is applicable to batteries discharging at relatively low rates. If properly done, the capacity of a battery for discharge may be increased by such a procedure.

In addition to the trickle-charge rates specified above, charging current must be provided to compensate for intermittent discharges which occur under service conditions. An excess of 10 per cent over the average discharge should be added for this purpose. Allen's[4] empirical formula for calculating the proper trickle-charge rate for any type of Edison battery, based on the higher limit mentioned previously, is as follows:

$$I = \frac{(C \times 0.16) + (D \times 1.10)}{24 - H}$$

In this equation I is the current in amperes, C the rated ampere-hour

4 *Electrical Engineers' Handbook,* p. 7-28, Pender and Del Mar, 1949.

capacity of the battery, D the average number of ampere-hours discharged per day, and H the aggregate time in hours of discharge. It should be noted however that H must be small in comparison with 24, the number of hours in a day. Otherwise, the charging rates that would be calculated by the formula can no longer be regarded as trickle-charging rates.

The Conditions for Floating

A battery continuously connected to a bus is said to float when the voltage of the charging line is slightly greater than the open-circuit voltage of the battery (see Table 39), and opposite in polarity. When a floating battery is connected to a line whose voltage is approximately equal to the open-circuit voltage of the battery, the battery will charge or discharge according as the fluctuations of voltage of the line rise above or fall below the battery voltage.

A battery which is properly floated upon a power line will automatically take care of the power required for a fluctuating load, and the battery will be maintained in the fully charged condition. Storage batteries are commonly installed in power stations and substations to insure an uninterrupted supply of current for the operation of control and protective equipment as well as for emergency lighting and other vital services. Such batteries often consist of 60 cells.

Batteries of the "sealed-in-cell" type of construction, with either pasted plates or Planté positives, are used universally in the United States today, instead of any of the "open-cell" construction type. They are in installations principally in services of the telephone companies, of central generating plants, and with various types of equipment for signal operations, emergency lighting, circuit-breaker control, etc. These batteries are seldom required to discharge very much of their capacity, and their charge is maintained by floating the battery across a very carefully voltage-controlled circuit, operating usually at not lower than 2.15 volts per cell nor higher than 2.18 volts. This voltage will seldom vary outside the limits of 0.01 volt from the desired mean volts per cell.

The unsatisfactory experience occasionally reported from Europe when attempts were made to use pasted plate batteries in float service has led to the nearly exclusive use there of Planté cells in such installations. The difficulty can probably be traced to lack of proper voltage regulation. The electric companies in the United States have paid meticulous attention to maintaining their power supply continuously and uninterruptedly for long periods of time, with very small fluctuations in line voltage and with practically no variation in the alternating-current frequency. The conversion equipment necessary for floating the batteries is located at the batteries and can be designed to reduce further any voltage variations to a minimum. It is possible, therefore, in the United States to use the pasted-plate batteries in float service over long periods of time, and many thousands of such batteries are being so employed.

Batteries containing calcium alloy grids are used largely by the telephone companies in float service. The current they require when floating fully charged is $\frac{1}{5}$ to $\frac{1}{8}$ that needed by batteries with lead-antimony grids under comparable conditions.

The amount of water that is necessarily added to each cell to maintain the proper level of electrolyte bears a definite relation to the amount of overcharge and may, therefore, be used as a check on the correctness of the charging rate. Too much water indicates a rate higher than necessary. Manufacturers usually specify the maximum needed.

Although nickel-iron batteries have a somewhat larger differential in voltage between charge and discharge, they are applied to floating services such as signal operations, alarm systems, and circuit-breaker control.

System-Governed Charging

The batteries charged in this way are, like the floated batteries, continuously connected to the electrical system. "System-governed," however, differs from floated methods in respect to the rather wide fluctuations between the state of full charge and partial discharge occurring more or less continuously. Charge and discharge are automatic and are governed by the schedule and adjustment of the system. A familiar example is the work of the automotive battery. It discharges to crank the engine, provide for ignition, lights, etc., until a certain engine speed is attained, when the control takes the load off the battery and transfers connections to a generator that charges it. The battery is then ready for the next discharge.

CHARGING AND DISCHARGING LEAD BATTERIES

Characteristics of Discharge

Voltage. When a storage battery of the lead type begins to discharge, there is an initial drop in voltage, which may be attributed in part to the ohmic resistance of the battery and in part to the sudden decrease in concentration of the acid in the pores of the plates, which reduces the potential of the plates. This abrupt drop in voltage is often followed by an almost equally abrupt rise in voltage. Such an effect may be seen at the beginning of the discharge curve in Fig. 79. No designation in English has been applied to this peculiar phenomenon, but in French it is referred to as the *coup de fouet* (stroke of a whip). This effect, which has been ascribed to various causes, is not always present. After the sudden decrease in concentration of the

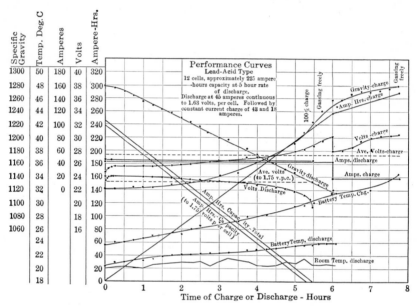

Fig. 79. Curves of discharge of a lead-acid battery, followed by charge by the constant-current method. Cells were of the motive-power type.

acid when discharge begins, the diffusion process is started and the stronger acid from outside the plates, diffusing into the pores of the plates, tends to restore the voltage.

In addition to this effect, there may also be a gradual rise in voltage during the early stages of discharge, particularly if the battery has stood for some days since being charged. This is attributed in part to the decreased resistance of electrolyte as the temperature is raised and also to the decrease in concentration toward 1.225 sp. gr., for which the resistivity is a minimum. During the greater part of the discharge, the voltage falls slowly from a value approximately equal to the open-circuit voltage to about 1.8 volts. The decrease in voltage is caused by a decrease in the plate potentials, as the rate of diffusion of the electrolyte fails to keep pace with the consumption of acid within the pores of the plates; to a lesser extent it may be attributed also to increasing resistivity of the electrolyte if the concentration falls below a specific gravity of 1.225. Near the end of the discharge, the curve drops more abruptly; this point is known as the knee of the curve. It signifies that exhaustion of the cell is approaching. The knee of the curve is less distinct in discharges at high rates than at low rates.

The average voltage during discharge is a factor of importance in determining the amount of energy delivered by the battery. The average voltage is determined from the time integral of the voltage curve from the beginning to the end of discharge. The average voltage varies with the rate at which the discharge takes place, with the type and construction of the cells, and also with the final voltage. Under normal conditions for motive-power batteries of the lead-acid type, the average voltages are approximately as follows:

Multiple of 6-Hour Rate	Volts per Cell
$\frac{1}{2}$	1.94 to 2.00
1	1.92 to 1.98
2	1.87 to 1.94
3	1.83 to 1.90
5	1.75 to 1.83
10	1.53 to 1.64

Wide separation of the plates, abundance of electrolyte, and relatively low rates of discharge afford discharge curves with the most nearly flat characteristic.

If the requirement for some specified service is a continuous discharge at a fixed current rate, the size of battery needed is easily determined in terms of the number of positive plates it should contain. However, if the duty cycle consists of discharges varying in amount or if the discharges are intermittent, the determination of the required capacity is more involved. Hoxie[5] has described a convenient graphical method for solving such problems and developed a formula of general applicability.

Temperature. The temperature of batteries of the lead-acid type rises slowly during discharge, but the effect is less than is observed when the same batteries are on charge at the same current. The reason for this has been pointed out in Chapter 4, where it was shown that there are two heat effects, one caused by the ohmic resistance and varying as the square of the current, and the other a reversible heat effect varying directly as the current. These two heat effects are opposed during discharge. The one resulting from the ohmic resist-

[5] E. A. Hoxie, Some discharge characteristics of lead-acid batteries, *Am. Inst. Elec. Engrs. Conference Paper*, June 15, 1953.

ance of the cell generally overbalances the second heat effect, so that the temperature of the battery rises. Since the irreversible heat effect varies as the square of the current, the temperature rise will be greater when the rate of discharge is increased.

Specific Gravity. The specific gravity of the electrolyte falls almost linearly during discharge, provided the discharge current is constant. The specific gravity of the electrolyte, therefore, affords a valuable means of checking the state of charge of the battery. If readings are taken for a time after the discharge has been stopped, it will be observed that there is a continued fall in specific gravity until the concentration of the dilute electrolyte in the pores of the plates has been equalized with the more concentrated acid outside.

Characteristics of Charge by Constant-Current Method

Voltage. When a storage battery is first put on charge, a sharp rise in the terminal voltage is noticed. This is probably to be ascribed to the sudden increase in concentration of the electrolyte as sulfuric acid is liberated in the pores of both the positive and negative plates. The diffusion processes then begin and check the rise in voltage.

The curve representing the terminal voltage of the cell continues to rise gradually to the point which is designated as 100 per cent charge in Fig. 79. By 100 per cent charge is meant a charge equal in ampere-hours to the preceding discharge. In Fig. 79 the 100 per cent point is found at $5\frac{3}{4}$ hours. The increase in the potential at the terminals during this period is caused by the increasing concentration of the acid within the cell. It cannot be accounted for on the ground of internal resistance, since it is well known that the internal resistance of a cell decreases during the charging period.

At the time the 100 per cent charging point is reached, or slightly before, a more abrupt rise in the voltage is noted. This is due only in part to the increasing concentration of the acid within the pores. At this time, most of the sulfate that is readily available for the action of the current has been broken down and the concentration of lead ions in the electrolyte has diminished. A consideration of the Nernst equation, relating the potentials of the plates to the ionic concentrations in the electrolyte, shows that as the number of divalent lead ions is diminished the voltage of the battery is increased. Therefore, when the lead sulfate on the plates becomes exhausted and the lead sulfate ions in the electrolyte are diminished, the plate potential rises to a point at which decomposition of water will take place. This is made evident by the formation of gas, including both oxygen and hydrogen, which escapes freely from the cell during the latter part of the charge. The gassing becomes free, as Fig. 79 indicates, at the end of about 6 hours. If the charging rate is reduced at this time, the voltage of the battery is lowered and the gassing is decreased.

Complications arise when charging batteries at extremely low temperature is attempted. Sometimes a battery will not accept charge, and it may be necessary to decrease the charging current to a value much below customary rates. The difficulty has been attributed to the slow rate of solution of the lead sulfate at these low temperatures. This results in a polarization that limits the battery's ability to utilize the current. Previous activities of the battery influence its behavior on charge. Excessive charging currents supplied to very cold batteries may give rise to peroxidized products, including ozone and Caro's acid, permonosulfuric acid, H_2SO_5.

Experiments on positive plates at low temperature and negatives at normal temperature, or vice versa, in the course of experiments down to $-50°$ C showed the effects[6] on positives to be slightly greater than on negatives, which seems rather surprising when compared with the results reported on page 224. When both positives and negatives were at the lowest temperatures the terminal charging voltage was 2.96 volts. The presence of ozone around the positive plate is said to increase its potential, but there is some question about which is the cause and which the effect.

Similarly cooling positive plates of nickel-iron batteries increases the potential at the end of charge more than for negatives. When both nickel and iron plates were at the lowest temperature, $-40°$ C, the final voltage was 2.04 volts per cell. No ozone was detected from the alkaline cells.

Gassing. The point at which gassing of a storage battery begins while on charge is determined by the voltage, but the quantity of gas depends on the portion of current that is not absorbed by the battery. The gases liberated are oxygen, evolved at the positive plates during charge, and hydrogen, evolved at the negative plates. It is not uncommon, however, for one plate to begin gassing before the other, and therefore the gas liberated from a storage cell, in the early stages at least, does not always correspond to the proportions of oxygen and hydrogen that form water. The gassing begins when the voltage at

[6] E. Briner and A. Yalda, Sur le fonctionnement aux basses températures de l'accumulateur au plomb et l'accumulateur au nickel-fer; values élevées des forces electromotrices en fin de charge, *Helv. Chim. Acta*, *25*, 416 (1942); see also their previous paper, *24*, 109 (1941).

262 Storage Batteries

the terminals of the battery has reached about 2.3 volts per cell. Table 53, obtained from a test, shows that during the early stages the gas is composed of almost equal parts of oxygen and hydrogen, but as the charge approaches the end, when the voltage has risen to 2.5 per cell, the gas has a composition of 2 parts hydrogen to 1 of oxygen.

Since oxygen and hydrogen unite with explosive violence to form water, it is necessary that certain precautions be taken to avoid accident. Open flames of any kind in a storage-battery room are not permissible, and suitable ventilation should be provided to prevent an accumulation of hydrogen. Four per cent of hydrogen in the atmosphere is dangerous, and the concentration should be kept much below this figure.

TABLE 53. EXPERIMENTAL DATA ON GASSING
(Per Cent by volume.)

Cell Voltage, v	Gassing	Composition Hydrogen, %	Oxygen, %
2.20	None	..	..
2.30	Slight	52	47
2.40	Normal	60	38
2.50	Hard	67	33

Other gases are not to be expected under normal conditions but may be liberated if certain impurities are present. Haring and Compton[7] detected small amounts of stibine in gases liberated during overcharge.

A certain minimum current density is required for the formation of stibine, and for this reason it is not normally detected until overcharge begins. At this stage of the charge the voltage across the cell terminals is about 2.43 to 2.45 volts. The passing off of stibine during charge is a useful way of eliminating accumulated antimony from the surface of the negative plates. Reactions for arsine are probably analogous to those of stibine. The amount of stibine formed is not necessarily proportional to the amount of antimony that is in the active material of the negative plates, since it is the freshly deposited antimony that is the more effective.

Carbon dioxide may be liberated, if organic matter is being oxidized.

[7] H. E. Haring and K. G. Compton, The generation of stibine by storage batteries, *Trans. Electrochem. Soc., 68,* 283 (1935).

Chlorine in relatively large amounts may be liberated, if sea water enters the battery. This comes from the positive plates, which are discharged by the action of the sodium chloride with the formation of lead sulfate and sodium sulfate.

Storage cells of the lead-acid type should not gas during discharge or when standing idle under ordinary conditions. There is, however, likely to be a slight liberation of hydrogen from the negative plates because of entrapped gas or local action that takes place when the cell is idle. It should always be presumed, therefore, that hydrogen is present within the space over the electrolyte, and precautions, necessary to avoid explosions, should be observed at all times. It is particularly dangerous to use a lighted match over the vent of a cell to see how high the liquid stands.

Specific Gravity. The changes in specific gravity of the electrolyte during the charging period are shown by a curve in Fig. 79. The most noticeable fact about this curve is the very slow rise during the first 2 hours. After this time the curve rises more rapidly, following in an approximate way the curve representing the ampere-hours of input. Toward the end of charge there is less lead sulfate available for conversion into sulfuric acid. The specific gravity reaches a maximum when no more sulfate is being converted into acid.

The fact that the specific gravity is slow to rise at the beginning of charge is worth more than passing notice. Three hours or more may elapse before any significant increase is noted. Even the so-called fast chargers (see page 253) may not produce a noticeable change within the time that impatient observers expect, and the battery is often erroneously declared not able to take a charge, to be worthless, or to be a fit subject for trying some rejuvenator.

Concentrated acid, as it is liberated from the pores of the plates, falls to the bottom of the cell. The rapid rise when the cell begins to gas freely is caused by the bubbles of gas stirring up the electrolyte. In a cell made as compactly as the motive-power type, for which performance curves are shown in Fig. 79, it is not possible for the acid to circulate throughout the cell as rapidly as it does in a cell of the stationary type with wider separation between the plates.

Tapering-Current Charge

In addition to charging batteries by the constant-current method, which has been described in the preceding section, it is possible to charge them by the modified constant-potential, or tapering-current, method. Characteristic curves are shown in Fig. 77, page 248.

Storage Batteries

Comparison of the Constant-Current and Constant-Potential Methods
of Charging

In Fig. 80, which is taken from an article by Woodbridge,[8] a comparison of the rate of charging by the constant-current and constant-potential methods and its relation to the state of charge and the gas liberated is shown. It has been stated that charging takes place more

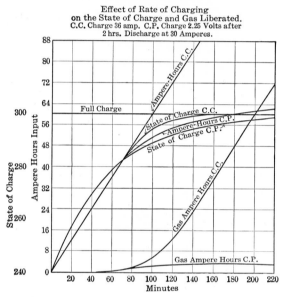

Fig. 80. Comparison of the constant-current and constant-potential methods of charging lead-acid batteries.

rapidly by the constant-potential method during the early stages of the charge, but Fig. 80 applies particularly to a comparison of the two methods when the charge is nearly complete. A storage battery of 300 ampere-hour capacity, fully charged, was twice discharged to the extent of 60 ampere-hours and put on charge by the constant-current and constant-potential methods successively. The curves show that the rate of charging by the constant-potential method is slightly greater than by the constant-current method, but that the curve representing the state of charge by the constant-current method lies above the state of charge by the constant-potential method, since the latter approaches full charge asymptotically. When the curve representing

[8] J. L. Woodbridge, Constant voltage charging and battery life, *Ry. Elec. Engr.*, *9*, 6 (1918).

the state of charge by the constant-current method begins to deviate from the ampere-hours actually put in by this method, it will be noted that a marked rise in the curve representing the ampere-hours wasted in gassing begins, and that this curve becomes parallel to the curve of ampere-hours input. The physical significance of the parallelism of these curves is that all of the ampere-hours supplied to the battery are eventually used in the decomposition of water. The curve representing the ampere-hours wasted in gassing by the constant-potential method shows that only a negligible part of the quantity of electricity passing through the cell is expended in this way. These curves also indicate that the constant-potential method of charging provides the maximum current input that the battery can absorb without waste of energy in producing gas.

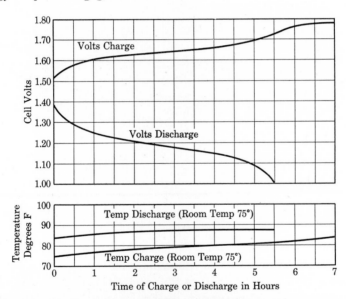

Fig. 81. Curves of charge and discharge of an Edison cell, constant current at the normal rate.

CHARGING AND DISCHARGING EDISON BATTERIES

The characteristic curves for charging and discharging an Edison storage battery are given in Figs. 81 and 82. The discharge curve resembles that for the lead battery, but the fall in voltage is somewhat more steep. The voltage curve on charge rises rapidly at the beginning and then more slowly until 5 hours have elapsed, after which the terminal voltage of the cell becomes constant at about 1.80 volts

per cell. No specific gravity readings are recorded, since the electrolyte in these cells does not change in concentration. The temperature
of the batteries rises markedly during both the charge and the discharge. In order to maintain Edison batteries in a state of maximum
activity, the charging rate should be maintained at an average of the
normal rate and should not be less than half the normal rate, but this
does not preclude the use of trickle charging if proper application as

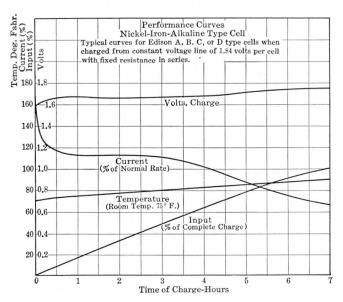

Fig. 82. Modified constant-potential charging curves of an Edison cell.

to size and type of cell and recommended rates of trickle charge are
followed. Edison batteries gas throughout the entire period of charge.

These cells may be charged by the constant-potential or tapering-
charge method. It is necessary to maintain a sufficient cathodic polarization during charging to liberate hydrogen, in order that the iron may
be properly activated.

Table 54 gives the average voltages for discharge of various types
of cell, and Table 55 gives terminal voltages for varying amounts of
discharge at various rates of discharge. The values given in these
tables are voltages at cell terminals with no allowance for connectors.

It has been found that the life of alkaline batteries is prolonged by
operating them at moderate temperatures. Charging them at high
temperatures is more detrimental than discharging them when hot, but
the latter may result in harmful reactions in extreme cases. At 50° C

TABLE 54. AVERAGE VOLTAGE, EDISON CELLS

(Computed for discharges of various types and sizes of cells in fully active condition at normal temperature to end voltages equivalent to 1.0 volt per cell at the normal rate. Normal rate for types A, B, C, and D is 5 hours. Based on cells giving 110 per cent of their rated capacity.)

Multiples of Normal Rate	Average Voltage
$\frac{1}{2}$	1.27
1 (normal)	1.23
2	1.13
3	1.04
4	0.94
5	0.85

(122° F) slight gas evolution occurs from cells on open circuit as a result of local action. The gas is mostly hydrogen, which is produced by the attack of the electrolyte on the iron of the negative plate. Charging the batteries when they are hot results in diminished capacity on the succeeding discharge. The largest immediate output is obtained by charging at a temperature of about 25° C (77° F) and discharging at 50° C (122° F), but it is wise to limit the upper temperature to about the same figure as for lead batteries, at most 46° C (115° F).

TABLE 55. EDISON CELL VOLTAGES, DISCHARGE

(The voltages are for various types and sizes of cells in fully active condition discharging at multiples of the normal rate after normal charges and at normal temperatures. Normal rate for types A, B, C, and D is 5 hours. Based on cells giving 110 per cent of their rated capacity.)

Per Cent of Rated Capacity Taken Out	Cell Voltages for Discharges Expressed in Terms of the Normal Rate					
	0.5	1 (Normal)	2	3	4	5
0*	1.51	1.47	1.39	1.31	1.23	1.14
0†	1.46	1.42	1.33	1.25	1.16	1.08
5	1.38	1.34	1.25	1.16	1.07	0.98
10	1.35	1.31	1.22	1.13	1.04	0.95
20	1.32	1.27	1.18	1.09	1.00	0.91
40	1.28	1.23	1.14	1.05	0.96	0.87
60	1.26	1.21	1.12	1.03	0.93	0.83
80	1.24	1.19	1.09	0.99	0.89	0.78
90	1.23	1.17	1.06	0.96	0.86	0.74
100	1.20	1.13	1.01	0.90	0.78	0.64
110	1.11	1.01	0.86	0.73	0.60	....

* Immediately after charge.
† After standing 6 hours.

The effect of temperature on the charging voltage of alkaline storage batteries (Edison type) is given in Table 56.

TABLE 56. EFFECT OF TEMPERATURE ON CHARGING VOLTAGE OF
EDISON BATTERIES AT NORMAL RATE OF CHARGE

Temperature		Average Voltage,	Maximum (Final) Voltage,
° C	° F	v	v
2	35	1.88	1.94
13	55	1.81	1.92
24	75	1.76	1.88
35	95	1.70	1.85
46	115	1.67	1.77

The knee of the discharge curve in Fig. 81 marks the end of the useful discharge of the battery. Beyond this a second stage of discharge may be observed in certain types of alkaline batteries (not including Edison batteries). Positive plates, except those containing graphite, do not contribute to this secondary discharge. Since the positive plate is normally the limiting factor, this second stage discharge is not of practical importance and should be avoided.

The operation of alkaline batteries can be controlled by voltage measurements on a pilot cell when discharging through a predetermined resistance. Comparisons of the readings are made with a table to determine the approximate state of charge of the battery. The instrument used is the "charge test fork."

The error in estimating the state of charge by such measurements is no greater than the error in determining the state of charge of lead batteries by specific-gravity readings. Table 57 gives the approximate

TABLE 57. EDISON CELL VOLTAGES, CHARGE

(The voltages are for cells charging at the normal rate.)

Percentage of Rated Capacity Remaining	Cell Voltage		
	35° F (2° C)	75° F (24° C)	115° F (46° C)
0	1.64	1.57	1.48
50	1.88	1.73	1.64
100	1.94	1.88	1.77

steady voltages per cell that may be expected when the battery has been on charge for a few minutes at the rate which is normal for the particular type of battery. The data in this table are reasonably consistent with the curves in Fig. 81 if interpolation is made for the temperature. Ampere-hour meters connected in circuit with the batteries are the most satisfactory means of terminating the charge. Lacking such instruments, however, recourse may be had to voltage measurements when the batteries are charging at specified rates and temperatures.

CHARGING AND DISCHARGING NICKEL-CADMIUM BATTERIES

Nickel-cadmium cells may be charged at constant current, constant potential, or modified constant potential. Evolution of gas begins at about 1.47 volts, usually at the end of $4\frac{1}{2}$ hours when charging is timed for completion in 7 hours. The final charging voltage is 1.75 volts per cell, and the required line voltage is 1.85 volts per cell. The limiting temperature for good operation is 115° F (45° C) measured in the electrolyte. Charging should be discontinued if this temperature is reached.

Some variation in the charging procedure is permissible. For example, a cell of 100 ampere-hour capacity charged at constant current normally for 7 hours at 20 amperes may be charged at 14 amperes for 10 hours or at 10 amperes for 14 hours. The floating charge voltage is 1.40 volts. Trickle-charge rates are adjusted at voltages between 1.40 and 1.45 per cell according to the need. Boosting, as applied to discharged cells, is permissible at the following rates:

5 times the normal 7-hour charge rate for 5–10 minutes
4 times the normal 7-hour charge rate for 10–15 minutes
3 times the normal 7-hour charge rate for 15–30 minutes

Discharging the cells at ordinary rates, 3 to 10 hours, should be terminated when the voltage per cell falls to 1.00 to 1.10 volts. At very low rates of discharge the final voltage should be 1.20 to 1.25 volts. Heavy discharges are permissible for short duration such as engine starting or emergency switch operation. At such times the terminal voltage will fall far below the limits mentioned above, but that will not be harmful if not long continued. In general the nickel-cadmium batteries are not intended for cycle service.

Characteristic curves of charge and discharge for cells of the type S are shown in Fig. 83. It will be noted that the general shape of the charge curve is more like that of lead-acid batteries than of the Edison type of alkaline cells, which is characterized by a steep rise in the terminal voltage in the early stages of charge. To determine the state of charge of a nickel-cadmium cell, it is necessary to measure the terminal voltage when the cell is discharging at a 5-hour rate. Comparison may then be made with the discharge curve in Fig. 83. The state of charge cannot be inferred from specific-gravity measurements of the electrolyte.

Curves showing operating characteristics of these cells are generally applicable to various sizes of cells, since they are based on the per-

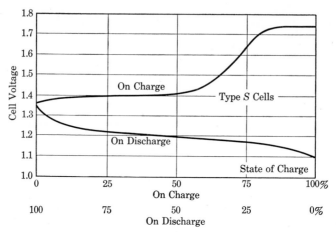

Fig. 83. Relation of voltage to state of charge of nickel-cadmium cells. A 7-hour charge current equals the normal 5-hour discharge current.

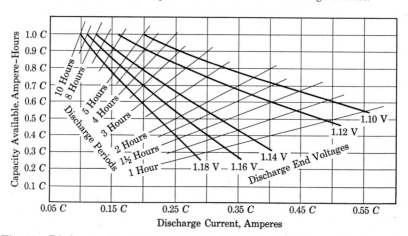

Fig. 84. Discharge characteristics of nickel-cadmium cells, type S. C equals rated capacity in ampere-hours at the 8-hour rate. The discharge current in amperes is expressed in terms of C.

centage of charge or discharge in Fig. 84 and on decimal fractions of C, the rated capacity of the cell in ampere-hours at the 8-hour rate. The discharge current in amperes is then numerically equal to the indicated fraction of the ampere-hours.

Figure 84 gives the capacity available at various rates of discharge to the indicated final voltages. These curves apply to the type S Nicad cells. Further data relative to the voltage characteristics at high rates of discharge for type S cells are given in Fig. 85. The average voltage on discharge at normal rates is about 1.2 volts per cell.

Batteries that are to be inactive for a few months should be given a full charge for 7 hours, the electrolyte checked for specific gravity

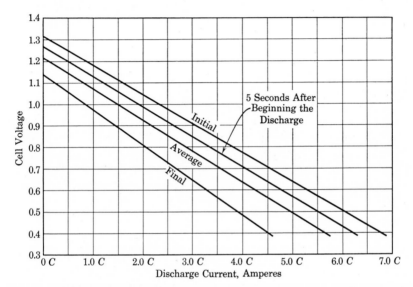

Fig. 85. High-rate discharge characteristics of nickel-cadmium cells, type S.

(to be at least 1.190), and several intercell connectors removed to eliminate stray currents. The battery can then be stored in a cool, dry place. For long periods of inactivity, trickle-charging at 1.40 volts per cell is recommended.

CHARGING EQUIPMENT

Except when direct-current bus bars of suitable voltage are provided for battery charging, the charging equipment will ordinarily include a motor-generator set, synchronous converter, or rectifier, to provide direct current at the proper voltage. Standard specifications for automatic battery-charging motor-generators, rectifiers, and panels are available.[9]

[9] *Recommended Specifications for Automatic Battery-Charging Equipment for Industrial Truck Motive-Power Service,* Electric Industrial Truck Association, 1949.

Selecting Equipment

It is not within the scope of this book to give a detailed description of motor-generator sets but rather to give such general information as may be required to plan the equipment in any particular case.

Battery Data. The batteries that are to be charged must be classified according to the number of batteries of each kind, the number of cells in each battery, the type of the battery, whether of the lead-acid or of the alkaline type, the capacity of the cells, and the purpose for which they are used. It is also necessary to know the maximum number of batteries which will be charged at one time, in order that the proper type and maximum capacity of the charging equipment may be chosen.

Charging Control. The rate of charging will ordinarily be determined by the type and size of the batteries. If it is necessary to give the cells boosting charges, this fact should also be taken into account in computing the maximum capacity required of the charging equipment.

The method of charging, whether constant-current or constant-potential, must be chosen in advance, since generators and auxiliary equipment of somewhat different characteristics are required for these methods. Generators for constant-current charging are ordinarily of the shunt-wound type with drooping characteristic. Hand regulation capable of reducing the terminal voltage of the machine is usually provided.

The trend in industry is to automatic control of battery charging. The modified constant-potential method is generally applicable to both lead-acid and nickel-iron batteries and is preferred because of its simplicity and freedom from the necessity of manual control. For charging one battery a shunt-wound type of generator, having suitable voltage characteristics, is used, but, if more than one battery is to be charged from the same generator at constant voltage, a flat-compounded generator is necessary. A voltage regulator is sometimes applied to the shunt field to provide uniform voltage regardless of temperature and load conditions. Each battery is connected to the charging source through a single fixed resistor. Batteries are removed from the line automatically as each finishes, and the generator is shut down automatically when the last battery is charged. In the event of power failure, the generator is disconnected from the batteries. This is quite necessary, but when the supply is restored the motor-generator is restarted and charging resumed.

For floating charge, the diverter pole generator, described by

Smith,[10] has found wide use, particularly in telephone offices and unattended substations. The generator maintains constant voltage from low loads to its full rated capacity, beyond which the voltage droops sharply. This transfers excessive current demands to the battery. Suitable voltage characteristics are obtained by placing the series windings on intermediate or diverter poles between the main poles and by the use of a magnetic bridge. By properly proportioning the shunt and series windings the flux in the armature can be made to vary with the load. The diverter poles and magnetic bridges also have an important function in stabilizing the machine in the event it should run as a motor during power interruption.

Similar characteristics can be obtained with regulated constant-potential rectifiers, which provide substantially constant voltage from small loads to full load and a drooping characteristic beyond full load.

Grouping Batteries. The small batteries, particularly those for starting and lighting service, may be grouped in several different ways during charging. The most advantageous grouping will depend upon the capacity and the number of cells in each battery. With low-voltage batteries of 6, 8, and 12 volts, there is a choice of connecting them in parallel, in series, or in a combination of series and parallel. It is practical to charge these batteries in series up to bus-bar voltages of 115 volts, but higher voltages are not recommended for use in service stations, where the floors are often wet and the insulation not of the best.

Sectional Panels. Charging equipment can be obtained in the form of sectional panels, which are designed for any practical conditions or method of charging. The sectional panels include incoming line sections, generator control sections, and charging sections with resistors both variable and fixed, voltmeters and ammeters, overload and underload circuit breakers, ampere-hour meters, reverse-current cutouts and fuses. If a proper selection of control panels is made, a convenient charging switchboard may be easily arranged. Several sections of control panels are shown in Fig. 118. These were designed originally for tests of aeronautical batteries, but they can be used also for automotive and similar, smaller types of batteries.

Rectifiers

A rectifier is essentially an electrical check valve which converts alternating current into unidirectional current by providing a low resistance path for the flow of current in one direction but interposing

[10] E. D. Smith, The diverter pole generator, *J. Am. Inst. Elec. Engrs.*, *48*, 11 (1928).

a high or infinite resistance when current would tend to flow in the other direction. Rectification may be accomplished by the use of mercury arcs, electrolytic valves, ionized gases, or synchronous switches, or by the flow of electrons across the junction of certain metals and semiconductors. The semiconductors include oxides and sulfides formed on the surface of copper, and selenium which is spread on a surface of iron or nickel. Whatever the method, a pulsating direct current is produced which can be used for battery charging without the necessity of "smoothing" it. The process of battery charging is governed by Faraday's law, and the electrochemical reactions which occur during an infinitesimal interval of time, dt, are, therefore, proportional to the quantity of electricity passing at that instant, that is, to I dt. For each half cycle the quantity of electricity flowing is equal to the average value of the current multiplied by the time that it flows. The current should be measured, therefore, with direct-current ammeters of the permanent-magnet moving-coil type. The magnitude of the current, as well as the time that the valve is open, is governed by relative values of impressed voltage, counterelectromotive force, and whatever capacitance, inductance, and resistance are present in the circuit.

If a single rectifying element is used, only one-half of the alternating wave is rectified, but two or more rectifying elements can usually be arranged to rectify both halves of the wave. This is called full-wave rectification. When half-wave rectification is provided the current cannot flow during one-half the cycle because it would have to pass in the high-resistance direction of the rectifier. Being stopped out, half the cycle is ineffective, but the efficiency is not materially reduced.

If the circuit is without appreciable inductance or capacitance, the rectified current is in phase with the impressed electromotive force and attains its maximum value once or twice each cycle, depending on whether half- or full-wave rectification is provided. This is illustrated in Fig. 86 by the tracing of an oscillogram shown for the simplest case of a circuit having no counterelectromotive force and negligible capacitance and inductance. The current begins to flow when the impressed voltage passes through zero value, and each maximum of current is in phase with the maximum of voltage. If a battery is being charged, however, the electromotive force of the battery, which is relatively constant, opposes the periodic impressed electromotive force, and current cannot flow until the impressed voltage has reached a value equal to the electromotive force of the battery, plus any other voltage losses. This is illustrated in Fig. 87. The valve opens at the point marked A and closes at the point marked B.

If the value of the impressed voltage is not greater than the counterelectromotive forces opposing it, no charging current can flow. However, if the rectifier and battery have appreciable capacitance, an

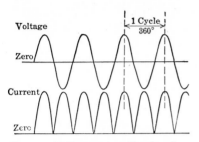

Fig. 86. Full-wave rectification in a
circuit having no counterelectromotive
force. Capacity and inductance are
negligible; hence the current is in
phase with the voltage.

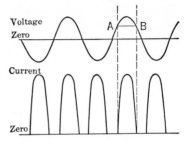

Fig. 87. Full-wave rectification in a
circuit having a counterelectromotive
force. The valve is open only from
A to B, that is, when the impressed
voltage exceeds the counterelectromotive force. If capacity and inductance are negligible, the current is in
phase with the voltage.

alternating current which leads the impressed electromotive force by
90° will flow through the battery, but this is entirely ineffective in
charging it. Such a condition is shown in Fig. 88.

Capacity currents, and leakage currents too (if the rectifying film
is imperfect), show negative loops in oscillograms of half-wave rectification. One may be distinguished
from the other, however, by the
fact that the capacity loop is leading
the impressed voltage, while the
leakage loop is always in phase with
it.

In the foregoing discussion, it has
been assumed that the circuit is
essentially non-inductive. The battery has no appreciable inductance,
but inductance may be added to the
circuit for purposes of regulation

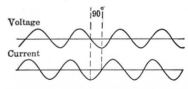

Fig. 88. Conditions when the impressed voltage is less than the
counterelectromotive force. No rectification occurs, but a capacity
current, leading the voltage by 90°,
flows through the battery.

or to extend the time that the valve is open. In Fig. 89, showing the
effect of inductance, it may be seen that the current lags behind the
impressed voltage. This oscillogram is for half-wave rectification.
Current in the positive direction begins after the impressed voltage has
attained a value predetermined by the counterelectromotive force.

The current increases slowly at first, by no means as rapidly as the impressed voltage. It attains its maximum value 90° later than the voltage, or somewhat less if there is appreciable capacitance in the circuit. As the impressed voltage decreases, the current lags and the current may still be flowing in the positive direction when the impressed voltage has passed through zero and become negative. This is because of the induced electromotive force, which depends on the inductance, L, and the rate of change of current, dI/dt. The valve is open longer, therefore, when inductance is present, unless the conditions of the circuit are complicated by large hysteresis losses in iron cores. Actually the voltage across the valve, E_R, which determines when it opens and closes is the algebraic sum of the impressed voltage, $E_0 \sin \omega t$, the counterelectromotive force of the battery, E_B, the induced electromotive force, $L(dI/dt)$, and the IR drop. That is, the voltage across the valve is

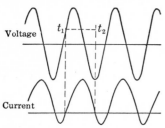

Fig. 89. Half-wave rectification in a circuit having counterelectromotive force and inductance. The valve is open from t_1 to t_2. The maximum of current lags behind the maximum of voltage. If an iron core is present, the hysteresis must be low to avoid loss.

$$E_R = E_0 \sin \omega t - E_B - L(dI/dt) - IR$$

Previous to the instant, t_1, that the valve opens

$$I = 0 \quad \text{and} \quad 0 < t < t_1$$

$$E_R = 0 \quad \text{and} \quad dI/dt = 0$$

but the impressed voltage and the counterelectromotive force are equal and opposite, whence, if a solution exists, $E_B < E_0$ and

$$\sin \omega t_1 = E_B/E_0$$

This determines in angular measure the time of opening the valve. To find the time of closing the valve is somewhat more difficult.

For this, $t_1 < t < t_2$, and in this interval $I > 0$

$$L(dI/dt) + RI = E_0 \sin \omega t - E_B$$

with the initial condition that

$$I = 0 \quad \text{when} \quad t = t_1$$

so that

$$I(t) = \frac{E_0}{R^2 + \omega^2 L^2} [R \sin \omega(t - t_1) - \omega L \cos \omega(t - t_1) + \omega L e^{-(R/L)(t-t_1)}]$$

$$- \frac{E_B}{R} [1 - e^{-(R/L)(t-t_1)}] \quad \text{for} \quad t_1 < t < t_2$$

where t_2 is to be determined as the next time in this interval that I vanishes, that is, $I(t_2) = 0$. Write

$$\tau = t - t_1 \quad \text{and} \quad \tan \alpha = \omega L / R$$

then

$$RI(t) = \frac{RE_0}{\sqrt{R^2 + \omega^2 L^2}} \sin (\omega \tau - \alpha) - E_B + \left(\frac{R \omega L E_0}{R^2 + \omega^2 L^2} + E_B \right) e^{-R\tau/L}$$

$$= E_0 \left[\cos \alpha \sin (\omega \tau - \alpha) - \frac{E_B}{E_0} + \left(\sin \alpha \cos \alpha + \frac{E_B}{E_0} \right) e^{-R\tau/L} \right]$$

For any particular case the values of E_B, E_0, L, and R are known so that it is not difficult to determine the value of τ which makes this expression for the current vanish. Using this and the value of t_1 determined previously, we have

$$t_2 = t_1 + \tau$$

Electrochemical Rectifiers. Aluminum, tantalum, and a few other metals possess the property of valve action when immersed in certain solutions. During the operation of the rectifier a film forms on the surface of these metals. This is permeable to hydrogen cations but not to anions, except those that act as "poisons" and destroy the film. The current can pass through the film on the valve electrode, but not from it, except at relatively high voltages when breakdown occurs. In addition to the valve electrode, each cell must contain another electrode to serve as an anode. This must be able to withstand the strong oxidizing conditions and must pass current in either direction. Lead, carbon, iron, chromium-steel, and iron-silicon alloys have been used for this purpose.

Tantalum rectifiers, used to some extent in railway signaling, consist of metallic tantalum as the cathode and lead or an alloy of lead as the anode in a solution of sulfuric acid, to which a small amount of ferrous sulfate is added. The specific gravity of the solution is about 1.250.

Metallic Rectifiers. *Copper Oxide Rectifiers.* The copper-cuprous oxide rectifier, which has developed from experiments made by Grondahl[11] in 1920, depends for its operation on the asymmetrical resistance

[11] L. O. Grondahl and P. H. Geiger, A new electronic rectifier, *Trans. Am. Inst. Elec. Engrs.*, *46*, 358 (1927).

at the junction of copper and a layer of cuprous oxide formed on its surface. That is, current can flow more freely in one direction than in the other. From oxide to copper the resistance is low, but from copper to oxide the resistance is many times greater. Such an arrangement provides a valuable rectifying unit which has no moving parts or chemical reactions.

Copper of a high degree of purity is partially oxidized in air at a temperature of 1000 to 1040° C. The latter temperature is slightly above the melting point of the oxide, 1025° C. Red cuprous oxide, Cu_2O, is formed on the surface of the copper as an adherent layer. This in turn is usually covered by a layer of black cupric oxide, CuO, which has a high resistance and which must be removed by being dissolved in a mixture of acids at a later stage of the process. The heat treatment of the elements after oxidation has an important effect on the operating characteristics of the rectifier. Slow cooling produces elements of relatively high resistance which are adapted to use in high-voltage rectifiers, since the elements are able individually to withstand 10 to 30 volts in the reverse direction and since the matter of their forward resistance is relatively unimportant. On the other hand, elements which are quenched have a lower resistance but should not be subjected to more than 6 volts. These elements have a high ratio of resistances in the high and low directions. They are well adapted to use in rectifiers carrying considerable current.

The oxidized copper is prepared in a variety of sizes and shapes. Some of these are buttons only a few hundredths of an inch in diameter, others are disks or "washers" 1 to $1\frac{1}{2}$ inches in diameter, and still others are rectangular plates. Contact to the oxidized surface is made in various ways, such as sheets of lead over a coating of colloidal graphite applied to the oxide surface. The elements are assembled as shown in Fig. 90, from the paper of Grondahl and Geiger. Ventilating fins are provided when necessary. These are interleaved with the elements. Without special ventilation a current

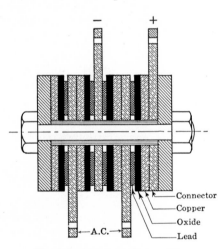

Fig. 90. Copper-oxide rectifier.

density of 0.07 ampere per cm^2 is reasonable. The current-carrying capacity may be increased by enlarging the area of contact (paralleling elements) and by forced ventilation. The life of the rectifier is adversely affected by high temperatures. Eighty degrees Centigrade is maximum, but for continuous operation temperatures should not exceed 50° C. Higher voltages applied to these rectifiers are sustained by placing additional elements in series.

The boundary area between copper and oxide has a capacitance of about 0.006 microfarad per cm^2. This is of some importance as the capacitive reactance must be large in comparison with the resistance component of the impedance in the low-resistance direction, otherwise rectification would be seriously impaired. At commercial frequencies the capacitive reactance is without detrimental effect on rectification and the power factor.

A wide variety of unit assemblies is available. These include single and double stacks with or without radiating fins. From these, rectifiers to meet any particular service condition of current or voltage can be built.

Selenium Rectifiers. The photoelectric properties of selenium have been known for many years, but its use as a rectifier[12] is comparatively recent. Like other metallic rectifiers it is assembled in series and multiple stacks to meet specified current and voltage requirements. Each cell consists of several layers: (1) a base plate and conducting member that makes contact with the next adjoining cell, (2) a layer of selenium, usually classed as a semiconductor, (3) a barrier layer between the selenium, and (4) the superposed counterelectrode. Rectification takes place at the barrier layer. The base plate is often nickel-plated aluminum. On it is sprinkled vitreous selenium with a small amount of a halogen to improve conductivity. Under pressure at a temperature between 100 and 150° C the selenium forms a film, and further heat treatment converts it to the so-called metallic state. The counterelectrode is a thin layer of a low-melting-point alloy which may be sprayed or evaporated onto the selenium. Current flow in the forward direction takes place from the selenium to the counterelectrode through the barrier layer which stops the current flow in the opposite direction. Selenium rectifiers are variously rated for ambient temperatures of 35 to 40° C; efficiencies range from 60 to 80 per cent, and the power factor is about 95 per cent. The manufacturer's ratings should not be exceeded.

[12] W. B. Roberts, Counterelectrodes for selenium rectifiers, *J. Electrochem. Soc.*, *97*, 181C (1950). Extensive bibliography.

Magnesium-Copper Sulfide Rectifiers. Another form of metallic rectifier makes use of copper sulfide,[13] which is also classed as a semiconductor. Rectifying action takes place at the junction of magnesium and the copper sulfide. When the latter is positive with respect to the magnesium, the current can flow in the low-resistance direction. In making these rectifiers, a zinc-copper alloy is usually employed in preference to pure copper. Initially the magnesium has a coating of oxide, and this is pressed against the copper coated with cupric sulfide. The rectifying film is formed by passing an alternating current through the cell, and this results in layers as follows: Mg-MgS-Cu_2S-CuS. The magnesium sulfide and the cuprous sulfide become locked together. The operating range is given as -70 to $+130°$ C, but for continuous operation it is better not to exceed more normal temperatures. Mobile units combined with TVR relays are made for charging lead batteries by the two-step method or for charging Edison batteries by the modified constant-potential method. Kotterman[14] has described these rectifiers as mobile chargers for railway passenger cars.

Gas-Filled Bulb Rectifiers. A combination of a hot and a cold electrode in a vacuum may act as a rectifier, because the hot electrode throws off electrons when it is charged negatively, and these, traversing the space to the other electrode under the influence of an electrostatic field, constitute an electric current. Space charges, depending on the relative rates of emission of electrons and their conduction of current, may occur and cause some loss of efficiency. The high-vacuum type of bulb is adapted to high voltages, but for low voltages and relatively large currents bulbs filled with an inert gas such as argon are better. The electrons streaming out from the hot electrode ionize the gas and permit considerable current to flow during the time that electrons are being emitted. On the other half cycle, neither electrode can emit electrons, and consequently no current can flow. Most of the current that passes through the bulb is carried by the ions from the gas. Rectifiers of this type are equipped with "Tungar" or "Rectigon" bulbs which were placed on the market about 1916. They are shown diagrammatically in Figs. 91 and 92. The cold anode of graphite is marked *a*, and the hot cathode of spirally wound tungsten wire is marked *c*. A small ring of magnesium wire around the stem of the anode, marked *b*, is included within the bulb to react with any gas (except the inert gas) that may be present and liable to impair the life

[13] Samuel Ruben, Magnesium-copper sulfide rectifier, *Trans. Electrochem. Soc.*, *87*, 275 (1945).

[14] C. A. Kotterman, Magnesium-copper sulfide rectifier battery charger for railway passenger cars, *Trans. Am. Inst. Elec. Engrs.*, *58*, 260 (1939).

or efficiency of the bulb. By a flashing process after the bulb is sealed, this magensium wire, called the "getter," is made to combine with the objectionable gases that may be present. The products of chemical combinations are deposited on the inner surface of the glass, giving it a mirror-like or sooty appearance. For this reason the bulbs are seldom clear. The gas pressure within the bulb may be from 0.01 millimeter to several centimeters, By using two tubes it is possible to rectify both halves of the wave. Some rectifiers of this type contain autotransformers connected as shown, but variations in the connections

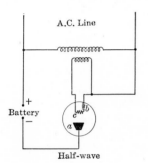
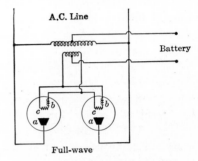

Fig. 91. Gas-filled bulb recti-   Fig. 92. Gas-filled bulb rectifiers.
fier.

will depend upon whether the secondary circuit is grounded or not. The small winding shown is a low-voltage, heavy-current coil supplying current to heat the filament.

Rectifiers of this type can be operated in parallel, if provided with suitable reactance. Two bulbs alone will not operate in parallel, since the one with the lower arc drop takes the load. The power factor is about 50 per cent.

The bulbs are mounted with transformers which are designed for specified voltages. Certain precautions should be observed in operating these rectifiers, to prevent damage to the bulb. The bulb, once started, will continue to operate without the cathode excitation, since the cathode can be kept hot by the bombardment of the positive ions. This does not eliminate the energy loss involved in heating the cathode, as the voltage drop across the arc increases. The arc tends to concentrate on a few turns of the filament when its normal excitation is cut off, and the filament will burn through at this point. It is not desirable, therefore, to operate the filament without its normal current, as the life of the bulb will be greatly shortened. If the battery voltage is 40 volts or more, both alternating-current and direct-current circuits should be opened simultaneously by the rectifier switch, to prevent a possible surge of current through the bulb, which may happen when the alternating-current circuit alone is broken, particularly if this occurs at the instant of peak voltage. When such a surge occurs, a large current from the battery discharges through the bulb in the reverse direction, destroying it.

The efficiency of these bulb rectifiers may amount to 60 or 70 per cent when charging batteries of nearly the maximum voltages for which they are designed, but the efficiency decreases to less than 20 per cent when used for batteries of only a few cells.

Eight or ten sizes of these bulbs are available, having direct-current ratings of 0.25 to 15 amperes. The voltages range from 7.5 to 250 volts. Usually the bulbs with the smaller current ratings are applied to trickle-charging batteries. Complete charging units comprise: (1) rectifier bulbs (one for half-wave or two for full-wave rectification); (2) transformers (autotransformers or two-winding, insulated transformers); (3) hand-operated selector switches for control of voltage and current to provide for charging any number of 3-cell batteries from 1 to 6, 12, or 24, according to the capacity and design of the charger; (4) a direct-current ammeter of the permanent-magnet, moving-coil type; and (5) switches, fuses, and leads for connection to the alternating-current supply and the direct-current load.

Self-Regulating Rectifiers for Floating Batteries. Fully automatic control of battery charging by rectifiers is a comparatively recent development which has found application in the maintenance of floating batteries such as those used for circuit-breaker control, telephones, and alarm and signal systems. Compared with cycle operation, floating the batteries increases their life and permits smaller sizes to be used. The constant-potential chargers are designed for an output which varies with the load demand, and the direct-current output is unaffected by considerable fluctuations in the alternating-current line voltage.

When the current demand is less than the rated output of the charger, all current is supplied to the direct-current circuit by the rectifier, and simultaneously a trickle-charge current maintains the battery in a fully charged condition. When the load exceeds the rating of the charger, the excess is supplied by the battery until the load falls to its normal value, less than the rated current of the rectifier. After this, the battery is quickly restored to a fully charged condition by a charging current, which for the time being is in excess of the trickle-charge rate.

The control units which provide the automatic features of these rectifiers are saturable core reactors. These are essentially choke coils

Operation

283

with two windings. An increase in the direct-current of one coil results in a proportionate decrease in impedance of the alternating-current coil. To prevent voltage fluctuations of the alternating-current line from interfering with the regulation of the direct-current load, an alternating-current stabilizer is added to the rectifier. This consists of a saturable transformer, reactor, and capacitor. It holds the alternating-current input constant to ±1 per cent, notwithstanding variations of ±15 per cent in the line.

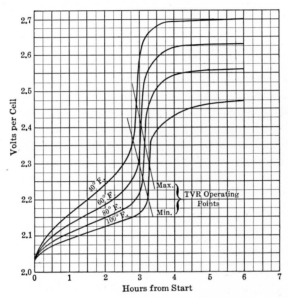

Fig. 93. Effect of temperature on charging characteristics of a lead-acid storage cell.

Voltage Relays

Volt-sensitive relays afford another means for automatically controlling the rate of charge supplied to a battery. As charging progresses, the voltage at the terminals of lead-acid storage batteries rises slowly until gassing begins, when a more abrupt rise of several tenths of a volt per cell occurs. At this point the battery is 90 per cent charged, or more, and charging should be completed at a lower rate. This abrupt rise in voltage is sufficient for operating a properly designed relay. However, the effect of variations in temperature of both battery and relay must be considered. For the same state of charge, the terminal voltage of the battery will be lower as the temperature is higher (see Fig. 93), but the operating voltage of an ordinary relay becomes higher as its temperature is increased. Such a relay is not well adapted to the purpose.

Inverse temperature compensation has been provided in the TVR relay so that its operating voltage is higher at lower temperatures by an amount which closely approximates the change in battery-charging voltage for a corresponding difference in temperature. This is accomplished by using a bimetallic strip which varies the air gap between the magnet frame and the armature. The relay may be set to operate at a desired voltage such as 2.37 volts per cell at 77° F when the battery is to be charged in 8 hours. The actual adjustment within a range of ±1 per cent is of no great importance as the battery voltage rises rapidly when gassing begins.

Figure 93 shows the volt-time characteristics of four similar batteries which, being fully charged initially, were discharged for 1 hour at the 8-hour rate and then recharged at one-third of this rate. The abrupt rise in voltage occurred after about 3 hours, as would be expected. The figure shows also the operating limits for a TVR relay.

Automatic intermittent charging is provided by Exide Model ES Control. By this equipment the battery is automatically placed on charge once every hour by contacts actuated by a synchronous clock, and each successive charge is terminated by a voltage relay when the voltage reaches 2.31 volts per cell at 77° F. A variety of conditions can be met by this type of control.

Ampere-Hour Meters

Since the ampere-hour is the unit commonly employed to measure the capacity of storage batteries, the ampere-hour meter is a convenient instrument for indicating the state of charge of a battery and for controlling the charge. Various types of ampere-hour meters have been developed, but the mercury-motor type is the most common form. This meter can be built as a rugged, dependable instrument which will keep in calibration and function satisfactorily even under severe conditions of service, such as may be encountered with mine locomotives, street trucks, or industrial tractors.

The mercury-motor type of meter consists of a copper disk immersed in mercury contained in a Bakelite chamber. The current is taken in through a contact ear, passes across the disk and then out through a contact ear at the opposite side. Permanent magnets create a field cutting the disk, and rotation is produced on the well-known principle of Faraday's disk. Inasmuch as the field of the magnet is constant, the torque exerted on the moving armature is proportional to the current flowing. The magnets are so situated that the pole tips are directly above and below the disk on opposite sides of the main shaft. When current is passed through the meter the disk rotates. This rotation induces eddy currents in the same disk, as it cuts the magnetic field, and damping is thereby accomplished.

The ampere-hour efficiency of a storage battery is less than 100 per cent, and therefore a greater number of ampere-hours are required to charge it than were discharged previously. It is possible to design the ampere-hour meter so that it will run slow during the charging period by a percentage that will be approximately equal to the difference between the efficiency of the battery under standard conditions and 100 per cent.

BATTERY REGULATION

In the preceding portion of this chapter, the voltage characteristics of the storage battery on charge and discharge have been discussed. The terminal voltage of the cells decreases during discharge and increases during the charging period. It is necessary to compensate for these changes when the battery is used to maintain the potential of bus bars in some power circuits. When the current flowing in the circuit is very small, this regulation is satisfactorily accomplished by the use of a regulating resistance. With cells of large capacity this would be a very wasteful and inconvenient method, and other methods must be employed. These may be classified as follows: end cells, counter cells, boosters, and methods of alternating-current regulation.

End-Cell Regulation. The falling voltage of a battery on discharge may be compensated for by adding additional cells from time to time. The name "end cells" comes naturally from the familiar arrangement of providing taps to each of a limited number of cells at the ends of a battery, so that any of these may be connected or disconnected at will without interrupting the flow of current from the battery to the bus bars. Figure 94 shows a simple diagram of such an arrangement for a three-wire circuit. If the battery is to be connected to the bus bars only during discharge or when floating on the line, a single end-cell switch at each end of the battery is all that is ordinarily required.

The use of end cells as a means of regulation is becoming more limited with changing conditions in central stations. Telephone central offices, however, employ several end cells on 24-volt batteries and proportionately more on batteries of higher voltage.

The successful regulation of voltage by the use of end cells requires well-designed switches. These switches must not interrupt the connection between the battery and the bus bars nor short-circuit any cell when the switch is being operated. The switch must be capable of carrying the maximum current of the battery. The switches must also be free from heating and sparking at the contacts.

The number of end cells required is calculated from the maximum voltage drop and the total number of cells in the main battery. In the simplest case, the initial discharge voltage may be assumed to be 2 volts per cell, decreasing as the cells discharge to 1.8 volts per cell, or 10 per cent. At least 1 end cell to each 10 cells in the main battery is therefore required. If the battery is to be charged while on the line, additional end-cell connections will be necessary. The cells will

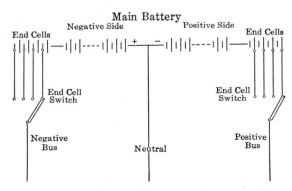

Fig. 94. Arrangement of end cells in a battery. For a two-wire circuit the end cells are at one end only.

increase to about 2.60 volts per cell, or 30 per cent, which, added to the 10 per cent required for discharge, makes a 40 per cent variation in voltage to be compensated for by the end cells. The actual number in any event can be calculated only when full data of the service are available. End cells are cut out on charge and in on discharge.

End cells are usually of the same size as those in the main battery, although they are seldom used as much. The last cell cut in on discharge obviously delivers only a small portion of the ampere-hours delivered by the cells in the main battery. They are consequently charged in a shorter time and must be cut out of the circuit when the charge is complete.

Counter Cells. Counter cells of the lead-acid type are provided with grids in place of the regular plates. They have very little capacity but develop a potential of 2.3 to 3.0 volts, depending on the current flowing, which opposes the potential of the main battery. The counter cells are placed at the ends of the battery, and the successive cells are cut out of the circuit as the voltage of the main battery falls, because their effect is subtractive instead of additive. Lead counter cells are practically obsolete, having been replaced by the alkaline type. The lead cells were subject to several operating difficulties, including a gradual increase in capacity, a high internal resistance after a period of idleness, and the formation of a scaly sediment which sometimes caused short circuits. The ideal counter cell would have no appreciable capacity, low internal resistance, unattackable electrodes, and ample space for electrolyte so that water would be required infrequently.

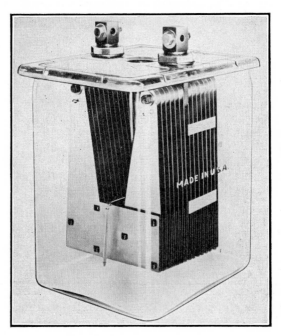

Fig. 95. Counterelectromotive-force cell, nickel-alkaline type NAK-30, continuous rating 60 amperes.

Alkaline counter cells fulfill these requirements quite satisfactorily. Each cell (see Fig. 95) consists of a glass jar in which are contained pure nickel electrodes in a treated solution of sodium hydroxide or stainless-steel electrodes in a similar solution. The solution is covered with a layer of oil. The electrode groups are identical and as the cells have no polarity can be used for either direction of current flow. Water is decomposed when current flows, oxygen being liberated at the anodic group and hydrogen at the cathodic group. Considerable water is required to replace that decomposed by the current, but several thousand ampere-hours can pass through the cells before they need watering. The layer of oil over the electrolyte prevents evaporation and protects the alkaline solution from the carbon dioxide in the air. The cells have covers similar to other storage cells.

The counterelectromotive force developed by the cells varies primarily with the current, but the absorption of carbon dioxide from the air (largely prevented by a film of oil on the surface of the electrolyte) tends to increase the counter emf. Voltage ratings are normally at an electrolyte temperature of 100° F. At lower temperatures the counter emf is slightly higher. The approximate cell voltages are given in Table 58.

TABLE 58. APPROXIMATE VOLTAGES OF COUNTER EMF CELLS
(Nickel or stainless-steel electrodes)

Per Cent of Continuous Maximum Current	All Cell Sizes	
	Intermittent Service Volts	Continuous Service Volts
10	1.63–1.72	1.66–1.82
25	1.73–1.80	1.83–2.03
50	1.78–1.88	1.98–2.18
75	1.82–1.94	2.07–2.26
100	1.85–1.99	2.13–2.32

The electrolyte should be renewed at intervals of 2 to 4 years.

Boosters. A third means of regulation is by the use of boosters. In the Standardization Rules of the American Institute of Electrical Engineers, a booster is defined as a generator inserted in series in a circuit to change the voltage. The booster is usually motor-driven, and then it is designated as a motor-booster. A wide variety of boosters is available for various purposes. These include (1) charging boosters used in connection with storage batteries, which discharge into a line and require the addition of a small voltage to the line voltage to complete the charge of the battery; (2) automatic regulating boosters for constant current or constant voltage, to take care of rapid fluctuations in the line; and (3) separately excited differential boosters. Boosters are still used for charging purposes, but automatic battery control by means of boosters has practically disappeared with the disuse of regulating batteries.

Alternating-Current Regulation. Although the storage battery is a direct-current apparatus it has found some application in the past to the regulation of alternating-current systems. If the distribution system is alternating-current, but the load direct-current, the application of the battery may be at the point where the load is required and may consist of the usual arrangement of batteries. If, however, the load is alternating-current, conversion apparatus, such as motor-generators or synchronous converters, is necessary in the battery circuit.

OPERATING CONDITIONS

Safety-Code Requirements

Rules relating to the installation of stationary batteries exceeding 50 volts are provided in the National Electrical Safety Code, Section 13.[15] This code is readily available by purchase at a nominal price from the Superintendent of Documents, Government Printing Office, Washington 25, D. C. It is not necessary, therefore, to reprint the rules here. Reference should be made to the code, but a word or two about the various rules is given here to indicate the scope.

Rule 130, *General*, applies to acid and alkaline batteries having nominal voltage in excess of 50 volts, sealed and unsealed cells, caution of fire hazard. *Rule* 131, *Isolation*, battery to be not accessible to other than qualified persons. *Rule* 132, *Ventilation*, requires provision for diffusion of gases with special provision for non-sealed batteries. Batteries of less than 50 volts come under this rule if capacity at 8-hour rate exceeds 5 kw-hr. *Rule* 133, *Insulation*, covers type of supports of cells in conducting and non-conducting containers. Special requirements for batteries exceeding 150 volts in rubber or 250 volts in glass. *Rule* 134, *Racks and Trays*, wood and metal racks and their treatment to prevent deterioration; similarly for trays. *Rule* 135, *Floors*, to be acid resistant. *Rule* 136, *Wiring in Battery Rooms*, to be in accordance with the code. *Rule* 137, *Guarding Live Parts in Battery Rooms*, current-carrying parts differing by a voltage exceeding 150 volts to be guarded. No bare conductors of 150 volts or more to ground in passageways unless guarded. Guarding to conform to code requirements. *Rule* 138, *Illumination for Battery Rooms Enclosing Batteries of the Non-Sealed Type*, specifies type of lamps and keyless sockets, prohibits open flames or incandescent resistors.

Care of Rooms

Aside from the proper installation of the batteries, the care of the rooms constitutes an important factor in their successful operation. Covered cells with sprayproof vents have so largely replaced the older type of open cells that the condition of battery rooms has been greatly improved. Many installations of batteries in rooms with other electrical equipment have been made successfully. Acid spray has been practically eliminated.

Ventilation. The objects in providing ventilation for storage-battery rooms and compartments are three. The first is to carry off the gases that are liberated during the charging period and to some

[15] *The National Electrical Safety Code,* 5th ed., Handbook H30, National Bureau of Standards, 1948.

extent by the negative plates at other times, since the hydrogen constituent of the gas forms an explosive mixture with the oxygen of the air. The second is to preserve the insulation of the installation by carrying off the acid spray (if any is present in the atmosphere), as this is destructive to woodwork, most kinds of paint, and insulating materials. The third objective of ventilation is to cool the batteries during the charging periods and when working under heavy loads. This is of particular importance for batteries in closely confined compartments, as on submarines and on certain industrial trucks. Forced ventilation is desirable wherever the natural draft is small or unreliable. As the gassing and maximum temperature occur near the end of the charge, it is customary in some installations to provide only moderate ventilation during the early stages of the charge and then to increase the ventilation after the battery has received about 75 per cent of its charge. Full ventilation should be continued for a considerable time after the completion of the charge. A 4 per cent mixture of hydrogen in the air is dangerous. The amount present at any place should not be allowed to exceed 2 per cent.

Insulation. In addition to the ventilation, certain other precautions are necessary to preserve the insulation. The supports for the tanks, now nearly obsolete, should be porcelain or glass insulators with oil cups. All metal except lead should be protected against acid spray. The overhead wiring should be carried on porcelain insulators which are easily accessible, so that they can be wiped off from time to time. The heads of the screws in the insulators should be covered with vaseline. In some large installations the copper bus bars and other cables are inclosed in a lead sheathing. The floors should be of waterproof material with drain at the lowest point so that they can be flushed with the hose. After the flushing, the floors should drain quickly and dry without water pockets.

Record Forms

Accurate records of the performance of the large stationary batteries and certain of the smaller batteries, such as motive-power and train-lighting batteries, are desirable. The records may indicate sources of trouble in the early stages, and they provide the necessary data for comparison of different makes and types as well as for making an accurate computation of the cost of operation. The forms to be used will naturally depend upon the service and to some extent upon local conditions, but the following have been in successful use and are given by way of illustration.

The first (Form 1) is applicable to the batteries in industrial trucks

STORAGE BATTERY RECORD FOR THE MONTH OF

Depot .... Battery of .... At .... cells. Type .... Date in service .... Battery No. ....

DAILY READINGS—DIRECTIONS: Start Charge as given in the Methods Bulletin recording time and amperes. Select a conveniently located cell as a pilot cell, and immediately after the daily charge is completed record the specific gravity and temperature of this cell. Once every week give battery an equalizing charge; replace evaporation with pure water before the charge is started. Note on reverse side of sheet any unusual condition or attention given the battery.

Monthly or Individual Cell Reading
Directions:
Once each month immediately after an Equalizing Charge, take and record Sp. Gr. of each cell in the battery.

Sp. Gr. Taken .... A.M. P.M.

Voltage Taken .... A.M. P.M.

Charging Amps. while read. Volt ....

Date ....

Cell	Sp. Gr.	Volts	Temp.
1			
2			
3			
22			
23			
24			

Date	BOOST						REGULAR CHARGE AT END OF DAY							
	START		FINISH				START OF CHARGE			END OF CHARGE				
	Time	Amp.	Time	Amp.	Sp. Gr.	Temp.	Time	Amp.	Sp. Gr.	Time	Amp.	Sp. Gr.	Temp.	
1														
2														
3														
4														
5														
6														
7														
8														
9														
29														
30														
31														

WEEKLY OR EQUALIZING CHARGE READINGS—DIRECTIONS: Once every week after the daily charge has been completed give the battery an equalizing charge (see Methods Bulletin) and record readings called for at half-hour intervals, below. Note any cells that do not gas as freely as the rest, on the reverse side of sheet.

DATE ....

Time	Amps.	Volts	Sp. Gr.	Temp.

DATE ....

Time	Amps.	Volts	Sp. Gr.	Temp.

WATER ADDED

REMARKS.

READINGS TAKEN BY

Form 1. Record sheet adapted for use with a truck or tractor battery.

and tractors. It was used at a large shipping depot during the Second War War. The second (Form 2) is for a battery in stationary service. The third (Form 3) is for a floating battery and includes the readings that should be made at the close of each equalizing charge. Form 4 is for Edison batteries.

Costs of Operation

It is often important to make an estimate of the cost of battery operation on trucks, tractors, or vehicles, and an outline of the general method is given below.

An estimate of the cost of operation is usually made for the purpose of determining the economy of a battery installation or of comparing one type of battery with another, to meet a given condition of service. In making such an estimate, the following items must be taken into consideration:

Equivalent batteries must be compared, that is, batteries having approximately the same watt-hour capacity at the service rate of discharge.

Interest must be paid on the investment, which is the cost of the initial battery and the renewals.

Depreciation is the amount of the capital investment to be written off per year and is equal to the total investment in the battery divided by the number of years for which the cost of operation is to be estimated. At the close of the period, the battery last in service may have a small scrap value. The amount of money set aside each year for depreciation is a sinking fund to amortize the debt.

The *watt-hour efficiency* of the battery is to be used in computing the costs of power for charging.

The *life* is the period of useful service, usually expressed in years.

The quantity of *electrolyte* is that required for renewal, or to replace losses by leakage, etc.

The quantity of *water* is that required to replace the losses by gassing on charge, that is, the quantity decomposed by the ampere-hours of charge exceeding 100 per cent of the previous discharge. It is, therefore, dependent upon ampere-hour efficiency of the battery.

The quantity of *power* required is that for charging, expressed in kilowatt-hours.

Cleaning for lead batteries includes renewal of wood separators, trays, and minor parts. This is seldom required now.

Labor is that required for normal maintenance. It depends on the size and kind of battery.

DISMANTLING AND ASSEMBLY OF LEAD-ACID TYPE

First, clean the top of the battery thoroughly.

Before starting to dismantle the battery, make a sketch or diagram showing the relative locations of cells, intercell connectors, terminals, and any other data necessary to insure the correct reassembly.

MONTHLY STORAGE BATTERY REPORT

Month of .... 19........

Plant of .... at .... Serial ....

.... consisting of .... cells, Type ....

Water was added to replace evaporation .... (date) .... Amount of water added ....

PILOT CELL READINGS   Cell No. ....

Are Specific Gravity Readings Corrected for Temperature?   Yes—No.

INDIVIDUAL CELL READINGS
(To be recorded once every month)

Temperature ....

Specific Gravity Readings Taken .... A. M. .... (Date)
P. .... (Date)

Voltage Readings Taken .... A. M.
P. ....

Buss Voltage .... Volts

Current While Taking Voltage Readings .... Amperes

DAILY RECORD | WEEKLY RECORD

Date	Time	Buss Volts	Pilot Cell Temp.	Pilot Sp. Gr.	Pilot Cell Volts
1					
2					
3					
4					
7					
30					
31					

Cell	Specific Gravity	Volts	Cell	Specific Gravity	Volts	Cell	Specific Gravity	Volts
1			51			101		
2								
32			82			132		
33			83			133		
34			84			134		
35			85			135		
36			86			136		
37			87			137		
38			88			138		
39			89			139		
40			90			140		
41			91			141		
42			92			142		
43			93			143		
44			94			144		
45			95			145		
46			96			146		
47			97			147		
48			98			148		
49			99			149		
50			100			150		

Temperature of Cells 1 .... 8 .... 21 .... 33 .... 47 ....
59 .... 75 .... 92 .... 102 .... 118 .... 125 .... 145 ....

81 | 131

REMARKS

READINGS TAKEN BY

Form 2.   Record sheet for a stationary battery.

STORAGE BATTERY MONTHLY REPORT
(Battery in Floating Service)

_____Company _____Station

Battery Consists of ____Cells, Type_____ Date Installed_____

19_____	OBSERVE AT SAME TIME DAILY					Cell	Volts	Hydrometer Reading	Electrol. Temp.	Remarks
YEAR	Bus Volts	Pilot Cell Hydrometer Reading	Day	Bus Volts	Pilot Cell Hydrometer Reading	1				
MONTH DAY						2				
1			16			3				
2			17			4				
3			18			5				
14			19			6				
15			30			17				
			31			18				

CELL VOLTAGE READINGS:

Observe once each month, 10 to 20 minutes after starting, equalizing charge. Read the hundredths as 2.29, 2.32. Use accurate voltmeter. If the difference between the highest reading and the lowest reading is more than .10 volts, take a check set of readings and record in adjacent extra column.

EQUALIZING CHARGE:

Give regularly once each month. One method for 60 cells is to raise bus to 140 volts for 8 to 24 hr. (preferably 24 hours.); then return to 129 volts for float.

CELL HYDROMETER READINGS:

Observe once every three months, 15 minutes to 24 hr. after completing equalizing charge. Record as read — do not correct for temperature.

CELL TEMPERATURE READINGS:

Observe at same time as hydrometer readings but only in 2 or 3 cells in each row.

ADDING WATER:-

Add water before starting equalizing charge or after completing hydrometer readings.

Date   Total Quantity

_____   _____Qts.

_____   _____

PILOT CELL:-No._____

19
20
37
38
39
40
41
42
43
44
45
46
47
48
49
50
51
52
53
54
55
56
57
58
59
60

Readings of all cells taken, date

_____

Signed:-

_____

Form 3.   Record sheet for a battery in floating service.

Circuit_____No. of Cells_____   Owner_____   Date_____On Charge_____M. Tested by_____

Kind {   Type {   _____   Address_____   Date_____Off Charge_____M. At_____

Electrolyte_____   Vehicle_____No._____   Date_____On Discharge_____M.

Date last renewed_____   Capacity or Model_____   Length of Charge____Hours Run No.___ Date_____

Clock Time	Lapsed Time	Cell Nos. A—H	Letters	1	2	3	4	5	6	7	8	9	10	11	62	63	64	65	66	67	68	69	70	Rate Amperes	Battery Voltage	Ampere-Hour Meter	Cell Temp.	Air Temp.	Av. Temp. Sol. for Sp. Gr. Rdg.
		Solution Level		1	2	3	4	5	6	7	8	9	10	11	12	63	64	65	66	67	68	69	70						
		Charged Specific Gravity																											

Form 4.   Record sheet for Edison batteries.

Fig. 96.   Removing an intercell connector by using a connector puller.

Fig. 97.   Drilling a post to permit removal of intercell connector.   The drill is
15/16 inch.   This is an alternative method to the use of a connector puller.

Removal of Connectors

Connector Puller. With the puller vertical, depress the plunger gradually until the connector is free from the post, Fig. 96. This method is quick and easy; it eliminates boring out but necessitates trimming the posts for reassembly of the cell.

Boring Out. If the puller is not used, bore out the connector, using a brace and bit. The bit may be either a twist drill or a wood bit. The wood bit is preferred and should be at least as large as the post, usually $\frac{5}{8}$ inch (1.6 cm) or $\frac{3}{4}$ inch (1.9 cm). Before boring, the bit should be centered carefully on the connector. The hole should be bored to a depth of about $\frac{3}{16}$ inch (0.5 cm). For posts reinforced with copper, a hollow or core drill called a "burn cutter" is available. Its use prevents damage to the copper cores. The filling plug should be in position while boring out, to prevent lead chips falling into the cell (see Fig. 97). When the hole has been bored to the proper depth, the connector will seem loose and the joint between the connector and the post can be seen. After boring out the connector, pry firmly but gently on the connector. Repeat this operation on the other side of the post, and continue until the connector is free. Care should be taken not to put pressure on the cover.

In all the above operations, care should be taken not to short-circuit the cell by allowing pliers or other tools to come in contact with both posts at the same time.

Removal of Cell

Grasp each post with a pair of pliers and pull vertically (see Fig. 98). If the jar sticks, a hot putty knife may be inserted around the edges. This will usually loosen small jars so that they may be pulled out easily. After the jar has been pulled out, rest it on the edge of the tray so that the pliers may be removed and the cell lifted by the hands to a position for the next operation.

In most instances cells will be too heavy or too tightly wedged in the case to be pulled by hand. To remove these an insulated cell-lifting device should be used with a block and tackle.

The sealing nuts, if used, should be removed by means of a special wrench (Fig. 99) unless the element and cover are to be removed together. If such a wrench is not available, use a pair of pliers, and only as a last resort use a monkey wrench or a pipe wrench. The sealing nuts are sometimes held in place by scoring the threads on the post with a prick punch.

Fig. 98. Removal of cell from the battery. If the jar sticks, a hot putty knife
may be inserted around the edges.

Fig. 99. Removing sealing nuts.

Removal of Element

With the filling plug out, blow the gases out of the cell. This is to reduce to a minimum the possibilities of an explosion. Warm the outside of the jar at the top with a flame and then insert a hot putty knife around the inner edge of the jar, melting out the compound to a

Fig. 100. Cutting out the sealing compound around the edge of the cover preparatory to removing the element.

depth of about ½ inch (see Fig. 100). This operation should be performed as rapidly as possible, as the compound cools quickly.

Place the jar on the floor with one foot on each side and pull the element upward while holding the jar with the feet (see Fig. 101), or use the device described above. When the element is nearly out, place it slightly out of plumb on the top of the jar and let it drain for about 5 minutes. The element should not remain out of the jar for

more than 15 minutes because of injurious heating. Should the negative plates begin to dry out and heat, sprinkle them with water until
they can be taken care of.

Fig. 101. Pulling the element from the jar, using chain and stirrup hold-downs
on the jar.

Removal of Separators

Place the element on the edge of a table or bench. When it is
necessary to use the separators again, it is desirable to use a special
tool called a "separator inserter." The broad side is used next to the
negative plate to loosen the separators so that they can be pushed and
pulled from between the plates. If the separators are not to be used
again, a putty knife may be inserted between each separator and the
negative plate to loosen them, and then they may be pushed from the
top and pulled from the bottom until free from the plates.

Separate the positive and negative groups of plates and soak each
of them in water for about 20 minutes. They are then ready to be
placed away on a shelf if the battery is to be stored.

Pour the electrolyte out of the jar and wash the jar with distilled water.

Assembly of a Cell

To install new separators, place the elements on edge as for removal of separators. Insert the new separators from the bottom. The smooth side of the wood separator is placed next to the negative plate, and the perforated rubber separator between it and the positive plate. (The Exide-Ironclad battery has only a smooth separator of wood or porous rubber between plates.)

With the element on edge and projecting slightly over the edge of the bench, place the jar over the plates. Then lift the jar and plates together to the floor and push the element into position. The electrolyte may be in the jar at this time, but it is preferable not to have it so. The inside edge of the jar at the top should be cleaned of compound and wiped with a cloth dampened with dilute ammonia or soda after the element has been inserted as above. It is necessary to have the top of the jar and cover free from all acid, as otherwise the compound will not adhere properly to the surfaces. Care should be taken that no ammonia or soda is allowed to get into the electrolyte.

Before putting on the cover, place the soft-rubber gaskets, if used, over the posts. The cover should be replaced while the jar is cold, unless it will not fit readily, in which event warm the outside of the jar with a flame until it becomes flexible enough to allow the cover to fit properly. Care should be taken not to burn the jar. Heat the compound and pour it into position, then trim with a hot putty knife until a level, smooth surface is obtained. Clean off all compound not needed to seal the cover. Replace the cell in the tray in the same position as it was in the beginning.

If there is electrolyte in the cell, first clean the posts with dilute ammonia or soda to neutralize the acid and allow them to become dry before doing anything further.

Trim each post, if a connector puller has been used, until its top is about $\frac{3}{16}$ inch below the top of the connector. This may be done by means of end-cutting pliers. Clean off the top and sides of the posts thoroughly. A file brush is good for cleaning the top, and gas pliers for cleaning the sides of the posts.

Always have the cover firmly in place and sealed before putting on the connector. If the intercell connector is not of the right length, it should be adjusted, and in no event should it be forced into position over the posts, as the jar or cover may be broken by strains.

Lead-Burning

This is a welding process for making a good mechanical and electrical connection between the plates and the strap and between the posts and the connector. Burning is done by means of acetylene and oxygen or by an electric arc. Use a reducing flame, as it is not desirable to oxidize the lead. Work with the tip of the inner blue flame and use a rotary motion, working from the center of the post upward

Fig. 102. Lead burning with an electric arc. Three cells of the battery supply current to a pointed carbon-rod electrode. Additional lead is supplied from an alloy "burning rod." Damp cloths protect adjacent parts and cover the vents.

and outward. The top of the post should be melted first and then fused to the wall of the hole in the connector; then lead from a piece of burning strip can be run in until the joint is flush with the top of the connector. Finish with a file and file brush. All parts must be thoroughly cleaned of dirt or foreign matter, as absolute cleanliness is necessary for successful work.

The electric-arc outfit consists of a carbon holder, with connecting cable and clamp and a carbon rod about ¼ inch in diameter. The battery on which work is to be done is usually used as a source of current. From 2 to 4 cells are required, according to their state of charge. Figure 102 illustrates this method of burning. The clamp is attached to an intercell connector a sufficient number of cells away to give the proper voltage. The carbon rod should be sharpened to a point and should project about 2 inches beyond the clamp. The carbon is first brought to a bright glow by contact with the post on which work is being done. The carbon is then worked with a rotary motion from the center of the post outward without drawing an arc. The carbon holder needs cooling occasionally by plunging it, with the carbon, into a pail of water. After a time the carbon may fail to work properly, owing to a film of lead oxide, which may be removed by a file or scraping with a knife. It is necessary that the operator use a pair of dark glasses to protect his eyes. If the battery is not available as the source of current, a 6-volt "starting" battery may be employed. In this case one terminal is connected to the clamp of the carbon holder and the other to the connection to be burned.

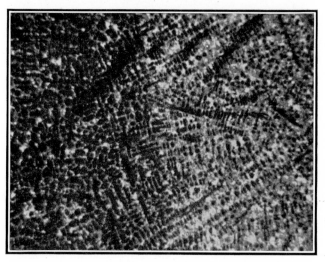

Fig. 103. Example of lead burning. A 4 per cent alloy has been burned to an 8 per cent alloy.

A device for lead-burning called the "pyrotip," operating on alternating current, is on the market. A small portable transformer reduces the 110-volt circuit to a suitable voltage. The carbon is large in diameter and terminates in a sharp point, which enables the operator to apply the heat at any place that he may desire. The alternating current may be passed through the cell without harm.

When a joint is well burned, the metal of one part fuses into the metal of the other part. This insures mechanical strength and good electrical conductivity. Imperfectly burned joints when examined under the microscope generally show void spaces, or dross. Figure 103 is a photomicrograph of two pieces of lead-antimony alloy showing complete union along the line of the burn. To the right, the alloy contains 8 per cent antimony, and to the left, 4 per cent.

SOURCES OF TROUBLE

Storage batteries large and small and in widely diverse services are subject to conditions that may lead to excellent service or through accident or abuse to something less than satisfactory. The heading "sources of trouble" implies faulty conditions which if recognized in time can often be corrected.[16] Many manufacturers have published detailed instructions for the proper care and maintenance of batteries. The Association of American Battery Manufacturers' Technical Service Manual deals with this subject in sections VII and XII. A book by Kretzschmar is entirely devoted to this subject. In the next sections many abnormal operating conditions will be described briefly.

Satisfactory life of batteries in service is measured in years. Those in cycle service are usually shorter lived than others in non-cycle applications. The reasons why batteries wear out are well recognized as a result of statistical studies in 1947 on thousands of batteries opened for examination at the end of their service life. Most of these were presumed to have given normal service. The most comprehensive data relate to automotive batteries. Grid corrosion of the positive plates then accounted for 42 per cent of the total but should be less now with the newer alloys. Shedding of the positive active material was a cause in 7 to 10 per cent of the batteries. Cracked partitions and leaking cases accounted for nearly one-third of the total. Buckled plates and short-circuited separators produced failure in 10 to 12 per cent of the batteries. Negative plate failures were about 10 per cent. Sulfation, which is the magic word of those who would seek to cure it by adding some of the common sulfates, appears, therefore, much less important. It is only one factor causing buckled plates and negative failures. At most, sulfation is a cause of failure in less than one-fifth of the batteries.

Overcharging

Overcharging produces corrosion of positive grids and excessive gassing, which loosens active material in the plates, particularly the positives. This material, sifting down between the separators and the plates, is deposited in the bottom of the jar as a fine brown sediment. Overcharging also increases the temperature of the battery and in some cases may carry it to excessive temperatures which are destructive

---

[16] F. E. Kretzschmar, *Die Krankheiten des Bleiakkumulators*, 3d ed., R. Oldenbourg, München, 1929. J. D. Huntsberger, Causes and remedies for troubles with lead-acid storage batteries, *Factory Management and Maintenance*, *107*, 143 (July 1949).

both to the plates and to the separators. Some cases of buckling of the plates are to be attributed to overcharging, although this is by no means the only cause. Overcharging, which is accompanied by excessive gassing, results in a needless loss of water, requiring constant attention to keep the cells filled to the proper level with electrolyte. Occasional overcharging is beneficial, but habitual overcharging decreases the period of useful service that the battery can give.

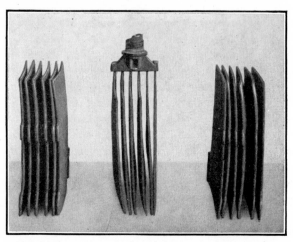

Fig. 104. Buckled plates. There are a number of causes for this, but perhaps the most common is undercharging.

Undercharging

Consistent undercharging of the battery results in a gradual running down of the cells. This is indicated by progressively lower values of the specific gravity readings and by a tendency of the plates to become somewhat lighter in color. The sediment deposited in the bottom of the jar when undercharging has been prolonged is usually a fine white powder, consisting principally of lead sulfate. Some of this material is deposited each time the cell is recharged. Consistent undercharging generally results in one or more of the cells becoming exhausted before the others, and some of these may become reversed by the other cells of the battery. When this occurs the most obvious remedy is to charge the battery until all the cells are again in normal condition. Equalizing charges should be a part of the maintenance schedule of stationary batteries. Insufficient charging is one of the most common causes of buckling of the plates (see Fig. 104). The lead sulfate occupies more space than the original material, and an excessive amount of it strains the plates.

Difficulties with the Charging System

Difficulties arising in the operation of starting and lighting batteries may in some cases be traced to difficulties with the charging system. The battery is generally held accountable for its own failures and those of the charging system also. Under normal conditions of operation of the car, the battery should begin to charge when a speed of about 12 to 15 miles per hour has been reached. Unless the generator cuts in at the proper speed, the battery will not receive its normal amount of charge. There is also the possibility of grounded circuits or partial grounds, which drain the battery and keep it in a more or less discharged condition. This can only be detected by testing the different parts of the circuit with a voltmeter. The charging system is always provided with some means of regulation that prevents the battery from receiving excessive currents when the car is being operated at high speeds.

Corroded Terminals

Corroded terminals may prevent a battery from delivering sufficient current for starting an automobile engine. This is because the products of the corrosion are non-conducting substances which form a layer between the terminal of the battery and the clamp connector connected to it. This film will not ordinarily interfere with the charging of the batteries or the use of the batteries for lighting lamps on the car but will make the resistance too great for the passage of the large currents required during the starting period. The products of corrosion should be removed and the parts cleaned with a dilute solution of ammonia to neutralize the acid and then should be covered with Vaseline.

Cracked or Broken Jars and Cases

Cracked or broken jars of batteries in trays can ordinarily be detected by leakage of electrolyte or, if the crack is slight, by the necessity of adding more water to the cell having a cracked jar than to the others. In such a case, the electrolyte of the cell will gradually become diluted, and this will result in diminishing the capacity of the cell. The obvious remedy is to replace the broken jar as soon as it is discovered.

Cracks in the partitions between cells in composition cases result in the gradual discharge of one or two cells in the battery, the leakage path between the cells being through the electrolyte. This occurs more frequently in batteries which are 18 months or more old. The first symptom noticed is usually the failure of the battery to hold charge.

The cells are in a state of unequal charge. Sometimes the position of the crack can be determined by running the corner of a hot putty knife along the top edge of the partition. Cases having this defect should be replaced or the battery discarded.

Glass-jar batteries which are cracked and leaking require prompt attention. It is usually necessary to remove the element from the jar and to immerse it in water in a non-metallic receptacle. The negative plates should not be allowed to become dry. If repairs cannot be completed within a week or two, the positive plates should be removed and dried. Negative plates should remain in water. When reassembled, new separators will be required.

Short Circuits

Short circuits within the cells may be caused by a breakdown of one or more of the separators between the positive and negative plates; by the excess accumulation of sediment in the bottom of the jars; or by the formation of a tree-like structure of lead from the negative to the positive plates. Treeing may result from two causes: (1) Sediment, brought to the top of the electrolyte by gas, settles on the top of the plates and bridges over the tops of the separators. This is sometimes called "mossing." (2) The presence of certain materials in the grids, such as cadmium, that cause the growth of trees at the side or bottom of the element. Pure lead grids have a tendency to grow trees from the negative to the positive plates. The presence of antimony in the grids, however, counteracts the tendency to treeing.

The evidences of short circuits within the battery are: (1) continued low readings of the specific gravity even though the battery is receiving a normal amount of charge; (2) rapid loss of capacity after a full charge, and (3) low open-circuit voltage. The remedy for this condition is to disassemble the cells, wash out the accumulated sediment, and replace the old separators with new ones.

Separator Failures

Wood and rubber separators are subject to mechanical failures as a result of strains and pressures of buckled plates or too tight packing of the element. They are subject to chemical action resulting in failure (1) by acid attack if the electrolyte is too concentrated or its temperature too high; (2) by oxidation in contact with the positive plate, which takes place very slowly in normal service but which may be accelerated by the presence of impurities in the electrolyte. Manganese and cobalt are typical.

Wood separators which have become thin and perforated should be dried and examined for a silvery appearance. If this is found, it is likely that manganese is the cause of their failure.

Wood separators which have become black are charred as a result of immersion in too strong or too hot acid. These separators have lost nearly all their mechanical strength. Pitting of separators may be caused by oxidation on the side next to the positive plate, or it may originate on the other side as a result of blisters on negative plates. Separators notched on the bottom are usually evidence of elements too loosely packed. The notches are made by the ribs in the bottom of the jars. Glass mats used without wood or microporous separators are subject to "treeing," which causes failure by short circuit.

Worn-Out Plates

Worn-out plates are ordinarily detected by a decrease in capacity of a battery which is receiving adequate charge. The specific gravity of the electrolyte may rise to the customary value, but the capacity on discharge is below normal. Eighty per cent of the rated capacity is usually taken as the end of service life. This loss of capacity may be more evident at high rates of discharge than at low rates. If the plates are examined, positives which have shed finely divided, active material more or less evenly from the whole plate are characteristic of long normal service. They are simply worn out. Corroded grids, loss of active material in lumps, and cracked frames are evidences of faulty charging arrangements. Positive plates which are hard like unglazed tile are heavily sulfated. For these a slow charge of as much as 100 hours may be necessary to restore them to use again. Negative plates which do not show a metallic streak when stroked with a knife are not in normal condition. "Sandy" negatives are usually the result of too concentrated acid electrolyte, too high temperatures, or undercharged condition. Prolonged charging may eliminate this sandy condition, but otherwise the plates are worn out.

Electrolyte below the Tops of the Plates

If the electrolyte is allowed to remain consistently below the top of the plates, an abnormal sulfation (see Fig. 105) takes place and the plates crumble. This is generally the result of neglecting to add the necessary water to replace evaporation. Unless the injurious effects have gone too far, it is sufficient to fill the cells with water to the proper level and allow the battery to continue its normal operation.

Electrolyte should not be added under any circumstances unless it is known that electrolyte has been lost from the cell. Whenever it is necessary to add acid to the cells, the acid should be in diluted form.

Fig. 105. Negative plate sulfated near the top, showing the effect of too low a level of the electrolyte.

Fig. 106. Positive plate showing loss of active material.

The cells should be placed on charge and the concentration of the electrolyte within the cells adjusted to the proper value before the conclusion of the charge.

Freezing

Sometimes freezing may be very destructive to the plates, but it is by no means certain that freezing will injure the battery. When freezing occurs, water crystals separate from the body of the electrolyte. These water crystals cause some expansion of the active materials of the plates, particularly if the pores of the plates are clogged with lead sulfate as a result of the discharged condition of the battery. The plates may become buckled, and the active material may be pushed out of the grids. Experiments with small plates in a fully charged condition carried to a temperature below the eutectic point, so that the entire mass of the electrolyte was solid, showed no destructive action on the plates. Under ordinary circumstances, however, freezing will only occur when batteries are in a discharged condition, and freezing then may destroy the plates.

Impurities in the Solution

Impurities in the solution may come from impurities initially present in the sulfuric acid, or in the plates, or from the container, or they may be added accidentally as, for example, by the use of impure water to replace losses caused by charging and evaporation. Storage batteries are particularly sensitive to certain kinds of impurities, and these have been described in Chapter 3. Perhaps the most common of the metallic impurities in the storage-battery electrolyte is iron. Iron produces discharge of the plates by being oxidized to the ferric condition at the positive plate and reduced to the ferrous condition at the negative plate. During this process the iron diffuses from one plate to the other, but it is not deposited upon either electrode. Some impurities in the electrolyte can be eliminated by pouring out the electrolyte and flushing the cells with distilled water, which in turn is also poured out. The cells are then filled with pure dilute electrolyte, which is adjusted to the proper specific gravity at the conclusion of a full charge.

Spalling of the Active Material

Blocks of the active material break out of the plate, as shown in Fig. 106. There are a number of possible causes for this effect, which is observed most often with positive plates but may occur with negatives also. Among the causes are the following: the active material may not have made good contact with the grid; the plate may not have been thoroughly dried; the plate may have been overpasted and the surface blown off by the formation of gas during charge; too high a charging rate; the active material may have been too hard and have cracked when the plate was dried or after formation; excessive sulfation may have cracked the active material. Usually the cause of the defect is to be found in the plate itself. The obvious remedy is to replace the defective plates with those of better quality.

Sulfation

The word "sulfation" has been used in several senses, and this has led to some confusion. In general it means the formation of lead sulfate on the surface and in the pores of the active material of the plates. Sulfate forms as a natural part of the process of discharge, and this fact is expressed by the chemical formula for the reactions discussed in Chapter 4. This sulfate is finely crystalline and easily reduced by the charging current. Sulfation in this sense is a necessary part of the operation of the battery and is not a source of trouble.

Lead sulfate is also formed as a result of local action or self-discharge of the plates. This is brought about by parasitic currents or by the action of the acid solution on the materials of the plates. The rate at which sulfation, in this sense, proceeds depends on the concentration and temperature of the electrolyte. The lead sulfate formed as a result of local action is easily reduced by the charging current, unless it is neglected.

The third and perhaps most common use of the word sulfation applies to the large crystals or crusts of lead sulfate that may form on the plates as a result of neglect or misuse. Excessive sulfation of this kind is difficult to reduce and may injure the plates. The active material of the positive plates which are sulfated is frequently light in color, and white spots of sulfate appear, but the color is not always a safe criterion. It may be quite dark, and the presence of excessive sulfate is revealed by a hard, rough surface and the gritty feeling of the material when rubbed between the fingers. Sulfated negatives likewise are hard, expanded, and gritty. They do not show a good metallic streak when stroked with a knife. Sulfation in this sense is the result of some form of abuse, as (1) allowing the battery to stand in a discharged condition for a considerable time, (2) neglecting to make repairs when evidence of trouble within the cells becomes apparent, (3) filling the cells with electrolyte when water should have been used, (4) operating the battery at excessive temperatures, (5) persistent undercharging. Excessive sulfation can be avoided with reasonable care, and it is doubtless true that the liability to sulfation of lead batteries has been exaggerated.

When a battery stands in a discharged state the crystals of sulfate grow larger and are more difficult to reduce by the charging current. Lead sulfate is sparingly soluble in sulfuric acid electrolyte, but the solubility increases at higher temperatures. Temperature fluctuations to which a battery may be subjected as it stands idle play an important part, therefore, in forming this hard sulfate. When the temperature of the battery is increased even slightly the smaller crystals of lead sulfate dissolve, and when the temperature falls again this lead sulfate crystallizes out slowly. As a result of such temperature cycles the large crystals grow at the expense of the smaller ones. The crystals on the positive plate are usually larger than those on the negative plate. This is probably because they grow more slowly. Sections of the active material may be pushed out of the grids and the positive grids fractured.

Various cures for sulfated batteries have been proposed. However, a simple and effective remedy for this condition is to pour out the electrolyte and fill the cells with water. After being allowed to stand for about an hour, the battery may be put on charge at a low rate of current, provided that the voltage at the terminals of the cells is less than 2.3 volts per cell. The resistance of the battery will be high at the start and the current initially small, but the current will increase as the sulphate is broken down if the voltage at the terminals is maintained. The cells will take the current as fast as they are capable of being charged, and the process becomes more or less automatic, but the temperature must be watched and the batteries cut off or the current decreased if the temperature reaches 43° C (110° F). The charging may also be done by the constant-current method at a low rate. The water that was put in the cells becomes a solution of sulfuric acid as the charge proceeds, and readings of the rising specific gravity can be made. If the final specific gravity obtained after prolonged charging becomes constant at too low a value, more electrolyte should be added. It not infrequently happens that the specific gravity of the electrolyte, initially water, will rise above the normal figure, say 1.280. This is clear evidence that acid has at some time been added to the cells improperly, that is, when they needed only water.

Grid Corrosion

This is one of the more common causes of battery failure. Long-continued overcharging causes oxidation of the positive grid structure, decreasing the cross section of the grid wires and eventually leading to collapse of the plate. The newer alloys should be less subject to this trouble. There are other causes of grid corrosion. Sometimes the active material becomes separated from the grid, and white patches of sulfate appear along the grid wires. Lead sulfate is no protection to the underlying metal. As an insulator it makes difficulty in charging the positive active material. Figure 107 shows a plate in advanced stages of sulfation along the lines of the grid. Attack by organic acids is another cause of failure. Of these acetic acid (see page 145) is perhaps the worst, but many organic acids produce similar effects. Inorganic acids such as nitric, hydrochloric, and perchloric acids attack grid metal. Some corroded brittle grids with soft positives and gritty negatives are attributed to electrolyte of excessive strength or above a safe operating temperature.

In aggravated cases the grid is converted to lead dioxide, an example of which is shown in Fig. 108. Fractures of the grid result from its weakened condition and from the excessive expansion of heavily sulfated active material.

Lander[17] studied the corrosion of lead in sulfuric acid solutions and found that below potentials for PbO_2 formation tetragonal PbO is produced on the lead. This is attacked chemically resulting in formation of $PbSO_4$. In a later paper he studied the corrosion and growth of positive grids in the lead-acid cell.

Reversal

Reversal may be caused by the overdischarge of a cell deficient in capacity when in series with others that have greater capacity or are more fully charged; generally, however, it is the result of charging a battery in the wrong direction, and then all cells are reversed. The reversal first becomes complete along the lines of the grid, but in the middle of each pellet some of the original brown active material can usually be seen. The active material becomes rough. A partially reversed plate contains both positive and negative active material and is subject to strong local action. Sometimes reversal of negative plates of the Planté type is resorted to in order to restore their capacity. The sponge lead of these plates shrinks and solidifies, the plate losing a large part of its capacity. By reversing it to a positive and then back to a negative the capacity can be greatly increased. Reversal of pasted-plate cells is not desirable.

[17] J. J. Lander, Anodic corrosion of lead in H_2SO_4 solutions, *J. Electrochem. Soc.*, *98*, 213 (1951); Effect of corrosion and growth on the life of positive grids in the lead-acid cell, *ibid.*, *99*, 467 (1952).

Fig. 107. Positive, showing sulfation along the line of the grid.

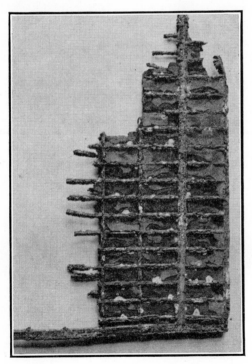

Fig. 108. Corroded grid of a positive plate, showing the effects of formation of the grid.

Growth of Positive Plates

Positive plates are subject to growth, but this condition should be considered abnormal. The cause of growth is usually the presence of organic acids, of which acetic acid is a typical example. This attacks unprotected portions of the grids causing oxidation of the metal.

Planté positives gradually shed active material as they are used, and the formation of new active material takes place. This is made from the underlying lead of the plate. The lead dioxide occupies more space than the lead from which it is formed, and expansion results. Plates that have grown several inches in length and a less amount in width are occasionally found. Figure 109 shows a positive plate that has grown.

Fig. 109. Growth of a Planté positive plate resulting in buckling.

Fig. 110. Shrinkage of a pasted negative plate. This is an abnormal condition.

Shrinkage of Negatives

The surface of the negative plate is in a state of flux during charge and discharge, and as a result the sponge lead tends to solidify, that is, to shrink (Fig. 110). This results in decreased capacity. To counteract this tendency, expanders are added to the pastes of the negative plates when they are made, as described in Chapter 2. The active material of Planté plates also shrinks, losing capacity. Various treatments have been devised to retard this effect, and reversal is sometimes resorted to, as noted in the section on reversal.

Explosions

The gases liberated during charge are hydrogen and oxygen, which explode with violence if a flame or spark ignites them. Figure 111 shows a double-compartment battery which exploded while on charge

Fig. 111. Storage cells of the double-chamber type, damaged by an explosion of gas.

because of a bad contact. The force of the explosion has blown out the side of the upper compartment, and the seams of the carrying case have opened up along the edges.

Static sparks are usually unexpected and sometimes result in explosions. Sparks are more likely to occur when the atmospheric humidity is low. Several authors have stated that such sparks do not occur if the humidity is above 60 per cent. A person walking on a woolen rug or on insulating floor material, paper running through a printing press, leather belts on motors, contacts of silk, woolen, or fur clothing, moving vehicles on rubber tires and other conditions may result in static charges and sparks. These may seem far removed from battery operation, but battery explosions have been traced to floor materials, leather belts, and silk parachutes. Some battery attendants are instructed to ground themselves by touching a grounded metal part before working on a battery.

Sparks from other causes, such as removing clips from battery terminals or accidental short circuits, can usually be avoided by following simple and obvious rules of safety. A charging circuit should be broken before connectors at battery terminals are removed.

The lower limit for explosive mixtures of hydrogen in air is 4.1 per cent, but for safety hydrogen should not exceed 2 per cent. The upper limit is 74 per cent. Maximum violence occurs at a mixture of 2 parts of hydrogen to 1 of oxygen, this being the composition of water.

The Bureau of Mines[18] has made extensive investigations of the hazards involved and the precautions to be taken in the use of mine locomotives. The hazard is in part due to the battery but involves also operation in "gaseous" atmospheres.

Ventilation of battery rooms is worthy of careful consideration. When all the charging current is being expended in the liberation of gas, each ampere-hour produces 0.418 liter of hydrogen gas, H_2. This streams upward and without complete mixing with the air of the room may give rise to higher concentrations in the upper part of the room than near the floor. This subject is covered in another Bureau of Mines paper.[19]

The discussion above relates to causes occurring outside the battery or cell, but explosions for which no probable cause can be assigned are not unknown. Some of these point clearly to the cause being within the cell. In one recent case the remaining parts of the cell revealed that corrosion of the positive terminal post, under the cover, had proceeded to the stage where only a small metallic contact remained.

[18] L. C. Ilsley, E. J. Gleim, and H. B. Brunot, Inspection and testing of mine-type electrical equipment for permissibility, U. S. Bur. Mines, Bull. 305 (1929). See also Bull. 306 and 313.

[19] G. W. Jones, John Campbell, R. E. Dillon, and O. B. Benson, Explosion hazards in storage-battery rooms, U. S. Bur. Mines, Tech. Paper 612 (1940).

No one was near the battery at the time of the explosion, but it occurred at the instant that a heavy load was thrown on the battery. It seems probable that the current burned through the corroded connection above the level of electrolyte and ignited the gas within the cell.

7

Resistance

RESISTANCE OF THE BATTERY AND ITS RELATION TO THE EXTERNAL CIRCUIT

The internal resistance of a storage cell is very small and for many purposes may be neglected entirely. When large currents are required, however, as when an emergency arises, or when a tractor begins to pull a heavy load, or in cranking an automobile engine, the resistance of the battery and its intercell connections becomes more important. In this chapter is given a simple discussion of the applications of Ohm's law and the resistance characteristics of storage batteries.

Every electrical circuit offers some opposition to the flow of electricity through it. This is called resistance, and the unit of resistance is called the ohm. For any direct-current circuit of which the resistance is constant, the current of electricity that flows is proportional to the voltage applied to it. This relation is expressed by equation (1) which is called *Ohm's Law:*

$$I = E/R \tag{1}$$

The current in amperes is represented by I, the voltage in volts by E, and the resistance in ohms by R.

The current flows through the external circuit from the lead dioxide plate to the sponge-lead plate, or in the alkaline battery it flows from the nickel oxide tubes through the external circuit to the pockets containing the iron; that is, the current always flows from the positive to the negative terminal. The current, however, does not begin with one plate and end with another but flows through the cell as well. The resistance of the circuit therefore includes not only the resistance of the external circuit but also the so-called internal resistance of the cell or battery. We may therefore write equation (1) in the following form:

$$I = E/(R' + b) \tag{2}$$

where R' is the resistance of the external circuit, b the resistance of the cell, E the total electromotive force of the cell or, as it is sometimes

318

called, the "open-circuit voltage." The value of b varies somewhat with the state of charge of the cell and other factors. It is highest when the cell is completely discharged. The current I is the same in all parts of the circuit.

Equation (2) may also be written as

$$E = IR' + Ib \tag{3}$$

That is, the total voltage drop is divided into two parts, one of which is the potential difference or voltage drop, IR', across the terminals of the resistance R' due to the current I flowing through it; the other part is within the cell itself. The quantity b is ordinarily so small in storage batteries that the product Ib may be neglected in comparison with IR'. If, however, the current I is very large, the product of Ib may amount to several tenths of a volt. The effect of this is to reduce the useful voltage of the cell.

The power developed in such a circuit is the rate of expenditure of the electrical energy, or in other words it is the rate of doing work. The work that the electric current does is proportional to the current, the voltage, and the time. It is expressed by a unit called the joule or the volt-coulomb, designated by W. If the time in seconds is expressed by t, the work done by the electric current is

$$W = IEt \tag{4}$$

If the circuit contains resistance only, the energy of the electric current will be converted into heat. By combining equations (1) and (4), we have

$$W = I^2Rt = E^2t/R \tag{5}$$

That is, the work done in heating the circuit of R ohms is given by equation (5), from which we may at once obtain the power expended by dividing the equation through by the time, which gives the rate at which the work is done. Letting P represent the total power expressed in watts, we have

$$P = W/t = IE = I^2R = E^2/R \tag{6}$$

We may consider the complete circuit as made of two parts as before, one external to the battery and the other the internal resistance of the battery itself. The power expended in each part of this circuit, considered as of resistance only, is obtained by combining equations (3) and (6):

$$P = P_1 + P_2 = I^2R' + I^2b = E^2/(R' + b) \tag{7}$$

The part I^2b is expended in heating the cell and is lost. It should be noted that this loss increases as the square of the current. This factor reduces, therefore, the watt-efficiency of the battery, particularly at high rates of discharge, and limits the useful power that the battery can give.

When several cells are connected in series, the resistance of the battery is the resistance of all the cells added together. If similar cells are connected in multiple, the resistance is reduced by a factor one divided by the number of rows in multiple. Ohm's law applied to a battery having s similar cells in series and p rows in multiple is therefore

$$I = \frac{sE}{R' + (sb/p)} \tag{8}$$

The circuit external to the battery absorbs part of the power which is designated as P_1, in equation (7). The voltage drop across the terminals of this part of the circuit will be E', which is less than the total value of E by the amount equal to Ib, which is the drop within the cell itself. If the current is I the power expended in this part of the circuit will be

$$P_1 = IE' \tag{9}$$

If this part of the circuit consists only of resistance, all of the energy will be expended in the form of heat and from equation (7) we have

$$IE' = I^2R'$$

Since the values of E and b are practically constant, the total power generated by the battery in a circuit consisting only of resistance will be small by equation (7) if the value of I is small but will increase as the current increases. The maximum current is obtained when the external resistance R' is made equal to zero as shown by equation (8). Here the power generated is a maximum, but it is all expended within the battery itself in the form of heat and no useful work is done. Between these two extreme conditions lies the maximum useful power delivered to the external circuit. The total power generated by the battery, equation (6), is $P = IE$. The power wasted in the battery (considered as a single cell) is by equation (7).

$$P_2 = I^2b$$

The power delivered to the external circuit is then

$$P - P_2 = P_1 = IE - I^2b = \frac{E^2R'}{(R' + b)^2} \tag{10}$$

The condition that P_1 shall be a maximum is found by differentiating it with respect to R' and equating to zero:

$$\frac{dP_1}{dR'} = \frac{(R' + b)^2 E^2 - 2E^2 R'(R' + b)}{(R' + b)^4} = 0$$

whence

$$(R' + b) - 2R' = 0$$

and

$$R' = b \tag{11}$$

That is, assuming E and b to be constant, the resistance of the external circuit which receives maximum power from the battery is equal to the internal resistance of the battery.

The external circuit may contain some apparatus for transforming electrical energy into something besides heat, as for example a motor to transform the electrical energy into mechanical energy. In this case

$$IE' \text{ is greater than } I^2 R'$$

$$IE' - I^2 R' = I(E' - IR') = IE'' \tag{12}$$

The factor $(E' - IR')$ is called the counterelectromotive force of the circuit. The rate of conversion of energy of the battery into mechanical energy, for given conditions, is equal to the expression $I(E' - IR')$. The useful mechanical energy obtained is equal to this multiplied by the mechanical efficiency of the motor.

The equation for the mechanical power is, therefore, (13) where C_1 is a constant.

$$P_m = C_1 IE'' \tag{13}$$

The counterelectromotive force is variable, depending on the speed and the magnetization. The current delivered by the battery is

$$I = \frac{E - E''}{R' + b} \tag{14}$$

Combining (13) and (14),

$$P_m = \frac{C_1(E - E'')E''}{(R' + b)} \tag{15}$$

From which the condition for maximum mechanical power is obtained:

$$\frac{dP_m}{dE''} = \frac{C_1(E - 2E'')}{(R' + b)} = 0$$

whence $E'' = E/2$ and

$$P_m = \frac{C_1 E^2}{4(R' + b)} \tag{16}$$

Equation (14) shows that the current delivered by the battery varies inversely as the battery resistance. The battery resistance, therefore, affects the torque and consequently the power that a given motor can exert, since the torque is a function of the armature current according to the equation:

$$\text{torque} = C_2 \phi I \tag{17}$$

where C_2 is a constant and ϕ the flux.

Similarly, the speed of rotation is given by the equation:

$$\text{speed} = C_3 (E''/\phi) \tag{18}$$

The product of equations (17) and (18) gives the expression for the mechanical power, equation (13), in which the constant depends on the units employed.

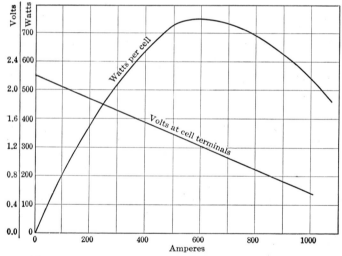

Fig. 112. Power available at various rates of discharge.

A battery is seldom required to deliver maximum power, but in emergencies or for short periods of time the relation of battery resistance and voltage to the outside circuit may be important.

Figure 112 shows curves, obtained experimentally, that illustrate the above statements. The battery used for these experiments was a small one, having a capacity of about 140 ampere-hours at the 5-hour rate of discharge. The voltage at the cell terminals falls to one-half its open-circuit value at about 600 amperes. This means that the drop in voltage because of the ohmic resistance of the cell is equal to the voltage drop in the external resistance when the cell discharges 600 amperes. As was shown above mathematically, this is the point of maximum power, and the curve marked "watts per cell" has a maximum at this point. The average internal resistance may be calculated from the slope of the voltage curve. It is found to be 0.0016 ohm. Other cells having different resistances would give results differing from these curves, but the principle is illustrated by this figure.

No specific statement as to the internal resistance of the many types and sizes of lead-acid batteries can be made. In general the larger sizes have the smaller resistance. Thus a small radio B battery has a resistance of about 0.1 ohm per cell; larger B batteries and small miscellaneous types have a resistance of a few hundredths of an ohm per cell; starting and lighting batteries and small motive-power batteries have a resistance of a few thousandths of an ohm per cell; larger motive power batteries have a resistance of less than 0.001 ohm per cell; very large motive-power batteries for locomotives have a resistance as low as 0.0001 ohm per cell.

FACTORS WHICH AFFECT THE RESISTANCE

The first and perhaps the most obvious source of resistance in the battery is the electrolyte. Accurate data on the resistivity of sulfuric acid solutions are available, and a table of values covering the range of concentrations in storage-battery practice is to be found in Chapter 3. Experiments by Morse and Sargent[1] on the resistance of storage cells have shown the variation with temperature to agree closely with the temperature-resistivity curves for the acid solution of corresponding specific gravity. As the temperature rises the resistance falls.

The presence of the separators adds a small amount to the internal resistance of the cells, particularly if perforated or slotted rubber separators are also employed. The resistance of separators cut from various kinds of wood is given in Table 8.

A third factor contributing to the resistance of the cell is the resistance of the active material and grids. Lead dioxide is classed as a semiconductor but is unique in being a better conductor than other oxides, more nearly approaching metals. Measurements of its resistivity in massive form by Palmaer[2] indicated 0.92×10^{-4} to

[1] H. W. Morse and L. W. Sargent, The internal resistance of the lead accumulator, *Proc. Am. Acad. Sci.*, *46*, 589–611 (1911).

[2] W. H. Palmaer, Über die elektrische Leitfähigkeit und einige anderer Eigenschaften des regulinische Bleisuperoxyde, *Z. Elektrochem.*, *29*, 415 (1923).

0.97×10^{-4} ohm-cm. Thomas[3] reported the resistivity at room temperature to be 1 to 3×10^{-4}, about equal to the resistivity of bismuth and mercury. In the porous form, as lead dioxide exists in the positive plate, Thomas reported the resistivity to be 74×10^{-4} ohm-cm. The resistivity of the sponge lead of the negative plate is given by Crennell and Lea[4] as 1.83×10^{-4} ohm-cm, increasing when in the discharged state by about 20 per cent. Comparing these figures with the resistivity of the grid metal (Table 3, page 16) for the 7 per cent alloy, 2.59×10^{-5}, it is apparent that the sponge lead of the negative plate is almost as good a conductor as the grid itself.

The resistance of the various parts of the cell has an important bearing on the distribution of current at the plates during discharge. Crennell and Lea have found that the current density in a cell having plates 77 cm high is not uniform over the surface. The greater part of the discharge comes from the upper portion, and, further, the current density in any particular place changes during the progress of the discharge. The current density at the top of the plate is initially high, but it decreases while the low current density at the bottom of the plate increases, and both approach a limiting value about equal to that maintained throughout the discharge at the middle of the plate.

Another factor affecting the resistance of a storage battery is the state of charge. Lead sulfate, which forms on the plates during discharge, is a non-conductor, and its presence increases the resistance to the passage of the electric current. The resistance of the lead-acid cell begins to increase slowly as soon as the discharge begins, and toward the end of the discharge the increase is much more rapid, reaching values from two to three times as great as the initial resistance. Morse and Sargent found that the characteristic resistance curves for Planté plates differ somewhat in shape from those of pasted plates. The curves for the Planté plates reach higher values, and the last rapid rise is preceded by a short region of almost constant values which these authors attribute to the distribution of active material on the plate.

During the first part of charge the internal resistance is high, but it falls gradually until gassing begins, when there may be a temporary rise in values because of polarization phenomena. The resistance may continue to fall for a time after the battery is disconnected from the charging line, because of the gradual equalization of the acid con-

[3] U. B. Thomas, Electrical conductivity of lead dioxide, *J. Electrochem. Soc.*, *94*, 42 (1948).

[4] J. T. Crennell and F. M. Lea, The distribution of current density in lead accumulators, *J. Inst. Elec. Engrs.*, London, *66*, 529 (1928).

centration and the dissipation of the gas layer on the active material.

The resistance of the alkaline batteries also increases during discharge and falls during charge, but the reasons for this are not the same as for the lead battery. The resistance of the alkaline batteries is higher than for the acid batteries of corresponding sizes, mainly because the resistivity of the electrolyte is greater.

During charging there is a back electromotive force that opposes the applied potential. The magnitude of this back electromotive force depends on the concentration of the acid in the pores of the plate and the prevalence of lead ions in the electrolyte. The difference between the impressed voltage and this back electromotive force is the effective electromotive force across the cell, and this, together with the resistance, determines the charging current that flows at any instant.

METHODS OF MEASURING THE RESISTANCE

The measurement of the resistance of storage batteries presents unusual difficulties, because of the small values and the complications arising at the surface of contact between the electrolyte and the electrodes. A bibliography on this subject may be found in the paper by Morse and Sargent.

Accidental contact resistances at binding posts and elsewhere in the measuring circuit may introduce errors. The values obtained by the use of alternating currents do not agree with those from direct-current measurements, apart from any consideration of polarization phenomena.

Direct-Current Methods

The numerous direct-current methods that have been proposed for measuring the internal resistance of batteries are based on the application of Ohm's law. The resistance of a cell is ordinarily defined by the equation

$$b = (E - E')/I \qquad (19)$$

The symbols have the same meaning as in the first part of this chapter. Preece[5] found many years ago that the values obtained by using the above equation vary with the values of I, the resistance becoming greater as the value of I was made smaller. This fact has been confirmed by many subsequent observers, and it is true for dry cells and most other forms of primary batteries as well as for storage cells. This effect is shown in Fig. 113. Whatever the cause may be, it is evident that b is not a true ohmic resistance, since it does not obey

[5] W. H. Preece, The charging of secondary batteries, *Electrician, 15,* 42 (1885).

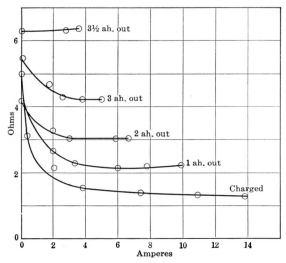

Fig. 113.   Resistance of a 12-cell, 3-ampere-hour battery at various rates of discharge.

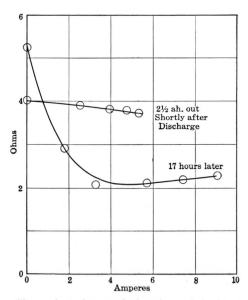

Fig. 114.   Change in resistance during the period of recuperation.

Ohm's law. The resistance varies, as is shown in Fig. 113, with the state of charge as well as with the current. A small battery was selected for these experiments as a matter of convenience. The recuperation of a battery after discharge results in a gain in voltage and a decrease in resistance, as Fig. 114 shows.

Alternating-Current Methods

In view of the difficulties of measuring the resistance of batteries by direct-current methods, the alternating-current methods, of which many have been proposed, deserve consideration, but only a few can be described here. It is desirable to prevent the battery from discharging through the bridge. Two methods have been proposed to accomplish this. The first and simplest is to connect two cells of like voltage in opposition and measure the combined resistance of the two. The second method is to put a capacitor in series with the cell to be measured.

Nernst and Haagn[6] proposed a method which permitted measurements to be made on one or more cells which might be discharging a direct current through a local non-inductive circuit or be on open circuit as desired. This method was somewhat improved by Dolezalek and Gahl,[7] who eliminated the error introduced by the traveling contact of the bridge wire.

The Wien bridge described by Grover,[8] although designed for the measurement of capacity, can be used for the measurement of battery resistance, and it is superior to either of the foregoing bridges. A diagram of this bridge is shown in Fig. 115. The arms are designated as A_1, A_2, A_3, A_4. The capacitor C_2 is of known value. The capacitor C_1 should be variable to compensate for the capacitance of the battery and must be calibrated. Capacitors of good quality are desirable in order that the correction for absorption may be negligible. The resistances R_3 and R_4 are equal, and R_2 is a small fixed resistance. R_1 is a small variable non-inductive resistance whose magnitude will depend on the resistance of the battery and the other resistances in the circuit. The condition for a balance of the bridge is:

$$A_2A_3 - A_1A_4 = 0 \qquad (20)$$

Since the capacitors usually have some absorption that may not be

[6] W. Nernst and E. Haagn, Methode zur Bestimmung des inneren Widerstandes galvanischer Zellen, Z. Elektrochem., 2, 493 (1896).

[7] F. Dolezalek and R. Gahl, Über den Widerstand von Bleiakkumulatoren und seine Verteilung auf die beiden Elektroden, Z. Elektrochem., 7, 429 and 437 (1901).

[8] F. W. Grover, Wien bridge (part of a paper on measurement of capacitance), Bull. Natl. Bur. Standards, 3, 378 (1907); Sci. Paper 64.

negligible, Grover represents the absorption as fictitious resistances ρ_1 and ρ_2, in series with the capacitors C_1 and C_2, respectively. Substituting the impedances of the various arms in equation (20) and

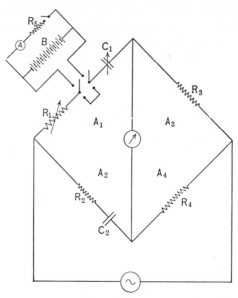

Fig. 115. Alternating-current bridge for measuring resistance of storage cells.

separating the real and imaginary parts, two equations are obtained, which are as follows:

$$\frac{\rho_2 + R_2}{\rho_1 + R_1} = \frac{R_4}{R_3} \tag{21}$$

and

$$\frac{C_2}{C_1} = \frac{R_3}{R_4} \tag{22}$$

from which

$$\frac{\rho_2 + R_2}{\rho_1 + R_1} = \frac{C_1}{C_2} \tag{23}$$

Assuming that the quantities ρ_1 and ρ_2 are negligible (if they are not, their values must be determined), R_1 is the only unknown quantity in the equation and its value is obtained from the solution of equation (23).

$$R_1 = R_2 C_2 / C_1 \tag{24}$$

When the battery is in the circuit, the resistance of the arm A_1 is $R_1 + b$ and the battery resistance is obtained from the difference of two measurements with the battery in and out of the bridge. The value of C_1 is different in the two measurements. If the battery is discharging through a local circuit of resistance R_5, the effect of this resistance as a shunt circuit must be allowed for in equation (24).

Another form of bridge suitable for the measurement of battery resistance, devised by the author, may be described as a resonance bridge in which the capacity reactance is balanced by an inductance and the capacitance of the battery is calculated from the change in the inductance when the battery is removed from the circuit. This bridge requires a very constant source of alternating current, free from troublesome harmonics. Some special apparatus, such as variable resistances and inductances, is also required.

When a capacitor is placed in series with a battery, the impedance of the circuit is given by the expression

$$\sqrt{b^2 + (1/\omega c)^2}$$

where $1/\omega c$ represents the capacity reactance of the capacitor and the battery combined. The resistance of the battery, squared, is extremely small as compared with the square of the capacity reactance. In order to eliminate the effect of the capacity, an inductance is tuned into the circuit so that

$$\omega L = 1/\omega c$$

Whence the impedance of the circuit

$$\sqrt{b^2 + \left(\omega L - \frac{1}{\omega c}\right)^2} = b$$

Balances of the bridge should be made with the battery both in and out of the circuit, and this is therefore a substitution method. To compensate for the battery, a low variable and non-inductive resistance is required. This was supplied by a straight copper wire that could be moved at will into or out of a copper-clad glass tube, filled with mercury. Since the storage cell or any other battery possesses a large electrostatic capacitance, the substitution of the mercury resistance for the battery requires a rebalancing of the inductances in the bridge. This was accomplished by the variable inductance, and from the change in this inductance the equivalent electrostatic capacitance of the battery may be computed. The cells may be on open circuit while being measured or may be discharging through a local non-inductive circuit at any desired rate. A diagram of the circuit is given in Fig. 116. R_1 and R_2 are non-inductive resistances of small values serving as the ratio arms of the bridge, R_3 is a variable resistance for balancing the

330 Storage Batteries

bridge, C is the capacitor to prevent the battery from discharging through the bridge. L is a variable inductance, B the battery to be measured, with its local circuit consisting of an ammeter and resistance. In place of the battery may be substituted a heavy link. R_4 is a non-inductive variable resistance to be used as a compensating resistance when the link replaces the battery. As a source of the alternating current a 5-watt electron tube, with loose coupling to the bridge, has been used. An amplifier has also been useful at times in the detecting circuit.

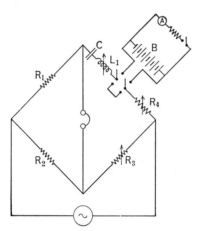

Fig. 116. Resonance bridge.

More recently Willihnganz[9] has described a modified alternating-current bridge for measuring battery resistances of the order of 0.004 ohm with an estimated accuracy of 2 per cent. The output of the bridge was amplified in three stages. He found the measured resistances were independent of the frequency over a wide range. Genin[10] published a paper giving an extended review of methods and results, including diagrams of many of the various bridges which have been proposed.

A question naturally arises in view of the above discussion: What is the physical meaning of the resistances as determined by direct- and alternating-current methods, and which is the true resistance? It seems likely that the alternating-current values are the true resistances, but they do not represent the effective values when the battery is being used for direct-current work. The resistance determined by direct-current measurements exceeds the resistance by alternating-current measurements. Since the resistance may vary with the current which flows through the battery, any measurement of resistance, to have a real significance, must be made under service conditions.

A comparison of the alternating- and direct-current measurements of resistance of a small radio B battery of 12 cells of the lead-acid type is given in Table 59. This shows that the voltmeter-ammeter method is too slow to catch the values obtained by the oscillograph, but prob-

---

[9] E. Willihnganz, A bridge for measuring storage-battery resistance, *Trans. Electrochem. Soc., 79,* 253 (1941).

[10] G. Genin, La résistance intérieur des accumulateurs au plomb et sa mésure, *Rev. gén. élec., 56,* 159 (1947).

TABLE 59. COMPARISON OF ALTERNATING- AND DIRECT-CURRENT
MEASUREMENTS OF THE RESISTANCE OF A STORAGE BATTERY
(Measurements were made on a small battery of 12 cells of about 3-ampere-
hour capacity. Values are expressed in ohms.)

State of Charge	D-C Resistance by Oscillograph Method		D-C Resistance by Voltmeter-Ammeter Method	A-C Resistance at 1000 Cycles
	Minimum	Constant		
Charged	1.00	1.08	1.29	0.99
Partly discharged, 2 amp-hr	2.00	2.23	3.2	1.18
Same, after standing 45 hr	1.52	1.60	....	1.03
Fully discharged, 3½ amp-hr	4.50	5.25	6.4	2.02

ably the voltmeter-ammeter readings come the nearest to representing
service conditions.

The resistance of various sizes and types of Edison batteries can
be computed easily from the data given in Fig. 117. These data are
on a single positive-plate basis. To obtain the resistance in ohms per
cell, divide the values read from the curve applying to the type of cell
by the number of positive plates in the cell.

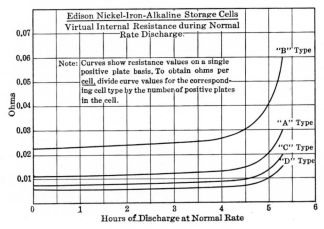

Fig. 117. Data from which the internal resistance of various types and sizes of
Edison storage cells can be calculated.

Similarly the internal resistance of type S, Nicad nickel-cadmium
cells can be approximated by dividing the value 0.15 by the rated
ampere-hour capacity of the cell. That is, a cell rated at 100 ampere-
hours' capacity would have an internal resistance of approximately
0.0015 ohm.

8

Efficiency

The efficiency of storage batteries is defined by the Standardization Rules of the American Institute of Electrical Engineers as follows:

36–300, Efficiency. The ratio of the output of a cell or battery to the input required to restore the initial state of charge under specified conditions of temperature, current rate, and final voltage.

36–301, Ampere-Hour Efficiency (Electrochemical Efficiency). The ratio of the ampere-hours' output to the ampere-hours of the recharge.

36–302, Volt Efficiency. The ratio of the average voltage during the discharge to the average voltage during the recharge.

36–303, Watt-Hour Efficiency (Energy Efficiency). The ratio of the watt-hours output to the watt-hours of the recharge.

The general expressions for efficiency will be considered first in this chapter, and then the conditions that affect the measurement of both the ampere-hour and the watt-hour efficiency.

GENERAL EXPRESSION FOR THE EFFICIENCY

If the current discharged by a battery at any moment be designated by I_1 amperes, the quantity of electricity delivered during an infinitesimal amount of time, dt, is $I_1\,dt$. For a longer period of time, represented by t_1 hours, the quantity of electricity expressed as ampere-hours is given by the integral:

$$\int_0^{t_1} I_1\,dt$$

Similarly, during the charging period, for which the current at any moment is I_2 and the length of the charge t_2, the total quantity of electricity passing through the battery is

$$\int_0^{t_2} I_2\,dt$$

The ampere-hour efficiency, according to the definition given above, is

332

the ratio of these two integrals:

$$\text{Ampere-hour efficiency} = \frac{\int_0^{t_1} I_1 \, dt}{\int_0^{t_2} I_2 \, dt}$$

If the current is kept constant during both the charge and the discharge, as is usual, this expression for the ampere-hour efficiency reduces to the simple ratio:

$$I_1 t_1 / I_2 t_2$$

The power delivered by a battery at any instant during its discharge is the product of the current and the terminal voltage. Expressing the latter in volts by E_1 and the current as above, the power in watts is $I_1 E_1$, and this multiplied by dt is equal to the energy delivered during the element of time. For a discharge lasting for a period of time t_1, the energy in watt-hours is represented by the integral:

$$\int_0^{t_1} I_1 E_1 \, dt$$

The corresponding expression for the energy received by the battery during the charging period is

$$\int_0^{t_2} I_2 E_2 \, dt$$

The energy efficiency is the ratio of these two integrals:

$$\text{Energy efficiency} = \frac{\int_0^{t_1} I_1 E_1 \, dt}{\int_0^{t_2} I_2 E_2 \, dt}$$

Either the current or the voltage may be kept constant during the charge and discharge. If the current is constant, the expression for the energy efficiency may be written

$$\frac{I_1 \int_0^{t_1} E_1 \, dt}{I_2 \int_0^{t_2} E_2 \, dt}$$

To obtain the integrals $\int_0^{t_1} E_1 \, dt$ and $\int_0^{t_2} E_2 \, dt$, the time curves of the values of E_1 and E_2 are drawn and integrated. The most convenient means of doing this is with a planimeter. Multiplying the first integral by the steady value of the discharge current I_1 and the second by the charging current I_2, the watt-hours delivered and received are obtained.

If both the current and the voltage vary, as when a battery discharges through a fixed resistance, the problem is somewhat more complicated but can be readily solved.

The current and time factors should be chosen with reference to the conditions of actual service when efficiency tests are to be made.

THE AMPERE-HOUR EFFICIENCY

The ampere-hour efficiency is of importance from the standpoint of the operation of storage batteries and is a step in the determination of the energy or watt-hour efficiency. The chemical reactions that occur during charging and discharging are reversible, and it is naturally to be expected, therefore, that under favorable conditions the ampere-hour efficiency should be only slightly less than 100 per cent.

The ampere-hour efficiency depends very largely upon the completeness of the recharge. A portion of the charging current is ordinarily wasted in producing gas, and this reduces the efficiency. There are also accidental factors that may lower the efficiency, such as (1) the self discharge of the plates, commonly called local action, (2) leakage of current because of faulty insulation either inside or outside the battery.

The completion of the charge is determined usually by continuing the charge at constant current until the voltage at the terminals of the battery reaches a maximum. Since this point can be determined only by carrying the charge beyond the point where the maximum is first reached, the efficiency will be low if calculated from the total amount of the charge. The ampere-hours put into the battery after the maximum is reached can be neglected, but in any event the error of an efficiency determination based on a single cycle of charge and discharge is likely to be large. It is common practice, therefore, to base efficiency determinations on a series of many cycles of charge and discharge in accordance with the normal operating conditions.

Temperature plays an important part in the measurement of efficiency, since the capacity of the battery is dependent very largely on the temperature, as has been shown in Chapter 5. Temperatures below normal result in diminished capacity and efficiency, but an increase in temperature may raise the calculated efficiency to over 100 per cent. This does not mean, however, that the battery can continue to furnish more ampere-hours than it receives during charge, as a few repetitions of the experiment would quickly show. The physical significance of a calculated efficiency in excess of 100 per cent is merely that the electrolyte diffuses more readily during discharge because of its decreased viscosity at the higher temperature, and therefore more of the active material of the plates may take part in the reaction than would otherwise be true. A standard temperature conforming to normal operating conditions should be specified when efficiency tests are to be made.

The rate of the discharge affects the capacity and, to some extent, the efficiency. The higher the rate of the discharge the lower is the delivered capacity, and a smaller number of ampere-hours are consequently required for recharging. The decrease in capacity of a battery because of increased rate of discharge is not accompanied by a corresponding decrease in efficiency. Tests on a group of vehicle batteries discharging at 45 and 90 amperes showed a decrease in ampere-hour capacity of 32 per cent at the higher rate as compared with the lower rate of discharge. Measurements of the ampere-hour efficiency of these batteries at the same rates gave 91 per cent at 45 amperes and 87 per cent at 90 amperes. The decrease in efficiency was only 4 per cent.

The following method for determining the maximum ampere-hour efficiency was suggested to the author by J. L. Woodbridge. A fully charged battery is discharged at some chosen current to a fixed cut-off voltage, careful measurement being made of the exact number of ampere-hours delivered. On the recharge the same number of ampere-hours are put back at the same current. A second discharge is then made to the same cut-off voltage as before. The efficiency of the battery is then calculated as the ratio of the ampere-hours delivered during the second discharge to the ampere-hours put in on the charge. After correcting for the slight increase in temperature as a result of the charge, the ampere-hour efficiency by this method has been found to be within the range 98 to 100 per cent. Such a measurement does not, however, represent the ordinary service conditions.

Nickel-iron batteries gas throughout nearly the entire period of charge and are subject to a rather large rate of loss of charge immediately after the charging period. The ampere-hour efficiency is therefore somewhat lower than for the lead-acid batteries. By shortening the charging period, efficiencies of 93 to 95 per cent can be obtained, but the output is then less than normal. The ampere-hour efficiency of a nickel-iron battery is about 82 per cent when discharged immediately after charge, but a 7-hour charge after a 5-hour discharge at the same rate would make the efficiency 72 per cent. However, the batteries ordinarily exceed the 5 hours, and this improves the efficiency.

THE WATT-HOUR EFFICIENCY

The watt-hour, or energy, efficiency is important because it shows the ability of the battery to return the energy that it has received. In this respect it is an important factor in determining the cost of operation. The watt-hour efficiency is affected by the same factors as the ampere-hour efficiency; in addition to these it is affected by the voltage relations on charge and discharge.

The internal resistance of a battery is small and variable, but it cannot be entirely neglected. Assuming the internal resistance to have a value b, the decrease of voltage during discharge at a current I_1 is equal to I_1b, and during charge there is a slight increase in voltage at the terminals which is equal to I_2b. The voltage on charge is always greater than the voltage on discharge. The part of this difference that is due to the resistance of the battery is equal to the sum of the two factors $I_1b + I_2b$.

By far the greater part of the difference between the voltage of the battery while charging and while discharging is due to polarization. There is some gas polarization, but primarily it is a concentration polarization in the case of the lead battery. The battery behaves as if it contained a very dilute electrolyte when it discharges and a concentrated electrolyte while charging. This is because the electrolyte in the pores of the plates becomes impoverished during the discharge and enriched during the charge. The relation of voltage to the acid concentration has been given on page 192.

The performance curves in Fig. 79 show that the average voltage during discharge at 45 amperes was 1.95 volts per cell and during charge it was 2.28 volts per cell. The ratio of these voltages, which is sometimes called the voltage-efficiency, is $1.95/2.28 = 85$ per cent. If the ampere-hour efficiency under such operating conditions could be 100 per cent, the energy efficiency could not exceed 85 per cent. The ampere-hour efficiency at 45 amperes discharge rate was 91 per cent, and the energy efficiency was therefore $0.91 \times 0.85 = 0.77$ or 77 per cent. A fair average for the energy efficiency of the lead battery under ordinary operating conditions is about 75 per cent.

The watt-hour efficiency of the nickel-iron cells at the normal rate of discharge ranges from 55 to 60 per cent. If discharged immediately after charge, the voltage efficiency is about 73 per cent and the watt-hour efficiency about 60 per cent. After standing for a day the lower figure is obtained.

It is apparent that, the better the equalization of acid within the lead cell, the less difference there will be between the average voltage of charge and discharge. High porosity of the plates and low viscosity of the electrolyte will help to accomplish this.

If the periods of charge and discharge are made very short, the concentration polarization becomes a relatively less important item. Highfield[1] gives the watt-hour efficiency of a central-station battery on lighting service during 12 months' operation as 74 per cent, but he reports the efficiency of a line battery on traction service, where the periods of charge and discharge alternated at frequent intervals, as 84 per cent. Hopkinson[2] made a series of experiments in short cycles of charge and discharge. By "time of the cycle" he means the sum of the time of discharge and the time of charge. Some of his results were as follows:

Time of the Cycle	Energy Efficiency
1 minute	96.5 per cent
10 minutes	93.6 per cent
30 minutes	92.0 per cent

Bailey[3] went a step further in this direction and made charges and discharges in cycles as short as $\frac{1}{60}$ second. He obtained energy efficiencies as high as 98.1 per cent. The difference between this and 100 per cent is probably to be attributed to losses caused by the ohmic resistance, because the concentration polarization must have been inappreciable at this frequency. Bailey's result is of no importance from an operating standpoint, but it is of very real interest theoretically in showing how perfectly reversible the storage cell is.

[1] J. S. Highfield, Storage batteries in electric power stations, *J. Inst. Elec. Engrs., London, 30,* 1070 (1901).

[2] B. Hopkinson, Losses of energy in accumulators, *Electrician, 48,* 211 (1901).

[3] B. F. Bailey, Maximum efficiency of a storage battery, *Elec. World, 47,* 829 (1906).

9

Testing Storage Batteries

The tests that are of primary importance are those to determine (1) the capacity of a battery at certain rates of discharge; (2) the ability of the battery to retain its charge over a period of time; (3) its ability to withstand vibration (portable types only); (4) the purity of the electrolyte; (5) its period of useful service, or life; and (6) its voltage characteristics. In this chapter is given a general discussion of the tests that apply to various types of batteries. Emphasis is laid on the nature and conditions of the tests, but fixed rules for making them are not prescribed.

The vital factors in formulating test procedures are: (1) the depth of the cycle; (2) the frequency of the cycle; and (3) the amount of overcharge. Tests of cycle life are primarily tests of the active material of the plates, but overcharge tests are tests of the grid material. In any event the test should be made to simulate service conditions.

CAPACITY TESTS

The manufacturer must determine by actual test the number of ampere-hours that any particular size and type of battery, which he makes, can deliver under specified conditions of discharge. Usually the capacity per positive plate is stated, and from this he can calculate a conservative ampere-hour rating for other batteries with these plates. The purchaser is interested in the capacity because he wishes to know also whether the battery in question can deliver sufficient energy to perform the service required of it.

The ratings made by manufacturers are usually conservative, and their batteries have some margin of excess capacity. Starting and lighting batteries and some others of the smaller types should ordinarily attain their full rated capacities on or before the third repeated cycle of charge and discharge. Stationary batteries and some motive-power batteries, however, may require as many as 12 cycles before attaining rated capacity.

In Chapter 5 the various factors that affect capacity are discussed in detail. The points to be observed in making capacity tests of any particular battery are: (1) rate of the discharge, (2) cut-off or final voltage, and (3) temperature.

Rate of Discharge

As the rate of the discharge is increased, the duration of the discharge is more than proportionally decreased (see page 214). It is important, therefore, that the current have a definite and constant value maintained throughout the test, except during intermittent tests. Storage batteries are commonly rated on the continuous discharge, "time" basis; for example, stationary batteries have a certain ampere-hour capacity at 8 hours, and starting and lighting batteries at 20 hours. Such ratings define the test current also. The ampere-hour capacity divided by the time fixes the test current. A starting and lighting battery having a rated capacity of 100 ampere-hours at the 20-hour rate would be tested at 5 amperes. So it is for any other time rating. Some tests of automotive and aircraft batteries are made at high current rates, such as 300 amperes, regardless of their size.

A difficulty arises in the testing laboratory when a group of batteries having slightly different ratings is to be tested. It is necessary first to determine whether the batteries are comparable and for the same kind of service. A fair average test current may then be chosen. If the batteries are to be tested in accordance with specifications which state the minimum required capacity, the problem is simplified. For example, if the minimum required capacity at the 5-hour rate is 72 ampere-hours, the test current is $\frac{72}{5} = 14.4$ amperes, although one or more of the batteries may exceed the 5-hour discharge period considerably.

Final Voltage

The "final" or "cut-off" voltage, as it is sometimes called, is the terminal closed-circuit voltage at which it is desirable to stop discharge. Batteries discharging at normal temperature and the 8-hour or longer rates have a cut-off voltage of 1.75 volts per cell; motive-power batteries, 1.75 volts at the 6-hour rate, etc. The cut-off voltages vary somewhat and have not been standardized except by general usage and in some cases by recognized specifications. In the absence of definite specifications, the cut-off voltages are generally determined by the shape of the discharge curve, stopping at the knee of the curve. The capacity that may be obtained beyond this point is small, and it is not economical to discharge the battery further.

When discharges are made at higher rates, the cut-off voltages are correspondingly lower. This is because of the increased voltage drop through the cell and the lower specific gravity of the electrolyte within the plates. The values for starting and lighting batteries are as follows:

Rate	Cut-off Voltage
20-hour, 80° F	1.75
20-minute, 80° F	1.50
300-ampere, 0° F	1.00

Temperature

The temperature of a battery under test is ordinarily specified as the initial temperature of the electrolyte, since this is under control of the laboratory. Standard temperatures for most portable batteries are 80° F or 25° C (77° F). The difference is small and may be neglected. The standard temperature for stationary batteries is usually 70° F. The final temperature, which is not under control, is really more important in its effect on capacity. Permissible limits of the ambient temperature should be stated. A battery should always be near the standard temperature when a test is made, to avoid large corrections. Corrections may be calculated from Table 49 in Chapter 5. The temperature correction varies with the rate of the discharge.

Accuracy

The accuracy of the measurements made in any determination of capacity deserves special mention. To attain an accuracy of 3 per cent or better, the current must be held constant to within close limits by regulation. Hand-operated rheostats are suitable for regulation if the steps are small, but carbon resistances afford the best means for continuous regulation. The ammeter chosen for the test should have a scale permitting readings to be made directly to 1 or 2 per cent of the test current, and it should be calibrated. The time is easily measured to 1 per cent, except in the case of high-rate discharges such as the 300-ampere test for automobile batteries or the 5-minute rate for airplane batteries, for which special precautions must be taken.

When the end of the discharge is approaching and the knee of the voltage curve is reached, the voltage may change so rapidly as to make it difficult to maintain the current constant at the time when it is most important to do so. To overcome this difficulty, recourse may be had to the use of "booster batteries." The battery under test is connected in series with a battery of larger capacity which serves to maintain the current and decrease the percentage change in voltage in the circuit.

For example, if the drop in voltage of the test battery is 0.2 volt per cell, or 10 per cent of the voltage of the cell, the use of 9 booster cells in series with this will reduce the voltage fluctuation to 1 per cent, provided the latter are of sufficient capacity. The proper number of booster cells to be used during test is determined mainly by convenience; even a few are of assistance, and ordinarily the ratio would not be as high as 9 to 1.

Records of the measurement of capacity should include the following: date of previous charge, date of the discharge, number of the cycle, initial specific gravity and temperature of each cell, initial open-circuit voltage of each cell, closed-circuit voltages at appropriate intervals with record of the time in each case (the intervals being shorter as the end of the experiment is approached), the cut-off voltage for the battery and the exact time that it is reached, and final specific gravities and temperatures for each cell. It is often worth while to repeat the final specific gravity and temperature measurements after several hours because equalization of the specific gravity of the electrolyte within the pores of the plate with that outside takes place slowly.

A new battery should be given several preliminary cycles of charge and discharge before determining its capacity. The capacity at any particular rate or temperature is influenced by the previous discharges (page 228), and therefore when changing to a different condition of discharge a preliminary cycle should precede the measured experiment.

Causes of Failure

Batteries which fail to meet capacity tests often require special attention, since it is always desirable for a testing laboratory to report the cause of failure as well as the fact. Experience is the best guide, but, if the observations have been carefully made, an analysis of them may suggest the cause of failure. The following items are suggested as a guide to the general procedure:

1. Compare the open-circuit voltages before discharge with Table 39. Appreciably low values may indicate internal discharge due to short circuits, a split separator, excessive sediment, or a similar cause.

2. Examine voltage readings throughout run and see if they progress regularly. Erratic readings may be an indication of bad contacts, loosely burned plates, internal short circuits. In the event of 2-volt changes, occasionally observed, the reversal of one cell is indicated.

3. Examine the gravity readings to see whether the range is normal for the type of cell under test. If the range is too small, and the plates are in good condition, the indication is that the electrolyte does

not circulate freely. Causes for this are high-resistance separators, separators without corrugations, wrong side of separator next to positive plate, or the addition to the electrolyte of some foreign substance which increases the viscosity. If the gravity range is abnormally great, the amount of electrolyte may be too small. Estimate the total amount of electrolyte if possible, and make use of Table 18, which will show the possible output for the amount and range.

4. The capacity of the cell is usually limited by the positive plate, but in some cases the negative plate may be deficient. The plate capacities can be determined by the use of the cadmium electrode (see page 235), which will indicate whether either plate is deficient.

5. By dismantling the cell or battery, and examining the plates and structural details, defects arising from manufacture or abuse are often revealed. Chemical tests of the electrolyte may reveal impurities that produce excessive local action.

Test panels for making automatically controlled tests of small types of batteries are shown in Fig. 118. The switches are operated by program machines, not shown in the illustration, within each of the panels.

TEST FOR RETENTION OF CHARGE

Besides determining the capacity of a battery, it is desirable to determine its ability to retain the charge over a considerable period of time. The Federal specifications[1] for automobile batteries require that a battery must not lose more than 25 per cent of its capacity during a standing period of 4 weeks. This is an average loss of about 1 per cent per day. The test is made as follows: After a careful determination of the capacity at a 20-hour rate, the battery is fully charged and allowed to stand on open circuit at a temperature of 70 to 80° F for a period of 4 weeks, after which it is discharged under the same conditions as before and the percentage loss in capacity is computed as the ratio of the decrease in ampere-hours delivered to the original capacity in ampere-hours.

The loss in capacity will depend on the specific gravity of the electrolyte, the temperature, and the purity of the electrolyte. The amount of loss allowed by the specifications is liberal and should cover any battery of reliable quality containing electrolyte not exceeding 1.280 in specific gravity. Lower specific gravities should result in smaller losses. On test of 17 different makes of batteries, the smallest

[1] *Federal Specification, Batteries: Storage, Vehicular, Ignition, Lighting, and Starting*, W-B-131e, May 4, 1953.

Fig. 118. Three of six battery panels installed at the National Bureau of Standards for the completely automatic control of charging and discharging storage batteries while on test. Each panel is provided with two motor-generators, operating in parallel when load is heavy, to supply charging current. Maximum capacity is 500 amperes at 30 volts, d-c, for each panel.

loss observed was 4 per cent and the largest 85 per cent; the average of the batteries that complied with the specification was 13 per cent.

One reason for the relatively small loss in capacity of some batteries is the fact that the negative plates at which the greatest local action occurs normally exceed the capacity of the positives, and it is possible, therefore, for them to be considerably affected without materially changing the battery capacity. A continuation of the test for an additional period may show greater losses.

The causes for loss in capacity while standing idle are local action and internal short circuits. Local action is accelerated by the presence of certain impurities in the electrolyte or on the plates. Antimony on the negative plates is a cause of local action, particularly in old batteries. Iron is a common impurity which can be eliminated in part at least by pouring out the electrolyte. Platinum, although rarely found, is exceedingly destructive even in amounts of only 1 part in 10,000,000. Among the causes of internal short circuits are excessive sediment, mossing, defective separators, metallic particles falling into the cells through the vents, and, rarely, porous sealing compound.

VIBRATION TEST

This test applies particularly to automobile and airplane batteries but may be used with other portable types to develop possible defects in lead-burning, sealing of terminal posts, vent-plug design, and shedding of active material.

For 2 hours the battery is subjected to a vibration consisting of a simple harmonic motion having a frequency of 1900 to 2100 cycles per minute through an amplitude of 0.045 to 0.050 inch (total displacement 0.09 to 0.10 inch) while the battery is discharging at the 20-hour rate. The temperature at time of test is specified as 70 to 90° F. Any departure from a simple harmonic motion changes the maximum acceleration and modifies the test.

The battery is fastened to the vibrating board by hold-down clamps. It must maintain a steady voltage and current during vibration. Fluctuations are generally due to plates breaking loose from the straps, as may happen if the lead-burning is not well done. Failure occurs most commonly at the terminal posts, which may become loose and allow the electrolyte to flood the top of the battery; vent plugs, if not provided with baffle plates, will usually allow electrolyte to escape. Excessive sediment may be produced. A battery of good design and workmanship should pass this test without difficulty. At the conclusion of the tests the cells are examined for broken plates, connectors, straps, and sediment in the bottom of the battery jars.

This test is considered superior to a bumping test, because the batteries in service are subjected to vibrations; also the test is easily specified in simple mathematical terms.

TESTS FOR PURITY OF THE ELECTROLYTE

Tests for the purity of the electrolyte form an important part of the complete test of a storage battery but cannot be made satisfactorily when suitable chemical laboratory facilities are lacking.

The presence of some of the impurities can be readily detected by inspection of the cells. Antimony and some of the more noble metals cause excessive gassing of the negative plates. This may be observed when the cells are on open circuit, if the cells are contained in glass jars, or a sample of the negative plates may be removed and placed in a tray containing a solution of pure sulfuric acid. Gassing will be somewhat in evidence even in a normal battery, particularly after the termination of a charging period. Negative plates which are contaminated will continue to gas until they are completely discharged. When manganese salts are present in the electrolyte they may ordinarily be detected by the characteristic permanganate color which appears at the positive plates when the cells are on charge. Hydrochloric acid and acetic acid, if present in any considerable amount, may be detected when the cells are on charge by the characteristic odors of chlorine and acetic acid, respectively.

Chemical methods for the detection of impurities in storage-battery electrolytes follow more or less the general procedure for analytical determinations, but they present the special difficulty that the impurities are present normally in small amounts in a fairly concentrated solution of a strong mineral acid.

Tests for the purity of sulfuric acid, both concentrated and diluted to 50 per cent are described in detail in the Federal specifications for sulfuric acid, O-S-801. This is obtainable from the Government Printing Office. The maximum limits for various impurities as given in the specification are included in Table 22 of Chapter 3. With some modifications the tests for these can be applied to samples of electrolyte removed from batteries. These tests, however, are of the type to determine whether the impurities in the test sample exceed the permissible limits of the specifications. Reference should be made to the complete specification for details of these tests.

Very briefly, the methods and reagents employed are as follows: For organic matter, charring when heated to fumes; for iron, ammonium thiocyanate; for substances oxidized by permanganate, potassium permanganate; for arsenic and antimony, Gutzeit test with modifications to distinguish between them; for manganese, potassium periodate; for nitrates, ferrous sulfate; for ammonium, Nessler's reagent; for chlorides, silver nitrate; for copper, ammonium hydroxide; for zinc, hydrogen sulfide; for selenium, iced sample overlayed with hydrochloric acid containing a little sodium sulfite—no red color to develop at zone of contact (selenium is found occasionally in acid from Oriental sources); for platinum, residue after evaporation is taken up in a little aqua regia and absorbed in asbestos paper, ignited, and while still hot tested in a stream of illuminating gas for incandescence.

A very useful method for volatile acids (acetic, formic, etc.) not included in the specifications was described by Craig.[2]

The choice of the methods for the detection of impurities in storage-battery solutions depends, first, upon the possibility of making the determinations quantitative, second, upon the accuracy which can be obtained by the method, and third, upon the presence or absence of certain impurities which may interfere with the results of the test. In order to determine the presence or absence of these impurities, preliminary tests are sometimes desirable before deciding upon the method of test to be used for the quantitative detection of any particular impurity.

Aside from chemical tests for the purity of the electrolyte, it is possible to determine the presence of impurities of the noble metals by spectroanalysis of material removed from the surface of negative plates. This method maks it possible to effect a concentration of the impurity, even when present in very small amounts, by scraping the surface of the negative plates. A quantitative estimate of the amount of impurities can be made from the relative intensities of the spectrum lines.

THE S.A.E. STANDARDS AND LIFE TEST

Standards of the Society of Automotive Engineers[3] apply to lead-acid batteries as used on motor vehicles, motor boats, tractors, and for automotive industrial applications for 6- and 12-volt requirements.

The sequence of tests prescribed is as follows: (1) conditioning charge at 20-hour rate to constant voltage and specific gravity; (2) discharges at 20-hour rate to 1.75 volts per cell; (3) discharges at 300 amperes 0° F to 1.00 volt per cell with recorded 5-second voltage readings[4]; (4) repeat (2); (5) repeat (4); (6) life tests.

The life test provides for all sizes of the batteries. Discharges of 1 hour at 40 amperes for a total of 40 ampere-hours are followed by recharges at approximately 10 amperes for 5 hours for a total of 50 ampere-hours. There are 4 cycles per day or 27 per week. The temperature of the center cell is maintained at 110° ± 5° F. One complete capacity test is made each week at 40 amperes to 1.70 volts per cell. When the capacity at 40-ampere rate drops below 40 per cent of the 20-hour rating, the test is considered completed. The S.A.E. standard specifies the required number of cycles for each of its listed batteries. Tests for cranking ability at low temperatures are made in

[2] Op. cit., p. 145.

[3] Soc. Automotive Engrs., Handbook, p. 770, 1953.

[4] For Diesel batteries record 30-second voltage instead of the 5-second voltage.

lieu of the capacity tests at 2, 5, 8, and each succeeding period of 3 weeks until the end of the test. These high-rate discharges at 300 amperes, 0° F, are counted in the number of cycles. The recharges after high-rate discharges are specified to be at 7.5 amperes for 1 hour for each minute of the previous discharge at 300 amperes.

An overcharge test is also specified. A charge of 990 ampere-hours is followed by a capacity test and represents 1 week's testing on a continuous basis or 1 complete life unit. The continuous charge of 990 ampere-hours is at the 9-ampere rate (battery in water bath) regardless of the size of the battery. This is followed by a standing period on open circuit for 48 hours in a thermostated water bath at 100° ± 5° F. The battery is then discharged at 300 amperes to 1.20 volts per cell, or a minimum time 30 seconds, whichever occurs first. Life units are repeated until failure occurs. The test is considered completed when the battery fails to meet the minimum voltage or time specified. Tentative requirements are given in the S.A.E. standard.

There are many details relating to the S.A.E. tests that are important, and reference should always be made to the Society's current standard. Past experience has shown that changes are made from year to year, and these changes quickly render obsolete information in more permanent book form. It is possible here to give only an outline of the general nature of the tests.

The S.A.E. standard gives dimensions of the batteries, the location of parts, and shape of the container (see Fig. 156 in Chapter 10); the minimum ampere-hour capacity at the 20-hour rate; the minimum time to 1.0 volt per cell at 300 amperes, 0° F; the minimum 5-second voltage during the 300-ampere test; the required life cycles and the overcharge life units. Batteries that have double insulation (retaining sheets of porous or perforated material between the positive plates and customary single separator) are allowed a deduction of 10 per cent from the time and voltage requirements but must give 15 per cent more life cycles.

At the conclusion of a life test the positive plates are examined for loss of active material, corrosion of the grid, color, buckling, texture of the active material, and breaks in the grids. The negative plates are examined for color, expansion or contraction, and texture of the active material, whether firm, spongy, or sandy. The amount of sediment in the bottom of the jars is estimated or weighed, and its color noted. The separators are examined for strength, degree of action of acid, and oxidizing effect of the positive plates, and particularly for splits or holes. From such an examination it should be possible to form an opinion as to the worth of the various parts.

VOLTAGE TESTS

The speed of a motor, the light from an incandescent lamp, and other services depend on the available voltage at the terminals of a battery under specified working conditions. These conditions will ordinarily indicate the kind of test to be made. One test, which has become recognized in the industry, is described in the S.A.E. *Handbook* and in the *Federal Specification*. These specifications provide for a voltage test at 0° F (−18° C), when the battery is discharging at the rate of 300 amperes. The measurement of voltage is made 5 seconds after beginning the discharge. To comply with the specification, the battery must have a voltage above the minimum specified for its type and size. This test is sometimes called the "5-second voltage test."

EFFICIENCY TESTS

The measurement of battery efficiencies has been discussed at length in Chapter 8, and little remains to be said about it here. Efficiency tests are of secondary importance, and quite often the results are an indication of the efficiency of the experimenter as well as that of the battery.

The ampere-hour efficiency is calculated as the ratio of the ampere-hours' output of the battery to the ampere-hours put in during charge, and is usually expressed as a percentage. Since the charging current, besides restoring the active materials of the plates, may produce gassing, the efficiency will be lowered in proportion as the quantity of gas produced is increased. By eliminating gassing altogether and correcting for temperature, it is possible to show that the current efficiency very closely approaches 100 per cent. This is seldom the normal condition of operation, however, as some gassing almost inevitably occurs. In making an efficiency test, therefore, it is necessary that the operating conditions be clearly defined.

The watt-hour or energy efficiency is of greater importance than the current efficiency. The watt-hour efficiency is equal to the ampere-hour efficiency multiplied by the ratio of the average voltage during discharge to the average voltage during charge. The average voltage is obtained from the time integral of the voltage curve.

The watt-hour efficiency is affected by the same factors as the ampere-hour efficiency and in addition by the factors that affect the voltage, namely, polarization, internal resistance, and, to a lesser extent, temperature.

OTHER TESTS

Tests of Intercell Connectors

A table is given in Chapter 10, showing the power losses in intercell connectors of motive-power batteries. For automotive batteries the voltage drop in intercell connectors should not exceed 20 millivolts per inch of distance between post centers when the battery is discharging 300 amperes, the temperature being 80° F. The test is made by "stabbing" the post centers with sharp metallic points connected to a suitable millivoltmeter through flexible leads while current is flowing from the battery at the prescribed rate.

The voltage of Edison batteries that are used for oil circuit-breaker installations and other purposes requiring heavy currents should be calculated on a basis allowing for voltage drop in the intercell connectors and jumpers. Voltages shown in Table 66 of Chapter 10 are the voltages per cell. From these values should be deducted 0.006 times the multiple of normal rate to obtain the effective voltage per cell at the battery terminals. Tests of any particular battery can be made to find whether it meets this condition.

Tests of Container Material

Test specimens (not including seams) are cut from the sides and partitions of the container as shown in Fig. 119. The specimens are immersed in water at 70° F (21° C) for at least 1 hour and then may be broken in a testing machine, provided this is not done until 16 hours after the specimens have been cut. According to the *Federal Specification*, the material from rubber cases of automotive batteries is subject to rejection if its tensile strength is less than 1300 pounds per square inch, or if the elongation (before rupture) is less than the following minimum requirements for corresponding tensile strength.

Tensile Strength, lb per sq in.	Elongation, %
1300	6
1350	5
1425	4
1525	3
1700	2
2000 or more	1

Rubber containers for motive-power, motorcycle, and some other types of batteries are usually subject to requirements of greater tensile strength.

Storage Batteries

Tests for acid absorption are made by measuring and weighing specimens (2 by 2 inches) before and after immersion in sulfuric acid solutions of 1.300 sp. gr. at 75° F (24° C). The containing vessels are placed in an oven at 155° F (68° C) for 7 days. When the specimens are removed from the solution, they are quickly rinsed in water, dried on the surface, and weighed. The material is subject to rejection if

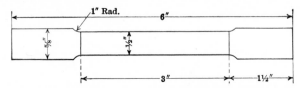

Fig. 119. Test specimen of container material for tensile-strength tests.

the increase in weight exceeds 1.5 per cent or if blisters or cracks are observed, or if the dimensions are increased 2 per cent. In this, as in other tests covered by the specifications to which reference is made, the full details should be obtained from the specification.

Tests of Glass Jars

Specifications for the quality of glass storage-battery jars may be found in the manual of the Signal Section of the Association of American Railways.

10

Present-Day Uses for Storage Batteries

Storage batteries find many uses. They fill the ever-present need for sources of power at times and places where other means are lacking, or because of their reliability they serve as safeguards to insure continuity of controls and service. The general public is most familiar with batteries to start motor cars, but more than these the less familiar types are many and vital to our present way of life.

Motive-power batteries are found on a variety of industrial trucks and tractors, mine locomotives, torpedoes, and submarines. Starter batteries provide the power to put internal-combustion engines in motion before they can operate on their own power. These include automobile, Diesel, marine, and some aircraft engines. Usually the batteries in these services provide lighting and ignition also. On the railroads track and signal circuits are actuated by batteries. Other batteries provide car lighting and air conditioning. Floating batteries are an important part of telephone central-office equipment, and many smaller batteries operate relays, message registers, vacuum tubes, coin-collecting devices, PBX boards, etc. Other types of floating batteries are a part of emergency lighting equipment in hospitals and places of public assembly. In power stations storage batteries serve as sources of auxiliary power and control, closing and tripping circuit breakers, operating oil switches. Other services include radio and television, laboratory experimental work, fire alarm systems, telegraph circuits, miners' cap lamps. These are but a part of the list which might be given. Only the more important applications will be described in the pages that follow.

TELEPHONE BATTERIES

Storage batteries form an important part of telephone power-plant equipment. The requirements of telephone service are very exacting, and much care and engineering skill have been devoted to perfecting the design, installation, and operation of these batteries. The large,

351

and most of the small, telephone central offices in cities are operated on what is called the common-battery system in distinction from smaller offices in rural communities where each telephone formerly had its own local battery of dry cells. The storage batteries with their associated generators at the central offices furnish current for transmitting, signaling, switching, and various miscellaneous services. In addition, the batteries serve as electrical noise absorbers, preventing crosstalk between simultaneous messages and reducing the circuit noise of motor-generators, rectifiers, or other equipment used to transform the incoming power supply to a form suitable for use on telephones. The storage battery is admirably adapted to this purpose by reason of its low resistance and impedance.

Besides the normal continuous uses for these batteries, they may be required in times of emergency to carry the entire load of the central office. Batteries with sufficient reserve capacity are provided, therefore, to permit operation of the central office to continue for some time if failure of the outside power supply should occur.

Historical Developments

Battery development in the telephone field has been stimulated by the rapid development in this system of intercommunication. Bell's first telephone apparently employed a primary battery to energize an electromagnet, but he discarded the battery shortly thereafter, preferring an instrument in which the feeble current induced in a coil of wire by a diaphragm of iron vibrating in the field of a permanent magnet served as the "talking current." The voice of the speaker was the source of power, and this has been estimated to be of the order of 1×10^{-5} watt. The receiver, which at this time was a duplicate of the transmitter, converted the varying electric current into corresponding vibrations of its diaphragm, thus producing sound waves heard by the listener. Early in 1878 the first commercial exchange was established in New Haven, Conn., without the use of batteries.

After the invention of the carbon transmitter, of which there are several forms, battery-operated circuits became indispensable. Various voltages were tried, but little was gained by increasing the voltage above limits which are familiar today. The phrase "24-volt talking battery," which is applied somewhat indiscriminately to batteries of 10, 11, or 12 cells, probably originated with an installation of 12 cells at Worcester, Mass., about 1896. About 1900, dry cells replaced other types of primary batteries in individual instruments, but the trend was strongly toward common-battery circuits in the larger communities. An early telephone power plant of 1893 is shown in Fig. 120.

Fig. 120. Early telephone power plant, August 1893. Four 4-volt batteries, two of them tapped at middle point; cells are of the Manchester box type; charging machines, ringing machines, and tone-generator are mounted on a wood bench. The "switchboard" above is really made of boards.

The evolution of the common-battery system has produced two types of central offices, manual and automatic-switching offices. Automatic-switching offices are often called dial offices; they may be further classified as (1) motor-driven panel offices, (2) step-by-step switch offices, and (3) the all-relay offices of which the crossbar system is one type. Obviously, any detailed description of these would be out of place here.

For many years, batteries of 11 cells have been standard in manual offices, with an additional battery of 11 cells connected in series with the first group to provide for toll calls and long-distance messages. The actual voltage of the 24-volt battery, however, ranged from 20 to 28 volts when no attempt was made to regulate the voltage. With the introduction of panel equipment about 1920 and the adoption of the continuous floating system of operation, 12-cell batteries again became common practice, and 1 emergency cell was added to hold up the voltage upon discharge. An additional battery of 11 cells in series with the first battery constituted the "48-volt battery" group, to which were added 3 emergency cells. The limits of voltage variation have been narrowed by the use of automatic regulators until 25.75 ± 0.25 volts is commonly found today in dial offices, although large repeater offices run a little lower, at approximately 23.5 volts. The regulation of the 48-volt battery is similarly held to 49.5 ± 0.5 volt.

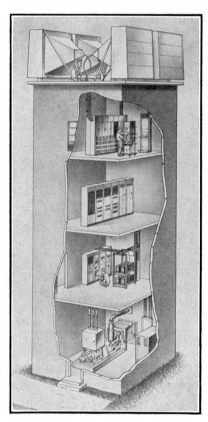

Fig. 121. Inside a typical radio-relay tower: Amplifying equipment is on the top floor, storage batteries and associated power apparatus on second and third floors, emergency power equipment on first floor. (Courtesy Bell Telephone Laboratories.)

More recently, the practice most commonly used in dial offices is to have one large battery of 23 cells with its floating generator, and with 4 emergency cells in reserve, to provide the major load, which is nominally at 48 volts. The 24-volt load is then provided through counter cells if small, or by a separate battery and floating generator if large. The battery may of course be made up of 2 or more strings of cells in parallel.

The rapidly multiplying achievements of basic and applied science since the mid thirties have produced changes in the telephone industry that are reflected in new and exacting battery requirements:

1. Long-distance telephone circuits of all types use large quantities of batteries at repeater stations to power amplifiers. These include open-wire conventional cable, coaxial cable, and the new radio-relay repeaters. In the older systems batteries supply filament and plate direct current to thousands of vacuum tubes. In coaxial systems, in emergencies, they drive motor-generators providing an a-c output which supplies repeaters through power-pack distribution circuits. The transcontinental radio-relay system connects with and supple-

ments previously existing facilities. The relay points, spaced approximately 30 miles apart, stretch across the continent carrying a microwave radio beam boosted at each successive tower and radiated to the next by 10-foot antennas. In a typical tower, Fig. 121, the amplifying equipment is on the top floor. Batteries of 12, 24, 130, and 250 volts, together with their rectifiers and other associated power apparatus, are on the floors below. Emergency equipment on the first floor consists of a pair of automatic-starting, gasoline-engine alternators to supply power in the event of failure of commercial alternating current. In very remote locations, where there is no commercial power available, these engines provide the primary source of power. The batteries must be absolutely reliable, charging automatically done, and maintenance limited to periodic visits of company personnel. Both speech and television are served.

2. Cross-country dialing is a new devlopment, permitting an individual to call directly any one of a vast number of stations in far-distant localities. This is an achievement still in its early stages but one to be reckoned with in changing and probably increasing the required battery capacity.

3. The development of storage batteries for floating service, in which ᴔhe grids of the plates are an alloy of lead hardened with 0.065 to 0.085 per cent calcium in place of the more usual lead-antimony grids. The gain is a battery of lessened self discharge, increased life, and floating-charge requirements reduced to about one-fifth of the usual amount. When gassing takes place during charge there can be no stibine because antimony is entirely absent.

4. Finally, the invention of the transistor, which can perform many of the services of vacuum tubes, has potential possibilities not yet fully appreciated. That it will have an effect on battery sizes, types, and usages seems quite certain. The power needed by transistors is much less than for vacuum tubes performing comparable service.

Power Supplies Needed

Although the 24- and 48-volt storage batteries (Fig. 122) are the principal ones in many telephone central offices, both manual and dial offices require a variety of other sources of power. These are mentioned later.

The primary sources of power are the large motor-generators, 2 or more being provided. At present the maximum size for a 24-volt generator is 1500 amperes and for the 48-volt generator 1200 amperes.

The main batteries are connected directly to a bus, to which are also connected the motor-generators. Normally the batteries float on this line, except in some very small offices when the load conditions are too light to warrant running the machines. The reserve capacity required of the battery in each case is estimated on the basis of load conditions and the probable reliability of the power supply. Although commer-

Fig. 122. Typical central-office installation showing cells of sizes F and H in hard-rubber sealed containers. (Courtesy Bell Telephone Laboratories.)

cial power services in duplicate are obtained where possible, in addition to these in many offices engine-driven generators are provided as stand-by protection, because times of emergency are usually periods of heavy load and continuous telephone service is of particular importance.

Direct-Current Loads

The amount of electrical energy required for the transmission of a single message is very small. About 35 ampere-seconds are required in a manual office to build up and restore connections, and about twice this amount for each 100 seconds of actual conversation. Similar figures for dial offices are small also. In the aggregate, however, the current requirements for an office of 10,000 lines or more may be surprisingly large, amounting to hundreds of amperes.

In manual offices the 24-volt battery carries a heavier load than the 48-volt battery, since the 48-volt battery is limited principally to the transmitter supply on long-distance or toll messages and to supplying PBX batteries over cable pairs. In automatic-switching offices, however, operation of the intricate mechanism constitutes a heavy load on the 48-volt battery, and this battery furnishes the greater part of the current. Approximate figures for certain multiunit offices of 17,000 to 20,000 lines illustrate this difference.

Type of Office	Battery	Heavy Load Amperes (Day)	Light Load Amperes (Night)
Manual	24-volt	300 to 500	50
Manual	48-volt	50	5
Automatic-switching	24-volt	250 to 300	100
Automatic-switching	48-volt	600 to 900	200

Power-plant loads are frequently greater than the above, particularly in large cities where several offices may be housed in one building and these offices may supply current for many private branch exchanges.

Peak loads in residential districts usually occur about 10:30 A.M. and again in the early evening. In business centers, the peak load is fairly well sustained during customary business hours, showing some decrease during the luncheon period. In long-distance repeater offices there is generally no distinct peak since most of the equipment operates 24 hours daily.

The size of cells in the main batteries will naturally depend on the traffic to be handled and whether the battery is assembled as a single series connection of cells or by paralleling strings of smaller cells. When parallel batteries are used they may be connected either through switches or by solid bus bars across the ends of the groups, and sometimes additional cross connections have been made within the groups to insure equalization of the load. Eight or more paralleled strings of cells are sometimes employed.

Open-Type Cells

The larger cells were formerly of the open lead-lined tank construction mounted on oil insulators. Open-type cells are now obsolete, but some installations of earlier years doubtless remain. The cells contained plates of the G or H sizes and ranged in capacity from 800 to 13,440 ampere-hours according to the local requirements. The largest cells contained as many as 85 plates.

Sealed-in Cells

Bell System telephone batteries are all of the pasted-plate type in sealed rubber or sealed plastic jars of polystyrene. The present tendency is toward increasing use of the plastic jars. Some independent companies, however, use batteries of the Manchex (see page 74) and Planté types in sealed glass jars.

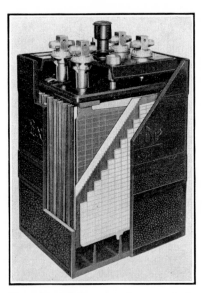

Fig. 123. Modern telephone central-office cell, type Exide EB and FB. The gas-escape vent is in the center of the sealed cover. In front of it is the filling funnel. The gas-collecting hood is the horizontal plate over the element. The grids are lead-calcium alloy.

Fig. 124. Gould-National calcium-grid type of stationary cell with explosion-proof ceramic cone, type FCW.

The standard Bell System batteries come in three size ranges: (1) 4 to 100 ampere-hour capacities. These are in units of 1, 2, or 3 cells in polystyrene containers. The grids are lead-antimony alloy except for the Plastical batteries which contain lead-calcium alloy. (2) Capacities of 180 to 1680 ampere-hours. These batteries contain plates of the E and F sizes, all having lead-calcium grids. The containers are plastic or hard-rubber, single-cell jars, except the smallest size, which comes in 2- or 3-cell units. (3) Large cells of 4000 to

7000 ampere-hour capacity. These cells contain plates of the H size and at present have grids of lead-antimony alloy. The tendency is, however, to change to lead-calcium grids.

All cells larger than 30 ampere-hours are supplied with explosion-proof vents. These are of several types. The Exide explosion-control

Fig. 125. Three-cell C and D battery with lead-calcium grids in a plastic case. The capacity of each cell is 180 ampere-hours. A plastic case of the same external dimensions for a single cell of 600 ampere-hours is the largest made now.

vent consists of a hood or dome extending below the surface of the electrolyte, Fig. 123. This collects the gas bubbles before they reach the surface and guides them into the neck of an inverted funnel, where at any one time there can be only a small amount of gas. If ignition should occur, this small amount of gas can produce no more than a harmless "pop." The Gould type of explosion-proof vent consists of a porous-ceramic cone, Fig. 124. This is permeable to the gases, allowing them to escape from within the cell, but on the principle of the Davy safety lamp the porous cone prevents the propagation of a flame to the interior of the cell. Vents on C and D cells, Fig. 125, are similar in principle but combine the funnel and vent, Fig. **126**. For the first two types of explosion-proof vents a separate filling tube extending below the surface of the electrolyte is necessary for taking hydrometer readings and for the addition of water without the necessity of removing the vents.

All lead-calcium cells have triple insulation, consisting of some combination of microporous-rubber separators, slotted retainers of hard rubber or plastic, and fiber-glass mats.

Steel racks are used instead of the impregnated wood formerly employed, since the attack by acid has been eliminated by the use of closed cells and by the precautions taken to make the vents effective against spray. The racks with batteries mounted on them can be located in rooms with other power-plant apparatus.

The specific gravity of electrolyte in most telephone batteries when they are fully charged is 1.205 to 1.225. The specific gravity being low, local action is reduced to a minimum and the dete-

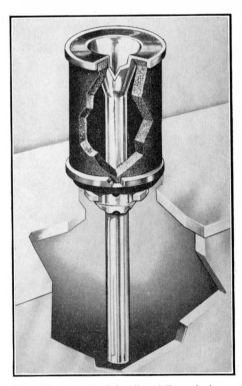

Fig. 126. Detail of the C and D explosion-proof vent. The gases escape through the porous stoneware. The central funnel serves both as a filling tube and for measurements of specific gravity. See Fig. 125 showing these vents in use on cells.

rioration of separators is also reduced. Use of double or triple insulation contributes to long life and freedom from trouble.

Operating Routines

Central offices operate on floating batteries which are, therefore, fully charged at all times except during an emergency discharge. The floating voltage is preferably 2.15 to 2.17 volts per cell. At these voltages the batteries absorb sufficient current to keep them in proper condition, and the higher figure obviates the necessity for periodic equalizing charges. The load of the office is really carried by the primary source of power, but the battery serves the double purpose of providing reserve power for emergencies and improving the transmission of speech.

Charging generators, which were formerly of a special type with smooth armatures and windings banded on the surface as a means of reducing the noise level, have now been largely superseded by the use of commercial-type generators with filters in the talking circuits. These consist of choke coils of high impedance and low direct-current resistance, together with electrolytic capacitors of high capacitance. Each electrolytic capacitor has 1200 microfarads capacitance or more at 24 volts, or approximately half this amount at 48 volts, and several capacitors in multiple are used. Similar but smaller capacitors are also available for 130 volts. Dry-type capacitors are used.

Batteries for Miscellaneous Uses

As indicated above, there are various miscellaneous batteries of small capacity which are required in addition to the main batteries. One of these is for operation of message registers in manual offices. Thirty-nine volts are required with an allowable variation of plus or minus 2 volts. For this purpose, a battery of 19 cells is provided and kept charged by floating continuously through an automatically controlled two-rate charging circuit from the 48-volt central-office battery. For operating coin-collecting devices on telephone lines, two batteries of 55 small storage cells each are provided at many of the larger central offices, while dry cells are used in small offices, the negative side of one battery and the positive side of the other battery being grounded. Connection being made to one or the other of these batteries, the operator is enabled to collect or return the coins that have been dropped into the slots at a pay-station telephone. These batteries seldom exceed 6 ampere-hours in capacity. Later it was found practicable to design a double-current generator, which in addition to 20-cycle and other signaling energy provides the currents for controlling coin boxes. Certain types of ringing systems commonly used have small storage batteries of 8, 16, or 22 cells to superimpose a direct-current voltage upon the alternating-current ringing voltage. This aids in tripping and improves the wave shape for ringing the bells.

In main offices where large groups of telephone repeaters are located for use in long-distance transmission, separate storage batteries and generators are provided primarily for the operation of the vacuum tubes. The current that heats the filaments is usually derived from a motor-generator set with automatic voltage control floating upon a

24-volt battery. The plate circuit needs 130 volts, and this is usually provided by a generator floating upon a storage battery. In the largest combined repeater, long-distance and telegraph installations, loads are heavy, approximating 5000 amperes at 24 volts and 500 amperes at 130 volts.

In telegraph offices, batteries for two systems are employed. For grounded telegraph these usually consist of 2 parts each of 61 cells, plus 8 emergency cells and 9 counterelectromotive-force cells, furnishing regulated 130 volts on the positive and negative sides of a neutral ground wire. For metallic telegraph 2 batteries of 16 cells each furnish plus and minus 34 volts. In both systems motor-generators or rectifiers float the batteries continuously.

Fig. 127. Twenty-four-cell counterelectromotive-force battery in right foreground, battery of sealed-tank cells in the background, 130-volt battery at extreme right.

Counter Cells and Emergency Cells

Counter cells of the nickel-alkaline and stainless-steel types have been employed for battery regulation within limits of their rating, Fig. 127. Thus in dial PBX equipment requiring regulated voltage, counter cells are introduced into the discharge circuit to keep the voltage at the distributing point below the upper limit during the time the battery is being charged. They are used in connection with plate voltages for repeater circuits and for telegraph circuits to prevent excessive voltages during charging of the batteries. They are also used to obtain lower-voltage current from higher-voltage battery, such as 24-volts from 48-volts. In these and various other uses counter cells may be in the circuit for brief periods or continuously.

The alkaline counter cells (Fig. 95 in Chapter 6) use several thin plates of nickel or stainless steel in a solution of treated sodium hydroxide. These counter cells have displaced the lead-acid type, because they do not build up objectionable capacity with continued use and can be short-circuited during switching without causing heavy currents to flow. They also have a longer life. The counterelectromotive force of these cells for various percentages of their rated current is given in Chapter 6, page 288. Counter cells are generally used in the ratio of 1 counter cell to from 7 to 12 active cells, depending upon the charging methods used and the voltage limits that must be maintained.

"Emergency cells" are regular storage cells arranged to be cut into the circuit automatically when the battery voltage is being reduced by an emergency discharge. They are kept fully charged during idle periods by small rectifiers which replace the internal losses.

Private Branch Exchanges

Private branch exchanges are large users of storage batteries. These exchanges vary greatly in size, from those having one trunk line and a few stations to those which have switchboards like large central offices and handle several thousand extensions. They are of two main types: (1) the private manual branch exchanges known as PBX, and (2) the private automatic branch exchange, sometimes known as PAX or dial PBX, since the telephones are equipped with dials. The community dial exchange, using equipment somewhat similar to a large dial PBX, is a small central office attended only at intervals. In these offices alarm circuits are provided to call an attendant from an attended office in the event of major trouble.

All these exchanges require storage-battery power for calls to the central office and for local messages. Voltages range from 14 to 26 volts for the smaller exchanges to 48 volts for the larger dial boards. Most of the private exchanges are too small to warrant a resident maintenance personnel, and, as a result, the maintenance of the local storage batteries, when provided, must be largely automatic. Small batteries are charged over cable pairs from the central office, but those in the

larger exchanges and in the dial type are charged locally by rectifiers or by motor-generators. This charging goes on during use so that
the batteries are relieved of a substantial part of the load.

In small PBX's the batteries are trickle-charged continuously, and
it has been found practicable to have a maintenance man visit the
exchange only at extended intervals when addition of water is needed,
while in the meantime the PBX attendant will call for a plant man if
a charge-indicator ball drops below a white line. Water-level charts
assist in making charging-rate readjustments.

Voltage control in conjunction with a two-rate floating-charge system has found application in automatic operation of these batteries.
Several kinds of temperature-compensated relays are available for the
purpose. The primary function of the relay is to select either of two
charging rates, high or low, depending on the state of charge of the
battery. If the battery is in need of considerable charging, the high
rate will be used first. As the voltage of the battery rises to approximately that of full charge, the relay operates to remove a short circuit
from part of the resistance in the circuit and the charging rate decreases
to a trickle-charge rate. Compensation for changing room temperatures is accomplished by a bimetallic strip which alters the tension of
the spring of the relay, thereby changing the voltage at which the
relay operates.

Grid-controlled tubes afford another means of close automatic voltage regulation of rectifiers continuously connected to storage batteries.
If only direct-current commercial power is available, small motor-
generators are provided.

The batteries are of the sealed-in type in plastic jars. A battery
plant widely used in small single-position PBX's has four 2-cell units
(Fig. 128) in a tall narrow cabinet matching the exchange switchboard to which it is attached at one end. The charge-indicator balls
in the top unit are conveniently in view of the attendant through an
opening in the cabinet.

Counter cells are provided, if necessary, in the larger dial PBX's
and these, together with a voltmeter relay and other control equipment,
maintain the bus voltage automatically within the proper range.
Automatic power plants to meet widely varying needs of private branch
exchanges and small central offices have been so skillfully designed and
constructed in compact units that relatively infrequent visits of the
maintainer are required. He can tell by the state of charge of the
battery and the amount of water it needs at the time of his visit
whether the installation is properly adjusted to the particular conditions of service.

Fig. 128.   Battery for a small private branch exchange.

BIBLIOGRAPHY

R. L. Young, Power plants for telephone offices, *Bell System Tech. J., 6,* 702 (1927). (Also *Bell Telephone System Tech. Monograph* B-276, November 1927.)

R. L. Young and R. L. Lunsford, Automatic power plants for telephone offices, *Trans. Am. Inst. Elec. Engrs., 50,* No. 3, 949, September 1931. (Also *Bell Telephone System Tech. Monograph* B-561, May 1931.)

G. L. Weller, Telephone power plants, *Bell Telephone Quarterly, 13,* 289 (1934).

R. P. Martin, Jr., The talking battery, *Bell Labs. Record, 15,* 138 (1937).

H. T. Langabeer, Supplying power to central offices, *Bell Labs. Record, 16,* 43 (1937).

U. B. Thomas, Lead-acid stationary batteries, *J. Electrochem. Soc., 99,* 238C (1952).

U. B. Thomas, An improved telephone battery, *Bell Labs. Record, 29,* 97 (1951).

C. H. Achenbach, Improving the service life of storage batteries, *Bell Labs. Record, 30,* 313 (1952).

W. L. Tierney, Radio-relay stations of the TD-2, *Bell Labs. Record, 30,* 327 (1952).

STORAGE-BATTERY APPLICATIONS IN RAILWAY SERVICE

Railway Signaling

Track and Signal Circuits.   A report of the Block Signal and Train Control Board of some years ago contained the statement: "Perhaps no single invention in the history of the development of railway transportation has contributed more toward safety and dispatch in that field than the track circuit. By this invention, simple in itself, the foundation was obtained for the development of practically every one of the intricate systems of railway block signaling in use today wherein the train is, under all conditions, continuously active in maintaining its own protection." The closed-track circuit was invented by Dr. William Robinson, August 20, 1872. This invention grew out of his previous invention of the open-track circuit, which proved to be unsatisfactory. The closed-track circuit differs from the open-track circuit in the essential particular that a small electric current flows continuously through a section of the track, and, when the track is clear, a signal indicates it to be so.

Signals are installed along railroad tracks to space trains properly and to instruct them when to proceed or stop. The distance between signals is dependent upon the character of the country, the concentration of traffic, and the allowable speed of trains in the particular locality. As a general average, a track circuit is about a mile long.

Almost every modern automatic-signaling system employed on the railroads involves some form of battery power for the various control, lighting, and operating circuits. Both primary and storage batteries are widely used for these important services. Railway signaling is one of the most exacting applications of storage batteries, because the highest degree of dependability is required and because the batteries are scattered along many miles of track.

Operation of automatic block-signaling equipment is usually controlled entirely by the movement of trains. This is accomplished through track circuits, the track being divided into sections which are electrically insulated from each other. Within each section adjoining rails are bonded together to provide a continuous low-resistance circuit. At one end of the circuit a "track battery" is located and at the other end a relay, which in turn controls a second or signal circuit (see Fig. 129). As long as no train is in the block, the relay is energized and the signal is held in the clear position. When a train enters, its wheels and axles shunt the primary circuit of the track relay, causing the armature to drop and thereby interrupt the signal circuit. The signal then changes to a restrictive indication, shown by a semaphore or by the position or color of lights. Broken rails or battery failure automatically result in a signal for trains to stop.

The voltage applied to track circuits is from $\frac{1}{2}$ to 2 volts. It is purposely kept low to minimize leakage of current from rail to rail across the ballast. The current flowing in the rails when the track is clear ranges from 200 to 500 milliamperes and, when occupied, from 500 milliamperes to 2 amperes, depending on the type of battery and the resistance in series with it to limit the current when a train enters the section. A single storage cell of 80 to 120 ampere-hours is generally sufficient for the track circuit, but if primary batteries are used the capacity is necessarily larger.

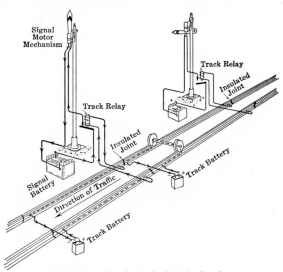

Fig. 129. Track and signal circuits.

Block signals are of three general types, described as semaphores, color lights, and position lights. Usually they operate on 5 to 12 cells. Semaphores require a steady current of about 40 milliamperes for holding the arm in the clear or caution position and 2 to 3 amperes for about 10 seconds to operate the motor when the signal arm changes from a restrictive to clear position. If semaphores are equipped with electric lamps, an additional current, which may be as much as 1.5 amperes, is needed. Color-light signals require from 0.5 to 2.25 amperes and position-light signals from 2 to 5 amperes, depending on the number and wattage of the lamps. Sometimes the lamps are lighted only when trains are approaching, in order to conserve power. Batteries for the signal circuits range from 80 to 300 ampere-hour capacity.

Although several methods of maintaining storage batteries in a charged condition can be used, the floating battery system (Fig. 130) is commonly employed at the present time. An alternating-current transmission line along the railroad's right of way furnishes the power,

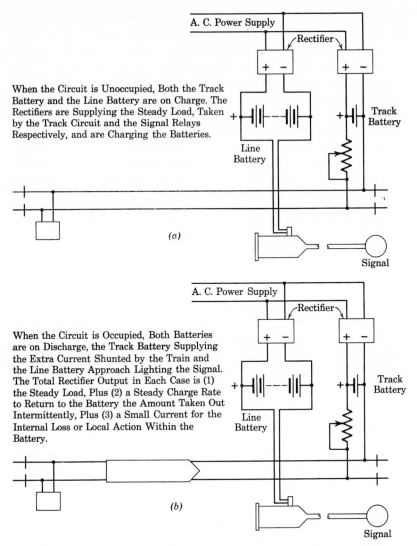

When the Circuit is Unoccupied, Both the Track Battery and the Line Battery are on Charge. The Rectifiers are Supplying the Steady Load, Taken by the Track Circuit and the Signal Relays Respectively, and are Charging the Batteries.

(a)

When the Circuit is Occupied, Both Batteries are on Discharge, the Track Battery Supplying the Extra Current Shunted by the Train and the Line Battery Approach Lighting the Signal. The Total Rectifier Output in Each Case is (1) the Steady Load, Plus (2) a Steady Charge Rate to Return to the Battery the Amount Taken Out Intermittently, Plus (3) a Small Current for the Internal Loss or Local Action Within the Battery.

(b)

Fig. 130. Floating-battery system for track and signal circuits.

and connection is made at each battery location through a transformer and rectifier. The output of the rectifier is adjusted to cover the average current required by the service, plus a small additional amount to compensate for local action in the battery.

At highway grade crossings, various automatic protective devices, including flashing lights, wigwags, stop-and-go signals, and barriers of several kinds, are employed to warn or stop vehicles when trains are approaching.   Power requirements for these range from **8** to **12** volts and from 1 to 20 amperes.   At many crossings these protective devices are operated by alternating-current power either directly or though rectifiers with a battery as a stand-by source of power.

Both lead-acid batteries and alkaline cells find use in various signal applications.   The type of cell and its capacity must in any event be determined by the character of the service.   A cell of the lead-acid type commonly used for railway signaling is shown in Fig. 23 (Chapter 2).   The nickel-iron cells are Edison cells of types A or B, type A being shown in Fig. 40.   Storage batteries of either type are required to meet specifications covering construction and performance of the Signal Section of the Association of American Railroads.

Interlocking Plants

Signaling facilities and switch mechanisms for governing the movements of trains at terminals, junctions, grade crossings, and drawbridges are usually interconnected and controlled through mechanical, electropneumatic, or all-electric interlocking plants in order to expedite traffic and to avoid conflicting routes and signal indications.   The interlocking machine is installed in a tower overlooking the tracks. Power for the control machine, switch-operating mechanisms, and sometimes for signals within the range of the plant is generally supplied by a storage battery of 10 to 120 volts, depending on the type of plant. The capacity of the batteries ranges from 80 to 280 ampere-hours. The batteries may be floated on a rectifier from an alternating-current line with provision for increased rates of charge following emergency discharges.   At some interlocking plants engine-driven generators are provided for battery charging.   The service requirements vary widely with the type of equipment and with the distance from the control tower.   The cells are of the same types as those used on signal circuits.

Centralized Traffic Control

One of the most important developments in automatic signaling is centralized traffic control.   As its name suggests, it is a method by which an operator in a convenient location can control the switching and movement of trains throughout a wide territory, extending many miles in either direction.   The operator sits before a panel board such as that shown in Fig. 131.   Pilot lights on the board indicate the location of trains, the position of switches, and the indications of signals throughout his territory.   By levers on the control board he can route trains, permitting fast trains to pass slower trains at passing sidings, or arrange for trains in single-track territory to meet and pass at sidings with a minimum of delay.   Trains move on signal indication without the necessity of written train orders.   This system increases track capacity and thereby postpones the necessity for increased trackage with increased traffic under older methods of operation.

Signaling and switch mechanisms are actuated through coded impulses for which power is supplied by storage batteries of relatively high voltage but small capacity.   The voltage used depends on the

Fig. 131.   Centralized traffic control board.   The dispatcher is operating a CTC machine, which controls 248 miles of main line between Hamlet, North Carolina, and Savannah, Georgia, on the Seaboard Air Line.   (Courtesy Union Switch and Signal Co. and the Association of American Railroads.)

length of the coded circuits.   At each remote switch, a relay tuned to its special code responds to its signal and actuates a secondary circuit controlling the mechanism.   The power supply for the signals may be the same as in any automatic block-signaling system, but switch mechanisms usually operate directly from a local storage battery of 100 to 150 ampere-hour capacity.

Car Retarders

In railroad classification yards, mixed car consignments are received with each incoming train.   The cars must be shifted and reclassified according to routes and trains.   A locomotive will do the shifting from a "hump" by allowing cars to glide from a running start on to the proper track, but the cars must be retarded at the proper time to avoid colliding with other cars. This was formerly done by hand brakes, but the use of electric or electropneumatic retarders and switches has done much to expedite the movement of cars through the yards. Car retarders are mechanical shoes along the rails which contact both faces of the car wheels and slow the cars down to the proper speed for traversing the yard tracks and for coupling to other cars. The retarders are controlled from a high tower overlooking the switching area. Power is furnished by storage batteries of 120 or 240 volts having an average capacity of 160 to 400 ampere-hours. The batteries are usually floated on rectifiers or motor-generators. The types of cell employed are the same as for signal service.

Multiple-Unit Cars

Mass transportation has been accomplished most effectively by electrified lines on which the cars are a modification of the standard railway coach. Each car, or car and trailer as a unit, has its own propulsion motors and other facilities, in contrast to a railway coach which depends on a locomotive for power. These cars operate in trains without a locomotive. In a sense they are trolley cars joined together and operated by one man at the control position at the head of the train, hence the name multiple-unit cars or MU cars. The number of cars in a train varies with the expected traffic density at the time of day.

The controller is essentially a switching device that operates the power switches by electromagnets. The control circuit is independent of the main propulsion current and is of lower voltage, that most commonly used being 32 volts. It provides the power supply to actuate motors and brake equipment. It supplies power also for door mechanisms, safety interlocks, marker lights, and emergency lights. Storage batteries are the indispensable source of power, protecting these vital services. The capacities needed vary from 50 to 300 ampere-hours. When lead-acid batteries are used, they are assembled in rubber jars and contain electrolyte of a relatively low specific gravity, 1.225. Nickel-iron alkaline batteries of the Edison type for 32-volt systems consist of 24 to 27 cells, depending on service requirements. Charging is accomplished by a motor-generator.

Switching Locomotives

In the field of heavy traction, several types of electric locomotives have been put in operation for switching service and as a substitute for steam locomotives in city streets where the use of steam locomotives is objectionable. Effective electrification of railroad yards and sidings thus becomes possible in many localities without the necessity of supplying trolley or third-rail systems.

Switching locomotives may be classified as (1) straight storage-battery locomotives for which the battery is the sole source of power; (2) combination battery and trolley (or third-rail) locomotives which may operate either on the battery or on external power; and (3) Diesel-electric locomotives, the most important now.

Diesel Locomotives for Railway Service

Originally applied in 1925 to switching service, Diesel-electric locomotives are now operating successfully also on main-line high-speed passenger service. Since 1934 the development of these has been rapid.

Most of the Diesel locomotives in railway service meet the transportation requirements with comparatively few sizes of engine units. These are 600, 660, 1000, 1350, 1500, and 2000 horsepower. Practically all locomotives in passenger, freight, and switching service use batteries of nominal voltages of 64 or 112 volts. The batteries are used for engine starting and control circuits. There are two general groups of batteries of 32 and 56 cells respectively. Engine starting is accomplished by an extra starting winding in the generator, which is directly connected to the Diesel engine. The generator operates for the time being as a motor. Current from the battery may range from a few hundred amperes to over 2000 amperes. A Diesel-electric locomotive is made up of 1, 2, 3, or 4 units. A 4-unit freight locomotive or a 3-unit passenger locomotive has about 6000 horsepower. Diesel locomotives have been replacing steam locomotives, but new types of locomotives are being developed, including those with steam turbines and gas turbines combined with electric drive. Diesel engines and their other applications are more fully described in this chapter in the section on Diesel-cranking batteries.

Car Lighting and Air Conditioning

From the earliest days when passenger cars ran only in the daytime and required no artificial light to the "modern" improvements when passengers brought their own candles, lighting of passenger trains developed rapidly. Oil lamps were introduced in 1850. Gas was first used in 1860. Electric lighting was first tried in 1872 on a sleeping car of the New York Central. A battery furnished the current, and it should be noted that this antedated the invention of pasted-plate batteries. The first trains to be entirely lighted by electricity were operated in 1887. Pintsch gas, however, continued to be the ordinary source of light from 1883 for many years. Brilliant lighting by incandescent lamps probably dates from one of the earliest attempts at lighting railway cars by electricity, made in 1881 with a straight storage-battery system. The Pennsylvania Railroad is said to have equipped eight parlor cars with batteries in 1885, and the Central Railroad of New Jersey installed the first axle-generator system of which there is record, in 1894. Electric car lighting is confined to passenger trains or trains made up of express, mail, or baggage cars.

The first passenger car to be equipped with fluorescent lights is said to have been on the New York Central in 1938, and the first train to be so lighted on the Burlington in 1939.

One of the important factors that has contributed to the rapid development of railway train lighting has been the introduction of metal-filament electric lamps. In the earlier days lamps were rated at 60 to 64 volts, but with the introduction of axle-generator systems and tungsten lamps this was reduced to 30 to 32 volts. This reduction was made possible because of decreased line drop and increased efficiency of the lamps. The voltage of train-lighting systems, with few exceptions, has been less than 110 volts because it is desirable to keep the number of cells in the battery as low as possible. In some foreign countries, voltages as low as 24 volts are found.

The traveling public has been educated to expect higher intensity of illumination than was previously acceptable. Four to six foot-candles which were once satisfactory have been increased to 8 to 12 foot-candles. To provide this without glare, indirect lighting or special light-diffusing fixtures have been used. Some roads have used spectrum lighting and other decorative effects. The effect has been to increase the demands on the battery for lighting and still more for air conditioning.

The straight storage system was the earliest system developed. It consisted of a battery on each car to provide current for the lights on that car irrespective of the other cars of the train. Charging was done at terminals or division points during the layover period. An advantage of the system is simplicity, but extensive charging equipment is required at terminals, cars must be "spotted" for charging, and they are unproductive of revenue while this is being done. The straight storage system is said to be the most expensive in spite of its simplicity and is seldom, if ever, used at the present time.

Head-end systems consist of a generator on the baggage car or locomotive, driven by a reciprocating engine or turbine supplied with steam, at reduced pressure, from the locomotive. The earliest forms did not include a battery. Failures were frequent and were necessarily caused each time the cars were parted or the locomotive uncoupled. For this reason, batteries were added with voltage regulators to avoid excessive voltages at the lamps when the batteries were charged. Head-end systems found their chief applications on trains having long runs without change in make-up, but the axle-generator system was more commonly employed. With the advent of streamlined trains operating as a unit, several installations of alternating-current or direct-current generators supplying current for lights and air conditioning have been reported. These are 110-volt systems.

The trend toward lighter passenger cars has been reflected in an effort to lessen the accessories load. This has given some impetus to the development of 110-volt systems of a-c supply which can be adapted to either a single car or to the head-end lighting system. Trial installations[1] were reported in 1948.

Axle-generator systems are most commonly used. Each car is provided with a battery and a generator to charge it. The generator is driven by the axle of the car. As used in the United States, the axle-generator systems are mostly of the 32-volt type. The essential parts are: First, the generator of the inclosed type, shunt wound and provided with some form of pole changer, unless of constant polarity, in order that the terminal voltage may have the same polarity irrespective of the direction in which the car may be traveling. Second, an automatic switch, usually of the solenoid type, to connect the generator with the lighting system and batteries when the car attains a predetermined speed. Third, a combined current and voltage regulator for the generator and a lamp regulator to hold the voltage at the lamp terminals to 30 to 32 volts. Fourth, the battery itself, consisting of 15 or 16 cells of the lead-acid type, or 24 to 25 cells of the alkaline or Edison type. The necessity for some regulating device may be seen from the fact that a 16-cell battery, when discharged to 1.8 volts per cell, has an effective voltage of 29, but the same battery near the conclusion of a charging period will have a terminal voltage of 2.5 to 2.6 volts per cell or 40 to 42 volts for the whole battery. Increasing the voltage from 29 volts at the end of discharge to 34 means doubling the illumination from the lamps. Regulators are normally adjusted to limit the voltage to 32 volts at the lamps. Some axle generators supply a constant current to the batteries and others charge the batteries at constant potential. When the train is standing the battery must carry the load, but charging commences when the train has reached a predetermined speed.

[1] H. H. Hauft, A-c air conditioning, *Ry. Mech. Engr.*, *122*, 147 (1948). L. B. Haddad, A-c power for passenger cars, *Ry. Mech. and Elec. Engr.*, *124*, 24 (1950).

Except for early experiments with "air-cooling" passenger cars, air conditioning began with the testing of mechanical air conditioning of sleeping cars by the Pullman Company and of coaches by the Baltimore and Ohio Railroad in 1927–1929. In 1931 this railroad put in operation the first completely air-conditioned passenger train. By 1953 there were 20,000 air-conditioned cars running on various railroads.

Before 1932, it was standard practice to use a 32-volt axle-generator system together with a storage battery whose size was determined by the electrical load and the protection period required in the opinion of each railroad electrical engineer. Railroad men thought in terms of 8 to 10 hours as the necssary protection periods during which the battery would be required to carry the load in the event of a failure of the generator equipment. It was common practice to provide 150- to 250-ampere-hour batteries on baggage cars, 200- to 400-ampere-hour batteries on mail cars, and 300- to 500-ampere-hour batteries on coaches, dining cars, and Pullmans. Since then the addition of air conditioning has so increased the load that protection periods of 10 hours are no longer considered, partly because the generating equipment has been developed to a higher state of perfection and partly because it was unwise to increase the battery to a size that would have provided equivalent protection periods. The changes in power requirements that began in 1932 as a result of air-conditioning cars were so marked that this date may be considered a beginning of a new era in the railroads' use of batteries. The electrical load, however, depends to a large extent on the type of air conditioning employed.

Air conditioning has been defined as the simultaneous control of all or at least the first three of the following factors affecting both physical and chemical conditions of the atmosphere within any structure. The factors include temperature, humidity, motion, distribution, dust, bacteria, odors, toxic gases, and ionization.

Three basic systems of cooling the air have found general application in railway passenger service. These are (1) mechanical-compression systems, sometimes subdivided into direct-mechanical (axle) and electromechanical (motor) systems; (2) steam ejector systems; and (3) ice-activated systems.

In the mechanical-compression systems, the refrigerant (dichlorodifluoromethane, commonly called "Freon") is in a liquid state under pressure on one side of an adjustable expansion valve. Through this the "Freon" passes to a much lower pressure, vaporizing and absorbing heat from its surroundings as it expands. The gas is compressed, which heats it, then from the compressor it is passed through a condenser which returns it to the liquid state, and the cycle is repeated. Air for the car is cooled by contact with the cold pipes.

The compressor represents the largest load, and it may be driven by an electric motor, by a gasoline engine, or by mechanical drive direct from the car axle. In the electromechanical systems (motor-driven compressors) the power requirement for cooling alone will vary

Fig. 132. A 2-cell unit of an Exide-Ironclad battery for railway-car-lighting and air-conditioning service. Cut away to show details of construction. Hinged vent plugs in center of each cell cover.

from 8 kilowatts for 5-ton units to 13 kilowatts for 7-ton units. The battery capacity can be supplied within a reasonable space, and it is common to find 32-volt batteries of 1000 to nearly 1300 ampere-hours installed on the cars (Figs. 132 and 133). However, the current requirements are large and the electrical conductors must be large. To decrease the expense and weight of the copper cables, some railroads are turning to 64-volt systems, which means doubling the number of cells in the battery and halving their capacity. One railroad is using a 110-volt direct-current system and a battery of 88 nickel-iron alkaline cells of 300-ampere-hour capacity.

In the direct-mechanical system the compressor is driven by a slip clutch from the car axle. Here the only additional electrical load is that required for circulating fans and pumps.

In the steam ejector system of cooling, water is the cooling medium. Live steam is passed through the ejector, and its velocity produces a vacuum in the evaporator tank which entrains the water vapor. The mixture of steam and water vapor then passes into a condenser and is liquefied. The cooling effect is brought about in the evaporator by

Fig. 133. Car-lighting and air-conditioning battery of the Edison nickel-iron-alkaline type in a roll-out cradle assembly.

part of the water boiling off at reduced pressure. The electrical load consists of motors to drive the condenser fan, cooling-coil fan, condenser pump, and cold-water pump. The total power consumption is about 4 kilowatts. The size of batteries applied to passenger cars using the steam ejector system is 600 to 1000 ampere-hours at 32 volts.

In ice-activated systems, the water which is cooled by ice is circulated through coils over which the air to be cooled passes. This is probably the cheapest in first cost, but it presents a serious servicing problem. Bunkers mounted under the car carry the large blocks of ice. Water is sprayed over these, and the cooled water is circulated through tubing with fins, which in turn cool the air. The increase in load on the battery over that required for car lighting is small.

In axle-generator systems (Fig. 134) when the car is traveling at

slow speed (below **16** to **18** miles per hour) the complete power requirements are taken from the storage battery.   As the train increases speed
a "cut-in" switch connects the generator to the electrical system for
supplying the electrical load plus charging current for the battery.

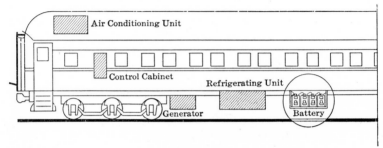

Fig. 134.   Diagram of an air-conditioning unit on a railway passenger car.

A 600-ampere-hour battery will take care of connected loads up to
100 amperes, and a 1000-ampere-hour battery for loads up to 250
amperes.   Lead-acid batteries can be fitted with hinged vent caps
which facilitate maintenance.   The caps are opened and closed by
using the nozzle of the cell-filler or watering gun.

    The capacity of the battery at the 8-hour rate in any particular
installation bears a fairly definite relation to the size of generator as

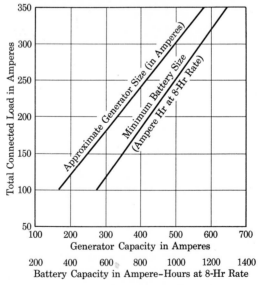

Fig. 135.   Relation of battery capacity to generator capacity for various
connected loads.

measured in amperes. Recommended sizes of battery and generator for various connected loads are illustrated in Fig. 135.

BIBLIOGRAPHY

American Railway Signaling Principles and Practices, Signal Section, Assoc. Am. Railroads, 1927–1935: Chapters V, "Batteries"; VII, "Direct-Current Track Circuits"; XV, "Automatic Block Signals"; XIX, "Electric Interlocking"; XXI, "Hump Yard Systems."

C. W. T. Stuart, *Car Lighting by Electricity,* Simmons-Boardman Publishing Corp., 1923.

Proceedings of the Association of Railway Electrical Engineers, 23 (1935); *24* (1936); *25* (1937).

Engineering Report on Air-Conditioning Railroad Passenger Cars, Division of Equipment Research, Assoc. Am. Railroads, 1937.

Invention of the Track Circuit, Signal Section, Am. Ry. Assoc., 1922.

Tracy. Railway car batteries, *Ry. Elec. Engr., 29,* 157 (1938).

Quiz on Railroads and Railroading, Association of American Railroads, Washington, 1951.

J. E. Gardner, Application of electricity for the auxiliaries of railroad trains, *Trans. Am. Inst. Elec. Engrs., 60,* 34 (1941).

G. T. Wilson, Railroad air conditioning, *Refrig. Eng., 45,* 323 (1943).

E. F. H. Grothe, Axle power for passenger cars, *Ry. Mech. Engr., 122,* 322 (1948).

H. C. Riggs, Lead-acid motive power and car-lighting batteries, *J. Electrochem. Soc., 99,* 236C (1952).

MARINE APPLICATIONS

Storage batteries find many uses on board ship. These range from the simplest applications on ignition circuits of internal-combustion engines of small craft to the complex circuits of the largest and finest ships. The batteries assure continuity of power for radio, lights, and essential auxiliaries. They carry the peak loads of pumps, hoists, and compressors. On naval vessels they provide for gun firing and for the propulsion of submarines when submerged.

Ordinarily the storage batteries float on the power line (Fig. 136). They are maintained, therefore, in a fully charged state ready for emergency service. The relation of battery voltage to line voltage, however, involves several interesting problems. The generator must have a drooping characteristic if the battery is to assume the load in excess of the generator capacity. This is especially necessary if the generator is driven by an internal-combustion engine, as such engines have relatively little overload capacity. When it is necessary to charge the battery, the generator voltage is raised from time to time in order to maintain the desired charging rate. This is essentially a constant current charge until the generator reaches a maximum voltage of 135 volts for a nominal 115-volt system. It provides an average voltage of

2.4 volts per cell for a 56-cell lead-acid battery. This is enough for a rapid charge of the battery, but it is too high for incandescent lamps, radio tubes, and gyrocompass. There must be a regulated portion of the voltage supply for these services. The unregulated portion of the bus is satisfactory for motor-driven auxiliaries which have largely replaced the steam-driven hoists and winches of some years ago. After the generator voltage has reached 135 volts, it is held at this point and the charge is continued as a constant-potential charge. The current through the battery then falls gradually until the charge is completed,

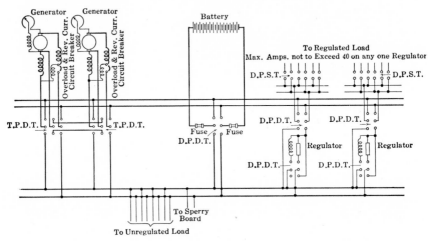

Fig. 136. Diagram of the Exide floating-battery system.

when the generator voltage is reduced to normal value and the battery again floats on the line. About 90 per cent of the charge can be given before the charging current begins to decrease. The time of charge up to this point is about 10 hours.

The electrical load on board ship is a fluctuating load, and the battery improves the conditions of operation by absorbing part of the fluctuations. A smaller generator than would otherwise be required can be used because the battery automatically carries the peak loads. The battery, on the other hand, must be large enough to carry the entire load for a specified time, which may be either to cover an emergency or to carry the night load on smaller ships, particularly yachts, if is desirable to stop the generators for reasons of economy or personal comfort.

Another method of operation for 110- to 120-volt marine requirements is by series-paralleling the two halves of the battery. Under normal operating conditions the generator supplies the ship's load and charges the two halves of the battery, but, if the generator voltage should fail, the two halves of the battery are automatically connected in series and assume the load. The principles of this system are illustrated in Fig. 137. Transfer of the load is accomplished by an automatic contactor when the line voltage falls below a predetermined setting. The contactor which connects the battery in series operates rapidly enough for the direct-current motor starting boxes in the load

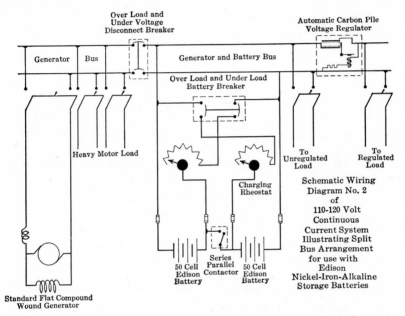

Fig. 137. Diagram of the Edison continuous-current system for marine use.

circuit not to be released and for no interruption in the current to occur. This system is used with Edison alkaline batteries. By dividing the battery, ample voltage for charging is provided, but when the batteries are in series the voltage may be too high for the lamp load and a voltage regulator must be provided. This is a carbon pile, subjected to pressure that varies inversely with the voltage. It is similar to voltage regulators used on the lighting circuits of railway cars. The heavy motor loads can be separated from the emergency and lighting loads by use of a disconnect circuit breaker that divides the bus. This has the advantage of relieving the generator of overload and of protecting the lamp load from large fluctuations. In full automatic control, loss of generator voltage causes the main disconnect contactor to open, which de-energizes the battery charge contactor, and the series-parallel contactor operates immediately to put the battery on discharge as one series-connected unit. The proper current for charging the batteries can be set on the manually operated charging rheostats. Normally 50 cells (Edison type) are used in each half of the battery.

Regulations prescribed by the Bureau of Marine Inspection and Navigation of the Department of Commerce for new vessels provide that two or more generating sets shall be installed so that, if one breaks down, there will be sufficient remaining capacity to carry the "peak sea load." Passenger vessels over 1600 gross tons are to be provided also with an independent emergency source of power. For the details and latest requirements of these rules, reference should always be made to official sources of information.

On ocean-going vessels a variety of types and sizes of batteries is to be found. The services that they perform include: general alarm, reserve power, Diesel starting, lifeboat engine starting and radio transmitter, radio direction finders and alarms, emergency transmitters, gyrocompass, fire alarms, and emergency lighting. These are usually under the supervision of licensed engineering personnel and are subject to specified programs of charging and maintenance.

Thirty-two- and 110-volt systems are available for auxiliary yachts and motor cruisers. The battery receives its charge from a generator operated by the main engine. However, if the main engine is seldom used, a small separate generator driven by some other means is generally supplied. The battery should have sufficient capacity to provide for several hours' continuous operation of the lights and all the auxiliaries that operate automatically, such as refrigerators and bilge pumps. Heavier loads, such as electric cooking, hoisting the anchor or sails, may be provided for either by installing a larger battery or by starting the main engines, after which the main generator supplies the necessary current. The service is intermittent, and some thought should be given to selecting convenient times to charge the battery adequately. The batteries are either 16 cells of lead-acid type or 25 cells of the Edison nickel-iron type for the 32-volt systems. At higher voltages, 55 or 60 lead cells or 100 cells of the Edison type may be needed.

With the development of Diesel engines for tugs and other small commercial craft, a need has arisen for electric power to replace steam formerly used for auxiliaries. Shaft-driven generators, somewhat like railway car-lighting equipment, have found increasing use. The smaller installations are usually 32-volt systems up to about 10 kilowatts. Larger units of 115 volts are employed in sizes up to 40 kilowatts or more. On many of the older installations it was the practice to call for two Diesel-engine generator sets of the same capacity, one being held in reserve, but more recently a shaft-driven generator of equal capacity has been used to replace one of the Diesel sets. When the main engines are operating, the auxiliary unit can be shut down.

The essential parts of a shaft system include: (1) A generator, the voltage of which will remain constant over a wide variation of the shaft speed. (2) A belt or chain drive from the propeller shaft, from which the system gets its name, "shaft-driven generator system." (3) A voltage regulator to maintain constant voltage and protect the generator from overload. (4) A reverse current relay to disconnect the battery and load from the generator when the terminal voltage of the generator is too low for battery charging. (5) A battery of the lead-acid or alkaline type which floats on the bus from the generator or which is charged in parallel halves. The battery assists the generator on peak loads, carries the emergency load, and provides for the lighting load at such times as the main engine and generator are shut down. (6) A lamp voltage regulator to protect the lighting units from voltages above their rating. On some installations a pole changer may be necessary to prevent reversal of polarity when the motion of the shaft is reversed.

The types of batteries used for marine service include various sizes of Exide-Ironclad batteries ranging from 57 to 1136 ampere-hours at the 8-hour rate. Five sizes of plates are available. Flat-plate batteries of the lead-acid type, assembled in rubber cases of 2 to 4 cells, are used also. These contain relatively thick plates, because space and weight limitations are of less importance than durability. Edison batteries in four sizes of plates are available in capacities from 18 to 900 ampere-hours at the 5-hour rate.

It is desirable that marine batteries should be rugged and durable enough to give long service. They are subject to continual vibration, and the safety of the vessel and its passengers may depend on their proper functioning under adverse conditions.

TRUCK AND TRACTOR BATTERIES

Economic Importance

Efficiency in handling materials depends partly on speed and the amount of goods that a truck can carry, but it depends also on how long a time must be spent in loading and unloading operations. There are practical limits to the speed of trucks from the standpoint of safety, and there may be limits to the amount transported at one time, depending on the rate of production, the weight, or the physical dimensions of the vehicle that are allowable. The earliest trucks and tractors were loaded and unloaded by hand. This often required considerable time, but it is characteristic of modern equipment that its measure of transportation service has been increased by the rapidity with which it can be loaded and unloaded by its own power.

Classification of Trucks, Tractors, and Locomotives

Industrial trucks were originally designed for use in railway terminals for the transportation of baggage. These trucks combined in a single unit the battery for motive power and the loading space. The

Fig. 138. Tractor hauling a train of modern castor-steer trailers which are carrying pallet loads. (Courtesy Mercury Manufacturing Co.)

next step was to divide the functions of these platform trucks. The load to be transported was placed upon trailers drawn by a tractor, which is a short, sturdy unit with rubber-tired wheels carrying only the battery and the driver.

Tractor-Trailer System. The tractor draws a train of trailers (Fig. 138). It may collect these at different points and may deliver them to various destinations.

Skid-Lift-Truck System. The platforms of these trucks may be raised, permitting them to pick up loaded skids. The truck can pick up its load in a few seconds and can deposit its load in an equally short time. The truck is provided with two motors, one for travel and the other for elevating the platform. A modification of this type of truck is found in the "high-lift" trucks, which can raise the load to a considerable height for stacking loaded skids in tiers. Sometimes the skids accompany the consignment of goods for quick unloading at their destination by similar lift trucks.

Electric Crane Trucks. Crane trucks, sometimes in combination with platforms, are adapted to general utility work. Modern crane trucks, without platforms, are adapted to heavy work, as illustrated in Fig. 139. This truck is equipped with tongs for lifting the loads.

Fig. 139. Electric crane truck equipped with tongs.

Pallet-Fork-Truck System. The pallet is essentially a skid platform with thin stringers instead of legs or a double platform with stringers between. This is picked up by the fork on the truck and tilted to a safe angle for stability while the load is being transported. This truck can lift the load to a considerable height for stacking goods in tiers (Fig. 140). The fork truck usually has three motors, one for travel, another for lifting the load, and a third for tilting the load. By taking the pallet with the goods into a car or into the hold of a ship, rehandling of the goods at point of shipment and at destination is avoided.

Motorized Hand Trucks. These small trucks, as the name implies, perform much the same work as the ordinary hand truck might, but

with the advantage of speed and saving of manpower, Fig. 141.
Power is derived from a storage battery of about 250 ampere-hour
capacity. This is less than the requirements for full-sized trucks and
tractors.

Fig. 140. A fork truck lifting a loaded pallet, stacking materials in tiers.

Other Systems. Special types of trucks have been developed for
handling materials of particular shapes, such as large rolls of paper,
coils of steel strip (Fig. 142), wire, or bundles of sheet metal. Decreased time for loading and unloading usually means that the truck
is operating a greater percentage of the time.

Mine Locomotives. A locomotive is distinguished from a tractor
by the fact that it runs on rails, while a tractor usually has tires and
a steering mechanism. Some rubber-tired propulsion units are called
shuttle cars. Mine locomotives are compact units of minimum height

Fig. 141. Motorized hand truck, powered by a storage battery, loading material into a freight car.

Fig. 142. A heavy-duty ram truck handling strip-steel coils.

which carry a battery for motive power (Fig. 143). They are capable of pulling a train of loaded trailers and are rapidly replacing other means of transportation, including the animals which were commonly used for transporting material within mines. No trolleys are needed, and the rails do not require bonding. With no voltage drop in a long line, the operating voltages can be lower than when a trolley is used.

Fig. 143. Mine locomotive operated by storage batteries. The battery is in the compartment beyond the operator, who is wearing a cap lamp for which a small battery is attached to his belt.

In gaseous atmospheres which are present in some mines the storage-battery locomotive is the safest form of motive power. Types approved by the U. S. Bureau of Mines are permissible in such atmospheres.

The Batteries

The batteries used for propelling trucks and tractors are of either the lead-acid or the nickel-iron type. Typical batteries for use in trucks are shown in Figs. 144 and 145.

For certain types of vehicle service, thin-plate batteries have been extensively used. The thin plates in these cells do not have a greater

Fig. 144. Exide-Ironclad battery for industrial tractors or locomotives. The cells are assembled in a steel tray.

Fig. 145. Cradle assembly of Edison 30 C8 battery for industrial trucks.

capacity per plate than the thick plates of the same length and width. The capacity of the thin plates is actually less, but because a larger number of these plates can be used within a jar of given size the capacity of the cell at the higher rates of discharge is considerably greater.

Connections between individual cells of a tray are made by intercell connectors, and similarly the connections between the trays are inter-tray connectors. These may be either burned or bolted to the terminals. Burned-on connectors are generally preferred, since the acid may creep in between the contact surfaces of the bolted connection and cause corrosion, which will destroy the joint electrically.

The intercell connectors may be either copper straps heavily lead-plated or solid connectors of lead-antimony alloy. The copper straps, which are sometimes subject to corrosion, have the advantage of flexibility and high conductivity. A loop in the middle of these straps allows for expansion and for a slight movement of the cells without danger of cracking the covers. The design of the intercell connectors is a matter of considerable importance, since the drop in voltage when the battery is subject to large current drains may become an appreciable part of the total available voltage of the battery. The power which is lost in the intercell connectors increases with the square of the current which the battery delivers. This power loss for low rates of discharge is entirely negligible, but at larger current rates such as may be required when a tractor is pulling a heavy load it may amount to $\frac{1}{2}$ horsepower or more. Table 60 shows the power loss for intercell

TABLE 60. POWER LOSS IN INTERCELL CONNECTORS

Type and Number of Cells	Discharge Current, amp	Drop in Voltage	I^2R Loss, w
Lead-acid batteries, 12 cells	45	0.053	2.38
(11 connectors)	90	0.106	9.5
	225	0.264	59.0
	450	0.528	237.6

connectors on a battery of excellent design. This table shows a hundredfold increase in the power loss as the discharge current is increased from the normal value of 45 amperes to 10 times the normal rate, or 450 amperes.

A wider variety of types and sizes of motor-power batteries is available now than in former years. It is obviously impossible to give detailed operating data for these, but their general characteristics may be found in Tables 61 and 62. The first of these applies to lead-

TABLE 61. DATA ON SOME TYPES OF BATTERIES FOR MOTIVE-POWER SERVICE

(The data are given on the basis of 21-plate cells, which contain 10 positive plates, to facilitate estimates of current and capacities of other cells of their respective types. Motive-power cells are often discharged at relatively high rates. By using the ampere-hour capacities at the 6- and 8-hour rates which are given, computations may be made of the capacity at any other rate by using the equation given on page 216. All cells have double or triple insulation.)

	Exide-Ironclad Cells				Exide Flat Plate	Gould-National Cells					C and D Cells	
	MVM-21	ML-21	TLM-21	MEH-21	MVA-21	XMZ-21 KMZ-21	XVLZ-21 KTZ-21	XLZ-21 KHZ-21	XEZ-21 KEZ-21	XRZ-21 KRZ-21	CDS-21	CHS-21
Number of positive plates per cell	10	10	10	10	10	10	10	10	10	10	10	10
Normal 6-hr discharge, amp	56.7	63.9	83.3	166.7	49.8	56.7	66.7	91.7	100	166.7	100	183.3
capacity, amp-hr	340	383	500	1000	299	340	400	550	600	1000	600	1100
capacity, w-hr	660	739	966	1910	553	666	780	1070	1170	1975	1170	2112
voltage, initial, v	2.04	2.03	2.02	2.01	2.04	2.03	2.03	2.015	2.04	2.06	2.04	2.03
average, v	1.94	1.93	1.93	1.91	1.94	1.96	1.95	1.94	1.95	1.975	1.95	1.92
final, v	1.76	1.75	1.74	1.70	1.76	1.75	1.75	1.75	1.75	1.75	1.75	1.75
Eight-hour discharge, amp	45	50.6	66.5	133.1	40.2	45	53.25	73.125	80	133.125	79.6	146.3
capacity, amp-hr	360	405	532	1065	321	360	426	585	640	1065	637	1170
capacity, w-hr	704	786	1034	2045	600	707	835	1140	1248	2095	1249	2258
Dimensions, length, in.	$8\frac{1}{8}$	$8\frac{1}{8}$	$8\frac{1}{8}$	$8\frac{1}{4}$	$6\frac{7}{8}$	$8\frac{1}{8}$	$8\frac{1}{8}$	$8\frac{1}{8}$	$8\frac{1}{8}$	$8\frac{1}{4}$	$8\frac{1}{8}$	$8\frac{1}{4}$
width, in.	$6\frac{1}{4}$	$6\frac{1}{4}$	$6\frac{1}{4}$	$8\frac{13}{16}$	$6\frac{1}{4}$	$6\frac{1}{4}$	$6\frac{1}{4}$	$6\frac{1}{4}$	$6\frac{1}{4}$	$8\frac{3}{4}$	$6\frac{1}{4}$	$8\frac{3}{4}$
height, in.	$15\frac{3}{16}$	$15\frac{11}{16}$	$20\frac{3}{8}$	$28\frac{5}{8}$	$15\frac{3}{16}$	$15\frac{1}{8}$	$15\frac{1}{2}$	$20\frac{3}{4}$	$21\frac{1}{4}$	$28\frac{5}{8}$	$20\frac{3}{4}$	$28\frac{1}{2}$
Charge rate, finishing rate, amp	22	23	25	50	22	17	20	27	30	50	27	50
Electrolyte, sp. gr.	1.280	1.280	1.280	1.280	1.250	1.280	1.280	1.280	1.280	1.280	1.280	1.280
weight per cell, lb	15.5	16.4	21.3	46	13.5	16	17	22	23	44	20.8	57.0
Weight, complete cell, lb*	69	74	95	189	60.5	86†	96†	123†	125†	228†	117	203
battery, per cell, lb*	76	81	104	229	69.5	90‡	99‡	130‡	133‡	....	132‡	247‡

* The weight of the battery per cell, frequently called "the trayed weight," varies considerably with the construction and the amount of reinforcing.
† Weight of battery, per cell, pounds when trayed in wood.
‡ Weight of battery, per cell, pounds when trayed in steel.

TABLE 62. DATA ON EDISON STORAGE BATTERIES FOR TRUCK, TRACTOR,
AND OTHER SERVICE

The electrical and physical characteristics are given for comparison on the
6-positive-plate size for the motive-power types of cell. Ratings are based
on 5-hour discharge for types A, B, C, and D; all to 1 volt per cell. Normal
charge is 7 hours for types A, B, C, and D.

	Type B6	Type A6	Type C6	Type D6
Number of positive plates	6	6	6	6
Normal discharge rate, amp	22.5	45.0	67.5	90.0
capacity, amp-hr	112.5	225.0	337.5	450.0
capacity, w-hr	135.0	270.0	405.0	540.0
voltage, initial, v	1.45	1.45	1.45	1.45
average, v	1.20	1.20	1.20	1.20
final, v	1.00	1.00	1.00	1.00
Normal charge rate, amp	22.5	45.0	67.5	90.0
average volts, 95° F	1.704	1.704	1.704	1.704
maximum volts, 95° F	1.845	1.845	1.845	1.845
Dimensions,* length, in.	3.85	3.85	3.85	3.79
width, in.	5.07	5.13	5.13	5.07
height, in.	8.81	13.41	19.16	25.83
Electrolyte,† lb per cell	2.55	4.12	5.99	8.30
Resistance, mean effective, ohms	0.004	0.002	0.0013	0.001
Weight of battery,‡ per cell, lb	13	22.4	34.4	45.6
Single-positive-plate data:				
normal discharge rate, amp	3.75	7.50	11.25	15.00
capacity, amp-hr	18.75	37.50	56.25	75.00
capacity, w-hr	22.50	45.00	67.50	90.00
voltages as above				

* Length and width dimensions are, respectively, the dimensions along and
across the tray axis.
† Electrolyte is for cells of normal height; "high" and "high-wide" cells
require more.
‡ Weights are for cells completely assembled in trays, including trays,
connectors, and electrolyte.

acid batteries having 21 plates per cell. This size of cell was arbitrarily chosen because each cell contains 10 positive plates, facilitating
computations of other sizes on a decimal basis. The various types
and kinds are not strictly comparable because of variations in size
and kind of plates, in their spacing, and in the separators employed.
These data were compiled from catalogs which should be consulted
for detailed information.

Table **62** applies to the nickel-iron type, Edison cells. All cells are of the sizes ending in 6, that is, each cell contains 6 positive plates, and the capacities of other cells can be calculated as fractions or multiples.

Figure **146** shows the variation of capacity of lead-acid batteries with time of discharge and the relation of power delivered to the discharge current. Similar curves for Edison batteries are given in

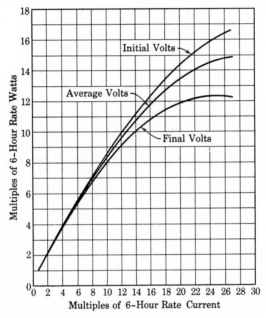

Fig. 146. Variation of watt output with the rate of discharge, type MVM Exide-Ironclad.

Fig. **147**. These apply to the type A. Values for other types can be estimated by recalling that the B plate has half the ampere-hour capacity of the A size, the C plate once and a half more, and the D plate twice as much.

The connections between the cells of Edison batteries consist of heavy copper wire swaged into steel lugs having an inside taper which fits the taper of the terminal posts of the cells. More of these connectors are necessarily used for the Edison battery than for the lead type, since 21 Edison cells correspond approximately in voltage to 12 cells of the lead type. In spite of this fact, however, the drop in voltage within the connectors exceeds that for the lead cells by only a slight amount. The Edison cells are assembled in hardwood crates which are provided with recessed hard-rubber buttons into which fit the bosses on the individual cells. They are therefore held rigidly in position with a slight space between each cell to provide the necessary insulation and ventilation.

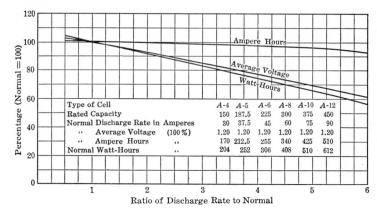

Fig. 147. Variation of capacity and power output of Edison cells with the rate of discharge.

The Service Which the Battery Has to Perform

The electric vehicle differs from the gasoline truck in the matter of starting torque. The torque of the motor on the electric truck increases as the speed of the truck decreases, that is to say, the torque is a maximum when the truck is beginning to move. The gasoline engine, on the other hand, delivers its maximum torque at high speeds. The performance of an electric truck as compared with a gasoline truck at *low speed* is therefore in favor of the electric truck.

The tractive effort of the truck is the force required to overcome the resistance of the vehicle and its load to motion. Drawbar-pull is a term applying to the trailing load and is usually expressed as pounds per ton of the total weight of the load. The drawbar-pull varies with the surface over which the vehicle moves, the grades it encounters, and the speed of the vehicle. Figure 148 shows the relation of the drawbar-pull to the battery current and voltage. From this it is evident that the drawbar-pull increases as the current increases.

The variation of the drawbar-pull with the surface over which the vehicle passes is shown in Table 63, which has been taken from an article by Pace,[2] who states that the drawbar-pull is increased 20 pounds per ton of load for each 1 per cent of upgrade. When the

[2] *Elec. World, 74,* 795 (1919).

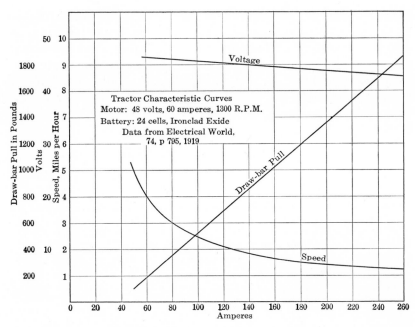

Fig. 148.   Operating curves of a tractor.

vehicle is going on a downgrade, however, an allowance of 20 pounds per ton may be made for each per cent of downgrade.

The service that a vehicle or tractor performs is measured in terms of ton-miles per hour of the "pay load," the ton-mile being a unit which corresponds to the transportation of 1 ton through a distance of 1 mile.   The quantity "ton-miles per hour" is therefore a measure of the transportation service which the truck or tractor can render. This varies with the amount of the load and the grade along which it

TABLE 63.   ROAD RESISTANCES ENCOUNTERED WITH TRAILERS

Type of Road Surface: Resistance lb per ton		Type of Road Surface: Resistance lb per ton	
Asphalt, hard	28	Snow, soft	66
Brick, smooth or cement floor	40	Gravel road	75
Brick, poor	57	Poor, macadam	75
Brick, glazed	47	Clay	200
Macadam	47	Sand road	275
Tarvia	47	Loose sand, 3 in. deep	330
Granite blocks	56	Concrete road	36
Wood blocks	40	Poor concrete	53
Snow, hard	50	Wood planking	43
Ice and snow	40	Wood planking, sticky surface	57

passes. It depends also on the size, the design, and the construction of the vehicle. A given tractor may have a large transportation capacity because it can haul a very heavy load at a low speed, but this would not be suitable for use where small or medium loads must be transported rapidly. Or again, two tractors may be capable of drawing equal loads, but one of these may exceed the other in ton-miles per hour because of greater speed. Sometimes high speed may be objectionable.

BIBLIOGRAPHY

Recommended Specifications for Automatic Battery-Charging Equipment for Industrial Truck Motive-Power Service, Electric Industrial Truck Association, Long Island City, N. Y., 1949.

S. K. Vaughan, How to determine industrial truck requirements, *Steel, 129,* 84 (1951).

M. W. Heinritz, Battery power for shuttle cars, *Coal Age, 55,* 91 (1950).

J. J. Buckley, Charging industrial truck batteries by a copper rectifier, *Gen. Elec. Rev., 50,* 27 (April 1947).

E. F. Grothe, Fundamentals of charging electric truck batteries, *Exide Ironclad Topics*, Winter ed., 1948.

Traction batteries, *Automobile Engr.*, Extra *33,* 466 (Nov. 4, 1943).

J. D. Huntsberger, Storage-battery charging, *Plant Eng., 3,* 43 (August 1949).

J. D. Huntsberger, Causes and remedies for troubles with lead-acid storage batteries (motive-power service), *Factory Management and Maintenance*, Data Sheet 159 (1953).

H. C. Riggs, Lead-acid motive-power and car-lighting batteries, *J. Electrochem. Soc., 99,* 236C (1952).

DIESEL CRANKING BATTERIES

Diesel engines belong to the class of internal-combustion engines. Like other engines of this class, they require a starting system to put them in motion before they can operate under their own power. Various systems for starting Diesel engines have been employed, including (1) electric starters, (2) compressed-air starters, (3) small gasoline engines connected for the moment through reduction gears, and (4) the introduction of explosive mixtures which can be fired in the cylinders. Electric starters are used nearly universally on the smaller types of engines, and modified forms of electric starters are employed on larger types, such as are found on Diesel locomotives. The voltage and current requirements of the battery exceed those for comparable gasoline engines.

The present Diesel engines are modifications of the type first built in 1892 by Rudolph Diesel. They operate on the principle that air compressed in the cylinders nearly adiabatically to about 500 pounds per square inch attains a temperature sufficient to cause ignition of the fuel, which is injected at the proper moment near the end of the compression stroke. Spark plugs to provide ignition are not necessary.

Diesel engines are an economical source of power because they burn relatively cheap grades of oil, but further economies are possible because they operate at high compression which increases the efficiency of using the fuel. Other fuels than oil might be used, and, in fact, Diesel proposed the use of coal dust.

Comparison of Diesel engines with the more familiar gasoline engines shows some points of similarity, but there are essential differences which distinguish one from the other, as shown in the following outline of four-stroke engines.

Stroke	Gasoline Engine	Diesel Engine
1. Suction stroke	Air and gasoline vapor enter cylinder together	Air alone enters cylinder
2. Compression stroke	Mixture is compressed to 90 to 150 lb per sq in. temperature 400 to 500° F	Air is compressed to 400 to 500 lb per sq in. 830 to 920° F
Beginning of combustion	Spark ignition near top of stroke, developing pressure of 200 to 400 lb per sq in.	Fuel injected near top of stroke, ignited by temperature of air in cylinder, developing pressure of 500 to 850 lb per sq in.
3. Expansion stroke	The gases burn	The gases burn
4. Exhaust stroke	The burned gases are expelled from cylinder	The burned gases are ex pelled from cylinder

Figure 149, which was furnished through the courtesy of H. C. Riggs, shows two oscillograms of the starting of internal-combustion engines. The first is for a 6-cylinder gasoline engine of 300-cubic-inch displacement, and the second is for a 6-cylinder Diesel engine of equivalent size. Each was cranked when the temperature of the lubricating oil was 32° F. The curves show the comparative starting times, currents, and voltages. The effect of compression is shown by maximums in the current curves and minimums in the voltage curve. On the basis of this oscillogram the following comparisons can be made.

	Gasoline Engine	Diesel Engine
Open-circuit battery voltage, v	6.3	25.2
Maximum breakaway current, amp	600	1200.0
Minimum voltage, v	4.2	15.0
Rolling current, amp	300	600.0
Cranking speed, rpm	90	180.0
Average voltage (rolling), v	5	18.0
Time to first fire, sec	1	7.0
Time to starter release, sec	2.6	10.5
Approximate watt-hours to start, w-hr	0.6	21.0

Diesel engines are usually equipped with a generator for charging the batteries when the engine is running. These are similar, more or less, to the generators on gasoline engines.

It is apparent from this outline that the starting problem is more difficult in the case of Diesel engines than gasoline engines. Higher

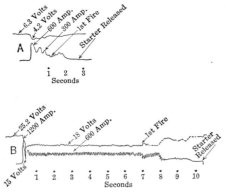

pressures in the cylinders must be overcome, but fortunately this is partly offset after the engine has begun to rotate by the expansion of air previously compressed in other cylinders. Because high compression is required in Diesel engines, the clearances between pistons and cylinder walls must be small, and friction is, therefore, a relatively large factor. Leingang[3] has estimated that 60 per cent of the power expended in starting the engine is used in overcoming fric-

Fig. 149. Comparative oscillograms of starting gasoline engine (A) and Diesel engine (B), each having 6 cylinders, 300-cubic-inch displacement, temperature 32° F.

tion between pistons and cylinder walls. The remaining 40 per cent is expended in overcoming compression and the friction of bearings and valve mechanisms. The Diesel engine must be turned fast enough to provide nearly adiabatic compression of the air. Otherwise the temperature attained on compression stroke may be too low to ignite the fuel. If the compression were strictly adiabatic, the temperature attained as a result of compression of the air to 500 pounds per square inch would be about 1050° F, but in ordinary practice the cooling effect of the cylinder walls lowers the temperature to about 900° F. High cranking speeds are necessary also to avoid losses of pressure resulting from

[3] *Diesel Power, 16,* 136 (1938).

"blow-by" past the piston heads. This effect is greater in the smaller sizes of engines.

Many sizes and kinds of Diesel engines are available for a variety of uses. The advent of small sizes and the decrease in weight per horsepower as a result of improved design and materials has stimulated the demand for Diesel engines. Some of their more important uses are: (1) portable power units for industrial service; (2) electric generating units including stationary and marine types; (3) marine propulsion and auxiliaries; (4) compressors and drilling equipment; (5) motive power on trucks, tractors, and buses; (6) industrial and railroad locomotives.

From the time of the first small Diesel-electric locomotive, tried experimentally in 1923 for switching service, to the beginning of 1950, when 35 per cent of all the railroad locomotives in the United States were Diesel-electric, the development has been extremely rapid. Coincident with this has been the battery development as a necessary part of the equipment.

The larger sizes usually operate at lower speeds. Diesel engines may be classified approximately according to speed as follows: Slow speed, up to 400 rpm; medium speed, 400 to 800 rpm; high speed, 800 to 2000 rpm. Cranking speeds are lower, of course, than these figures but must be high enough to produce ignition.

Climatic conditions are important. Diesel engines in transportation service, and perhaps other services as well, may be exposed to temperatures of 0° F or below. Obviously the engines are more difficult to start under such conditions, and much depends on the viscosity of the lubricating oil. To meet this difficulty the engines are kept warm by heating the circulating water. On the railways the temperature of the water is said to be above 125° F when starting. With proper preparation the engine should start in not more than 20 seconds.

Some of the larger Diesel engines that are directly connected to generators are started by the use of auxiliary field windings in the generators, which for the time being operate as motors, current being supplied by a battery of 32 or 56 cells. On engines of 1000 horsepower the breakaway current to be supplied by the battery may be as high as 2000 amperes and the rolling current 1200 amperes. When the engine has attained a predetermined speed, the starter circuit is opened.

Recent applications of high-speed Diesel engines to bus transportation and other services have necessitated the development of 6- to 12-volt starting systems somewhat analogous to those used on gasoline engines. The electric motors used for Diesel-engine starting are of the series-field type for which the torque increases with the current and the speed depends on the voltage. Starting motors for engines up to 300 horsepower are relatively small, and they are usually mounted on the side of the engine.

The starting requirements and how to determine them have been the subject of many papers. The battery engineer needs to know: (1) The maximum torque during the first revolution of the engine. This begins with the breakaway torque, and that depends on overcoming inertia, on the drag exerted by oil of high viscosity when the engine is cold and continues with compression of air in the cylinders. Sometimes, however, the maximum compression may not occur until near-completion of the first or second compression strokes. (2) The torque at rolling speed when the engine should be ready to fire. (3) The firing speed. (4) The lowest voltage that will give firing speed after a specified time. (5) The lowest temperature to which the engine is likely to be exposed. (6) A defined cranking or duty cycle.

Perhaps the most common cranking cycle specifies 6 consecutive cranking periods of 15 seconds each, spaced 15 seconds apart, without the cranking speed falling below the required firing speed. A battery that can furnish 1½ minutes of continuous cranking at not less than specified speed and temperature should be able to meet the intermittent cycle easily.

The relationship of speed of cranking to battery voltage for a motorized generator operating as a starting motor for a Diesel engine at 32° F is shown by a group of curves in Fig. 150. This illustration

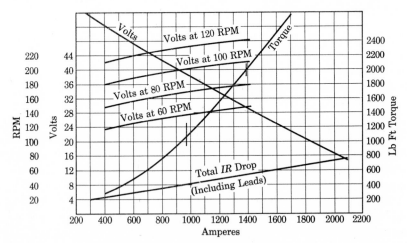

Fig. 150. Curves for a motorized generator operating as a starting motor for a Diesel engine, battery temperature 32° F.

is from a paper by Hoxie (see the Bibliography, page 403) and is recommended by him as a means of estimating the voltage at specified speeds of rotation. By superposing a curve indicating initial or 5-second voltage of a particular battery on the starting motor characteristic curves, the maximum or stalled torque using this battery is indicated by the intersection of the battery voltage curve and the

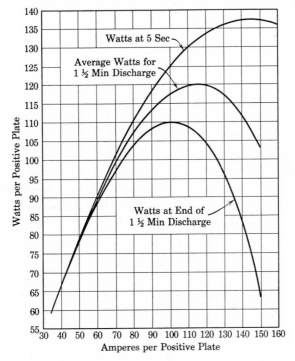

Fig. 151. Relation of watt output per positive plate to the current drain per positive plate. Battery of the type MVAHT, Exide-Ironclad. Temperature 32° F.

line marked total IR drop. In Fig. 150 this is shown at a current of 2070 amperes. Since the inductance of the starting motor prevents an instantaneous build-up of the current and the rotation generates a counterelectromotive force, the calculated stalled current will seldom be encountered in practice.

The effect of temperature is important particularly at high rates of discharge. Reduction in capacity at cranking rates of discharge and at low temperatures is much greater proportionately than at low rates such as the 6- or 8-hour rates. It is better in dealing with these starting batteries, therefore, to express the reduction in relative ca-

pacity in terms of the percentage of current obtainable for the specified
time and final voltage.

Many of the smaller Diesel engines are cranked by ordinary 3- or
6-cell batteries such as might serve on automobiles with gasoline
engines. For intermediate sizes of engines 6- to 12-cell batteries of
the heavy-duty type are employed. These have thicker plates and

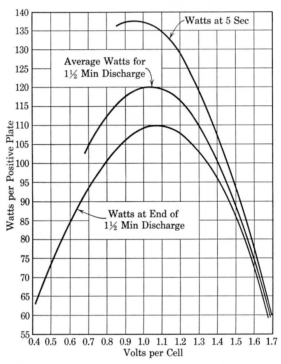

Fig. 152. Relation of the watt output per positive plate to the terminal volts
per cell. Temperature 32° F, type MVAHT, Exide-Ironclad.

separators. Larger Diesels used in marine service employ batteries
of 12 to 64 cells. In this they are similar to those on locomotives.

The initial watt output of the battery when discharge begins
depends on the rate of discharge. If the rate of discharge is increased,
the watt output will normally increase until a maximum is reached
(see Fig. 112) beyond which any increase in the current results in
a diminishing watt output. As discharge is prolonged the peak of
the watt output occurs at successively smaller currents and at higher
voltages. As an example of a battery discharging at 32° F, curves
for an Exide-Ironclad battery of the type MVAHT are given in Figs.

151 and 152. The former shows the current per positive plate and the latter the volts per cell plotted against the watts delivered per positive plate.

BIBLIOGRAPHY

W. C. Leingang, Selection of cranking batteries for light-weight high-speed Diesels, *Diesel Power, 16,* 136 (1938).

Diesel Engineering Handbook, ed. by L. H. Morrison, Diesel Publications, 1939–1940.

J. C. Davidson and R. Lamborn, Investigation of the starting requirements of a 660-hp locomotive Diesel engine, *Trans. Am. Inst. Elec. Engrs., 60,* 31 (1941).

A. O. Ridgely, Batteries in Diesel service, *Ry. Mech. Engr., 122,* 160 (1948).

George W. Grupp, Diesels in the railroad industry, *Diesel Power, 28,* 40 (September 1950); Diesels in highway transportation, *ibid., 29,* 36 (February 1951); Diesels in the mining industry, *ibid., 29,* 34 (March 1951).

K. A. Vaughan, How to select starter batteries, *Ry. Mech. Engr., 123,* 796 (1949).

Anon., How to select correct Diesel starting battery, *Elec. World, 133,* 72 (Jan. 2, 1950).

F. H. Brehob, Starting requirements for locomotive engines, *Ry. Mech. Engr., 124,* 381 (1950).

E. A. Hoxie, Matching batteries and starting equipment, *Diesel Power, 31,* 52 (March 1953).

S. K. Lessey and K. A. Vaughan, Diesel-electric locomotive batteries, *Ry. Mech. and Elec. Engr., 125,* 97 (December 1951); *126,* 76 (January 1952); *126,* 101 (February 1952).

Nickel-Cadmium Starter Batteries for Diesel Engines

Increased use of Diesel engines has greatly stimulated the demand for heavy-duty batteries for starting them. The lead-acid batteries are doubtless in greatest use for this service, but nickel-cadmium batteries have also been used both in this country and abroad. These alkaline batteries are applied to various services of which starting Diesels on heavy automotive trucks, marine engines, and railroad locomotives are illustrative.

For use with marine engines, the Nicad batteries are assembled in unit trays of 5 cells. A 12-volt battery consists of 2 trays, 10 cells. Similarly 24- and 32-volt batteries are made up of 4 and 5 trays, 20 and 25 cells respectively. A 5-cell unit tray is illustrated in Fig. 153.

Heavier batteries for starting large engines such as those on locomotives are assembled in 4-cell unit trays. They are charged by generators driven by the engines when running. The recommended voltage regulation is given in Table 64.

Fig. 153. Typical Nicad 5-cell battery, 6 volts.

TABLE 64. HEAVY-DUTY NICKEL-CADMIUM BATTERIES FOR
ENGINE STARTING

Nominal Battery Voltage, v	No. of Cells	Voltage Regulator Setting, v	Trickle Chargers, If Used,* v
24	20	29–30	28–29
32	24	35–36	34–35
64	48	70–72	68–70

* Trickle-charging is recommended in stationary applications or where the engine is run too infrequently to keep the battery charged.

Current and capacity ratings for heavy-duty starting batteries of the type THR are given in Table 65.

SWITCHGEAR CONTROL BATTERIES

Battery installations in power stations and substations are principally for insuring uninterrupted supply of currrent for operating remote control and automatic protective apparatus (Fig. 154). The services include closing electrically operated circuit breakers, operating

TABLE 65. ENGINE STARTING BATTERIES, TYPE THR

Cell Type No.	Capacities			Charge Current 7-Hour Rate, amp	Overall Dimensions of Battery Trays, in.					Approx. Weight per Cell, lb	
	Amp-Hr 8-Hour Rate to 1.10 Volts per Cell	Amp			Width	Height	Length of 1 Tray Containing			Net	Shipping
		8-Hour Rate to 1.10 Volts per Cell	5-Sec Rate to 0.66 Volts per Cell				2 Cells	4 Cells	6 Cells		
THR30	142	17.75	1700	28.4	8	$17\frac{1}{4}$	$10\frac{1}{4}$	$19\frac{1}{4}$	$28\frac{3}{8}$	38	45
THR44	210	26.25	2533	42.0	11	$17\frac{1}{4}$	$10\frac{1}{4}$	$19\frac{1}{4}$	$28\frac{3}{8}$	56	66

field rheostats and valves, and providing emergency lighting and in some installations for telephone communication.

The character of the service required of batteries in power stations changed during the passing years, and this has brought about a change

Fig. 154. Control bus battery, type 60 DME-17, installed in a station of the Philadelphia Electric Co

in the type of battery and the method of its operation. In the early days of electrical engineering, central stations were confronted with heavy loads at certain times of the day and light loads at other times. The load factor was low, and efforts were made to improve the efficiency of operation by the use of batteries. The first central-station battery in this country is said to have been installed in 1885, and in the years that followed many battery installations were made. The primary purpose of these batteries was to assist in carrying the heavy, or "peak," loads, the battery receiving its store of energy at such times as the generating equipment could carry the external load and charge the battery too. In some installations, when the minimum load was too light for efficient operation of the generators, the battery was used to carry the entire load for part of the time. The batteries were charged and discharged daily. The necessity for the storage battery in central stations, as it existed some years ago, has vanished, but at the same time a new demand for central-station batteries arose.

Experience in the operation of electrical switching and protective apparatus has led to the conclusion that the storage battery is the most dependable source of power where direct current is required for operating such equipment in manual and automatic stations.

Standard control voltages for closing and tripping electrically operated circuit breakers are specified in standards of the American Standards Association and other organizations. Mechanisms nominally rated for 125 volts have ranges of 90 to 130 volts for closing and 70 to 140 volts for tripping. The maximum of 140 volts is practically fixed by limitations of lamps, coils of relays, and other equipment. The lower limits of these voltage ranges have been questioned. Switch operation occurs in a fraction of a second, but battery capacity is usually based on a momentary load for a period of 1 minute to a final voltage of 1.75 volts per cell. Measurements reported by Hoxie,[4] using a cathode-ray oscilloscope, yielded interesting information about the actual service these batteries perform.

Experimenting first with a non-inductive load representing the battery, Hoxie found: (1) A rather large instantaneous drop in voltage representing the IR drop where R is the momentary internal resistance of the battery. (2) A rapid decrease in voltage lasting about 0.1 second, the time depending on charging conditions existing at the time. (3) A less rapid drop to a fairly stabilized voltage. This occupied about 15 seconds and probably represents the effect of diffusing electrolyte. Beyond this point the effects produced are small. The whole phenomena occur in much less than 1 minute.

In actual practice a delay occurs in building up the current because of inductive parts of the circuit. Hoxie points out that the current build-up in the coil of a closing solenoid is complicated. The battery current at any instant is influenced by the inductance of the circuit, but the inductance itself is changing. To a less extent the current is affected by the changing battery voltage. Oscillograms obtained with

4 E. A. Hoxie, Application of storage batteries to the control of switchgear, *Trans. Am. Inst. Elec. Engrs., 66,* 1561 (1947).

five typical solenoids show that there was no large instantaneous flow of current when the circuit was closed and as a result no large IR drop. In effect he found that the voltage at the battery terminals that is available for closing circuit breakers is appreciably higher than the voltage at the battery terminals following a continuous discharge at the 1-minute rate. This is fortunate. The ampere-second capacity of the battery becomes more important than its rating for longer time. However, the batteries are usually rated for 1, 8, and 72 hours as well as the short-time ratings of both current and capacity at 1 minute.

The commonly preferred method of operating these batteries is to float them continuously at 2.15 volts per cell or 129 volts for a 60-cell lead-acid battery, with occasional equalizing charges at 2.33 volts per cell or 140 volts for the battery. As an alternative, when carefully controlled floating is not possible, the battery may be placed on charge automatically once each hour and disconnected by a relay when its voltage reaches 2.31 to 2.33 volts per cell. Either method keeps the battery practically fully charged.

There are practical advantages to full-float operation. The personal element is eliminated, it is applicable in both attended and unattended stations, maintenance is reduced, and the life of the batteries is extended. Generators are provided with drooping characteristic, or flat compounded up to full load and drooping beyond this point. The battery then takes the load when the need arises.

In addition to the generators mentioned above, charging the batteries is also accomplished by floating them on constant voltage circuits supplied by current from rectifiers. Usually the load is less than 12 amperes, but Phano chargers can take care of loads up to 25 amperes. Other rectifiers of the dry disk type are possible for similar uses.

The fact that the batteries are floated means that the battery has very little work to do under normal conditions. It merely supplies momentary current for switch operation while the generator or rectifier supplies the steady demand and keeps the battery fully charged. The characteristics of the generator or rectifier must be such that they are protected when heavy discharges are needed. At such times the battery must take the load.

In estimating the size of battery required for any particular purpose, the engineer is at once confronted by the widely different services that these batteries perform. Current needed for the various low-rate services may be estimated and the proper number of positive plates of a specified size decided upon. Then the heavy current demands are added. Normally a switching operation requires a large current for 1 second or less, but the number of probable simultaneous switching operations must also be considered in estimating the current demand. The number of positive plates to deliver the required number of amperes at the 1-minute rate is determined, and the total number required for the battery is the sum of the two estimates.

In previous parts of this book, the fact has been emphasized that the ampere-hour capacity of a battery is proportional to the number of positive plates at a specified time rate of discharge. At extremely high rates, however, the voltage drop in these cells caused by the internal resistance of the cells becomes appreciable and the proportionality no longer exists. The current-carrying parts of the cell might be made heavier, but this is not always desirable or economical. The ampere ratings, like the ampere-hour capacities, are not strictly proportional. For example, a flat-plate cell, having 3 positive plates of the E size, has a discharge rate in amperes at the 1-minute rate of 288 amperes, but a similar cell, having twice as many positive plates, has a rating of 554 amperes, or less than twice as much.

Planté batteries, including cells with Manchester positives (Fig. 23), are used for this purpose, but many pasted-plate batteries are used also. The service life is normally greater in full-float service than in cycle service, but continuous floating on the line tends to form (peroxidize) the grids of pasted-plate batteries. This ultimately results in terminating the service life of the cells, and the effect is greater if the average floating voltage exceeds that specified. To meet this situation, manufacturers have developed various types of batteries. (1) In certain small types of batteries the cells at one time contained a greatly increased ratio of positive active material to negative active material. This was accomplished by using 2 positive plates and 1 negative plate in 3-plate cells, providing low current density at the positive plates. Grid formation was retarded. The higher current density at the negative plates was beneficial in keeping them fully charged and active. (2) Batteries containing reinforced grids (see Fig. 9) have been developed for floating service under the trade name of Floté (Fig. 22). The grids have heavy members interspersed among those of more nearly the conventional size. The mass of metal is embedded in the active material. (3) Planté plates, which are well adapted to floating service, are combined with pasted negatives (see Fig. 30). The negatives are not subjected to the corroding action which affects the positives.

All of these are batteries of the glass-jar type. They have covers which effectively close them, and the vents are designed to trap the spray and return it to the electrolyte within the cell. These batteries may be installed in the same room with other apparatus.

Alkaline batteries are used also for circuit-breaker control and emergency-lighting service as required by central-station or substation operation. They find their widest acceptance on low-voltage tripping circuits where 19-cell batteries are recommended for nominal 24-volt systems, 38 cells for nominal 48-volt systems, and 88 cells for 120-volt systems. These batteries are so applied that they are trickle-charged from normal power sources within the maximum voltage ranges of circuit-breaker equipment. For 120-volt systems the maximum is 140 volts, and for other systems values are correspondingly above the nominal-system voltage. The formula for computing the proper trickle charge for Edison batteries is given in Chapter 6. Discharge voltage data on cells of the A type are given in Table 66.

TABLE 66. CELL VOLTAGE AFTER 15 SECONDS' DISCHARGE AT INDICATED MULTIPLES OF NORMAL RATE, TYPE A CELL

Condition of Cell	Multiples of the Normal Rate					
	1	2	3	4	5	6
Fully charged	1.38	1.29	1.20	1.11	1.02	0.93
50% rated capacity out	1.21	1.12	1.04	0.97	0.87	0.79
100% rated capacity out	1.08	0.97	0.85	0.74	0.63	0.52

BIBLIOGRAPHY

P. Torchio, Thin-plate batteries for reserve service, *Elec. World, 68,* 77 (1916).

J. L. Woodbridge, Storage-battery practice in central-station service, *Elec. Rev., 77,* 674 (1920).

E. A. Hoxie, Control storage batteries, *Gen. Elec. Rev., 29,* 232 (1926).

"Batteries for Control Bus Operation," *Bull.* 204, Electric Storage Battery Co., 1936.

W. E. Holland, Station battery selection, *Elec. World, 105,* 2549 (1935).

R. A. Harvey, Automatic control of lead-acid battery-charging equipment, *Proc. Inst. Elec. Engrs., 96,* 607 (1949).

E. R. Sanderson, Storage batteries in generating stations and substations, *Elec. Rev., 142,* 119 (1950).

E. Barrow and P. W. Wichersham, Nickel-cadmium battery installed in substation, *Elec. World, 135,* 72 (Jan. 1, 1951).

E. A. Hoxie, The application of storage batteries to the control of switchgear, *Trans. Am. Inst. Elec. Engrs., 66,* 1561 (1947).

I. F. Freed, Storage batteries and full-floating operation, *Elec. World,* p. 92 of Feb. 16 and p. 70 of Mar. 2, 1946.

EMERGENCY-LIGHTING BATTERIES

Increased use is being made of storage batteries as a source of power for emergency lighting in buildings. Compact, self-contained units, including batteries and provision for automatically charging them, have been developed to meet increased needs of this service. Notwithstanding the present high standard of reliability of central-station power, occasional interruptions of service occur as a result of storms, fire, floods, traffic crashes, short circuits and other causes. Failures are generally localized and beyond the control of the central station, but the public of today is less tolerant of interruptions and the effects may be serious. Continuous light lessens the liability to accidents, loss of life, panic, theft, property damage, and law suits. Such a list suggests the places where emergency lighting is particularly applicable: places of public assembly, banks, hospitals, large stores, power and industrial plants, schools, hotels, and ships.

Individual emergency-lighting systems, which may be installed anywhere that power is normally available, do not necessarily provide for the full lighting load but rather for lights in particular locations where dangerous conditions could arise suddenly and without warning in the event of power failure. The danger zones are typically assembly halls, exits, operating rooms, bank vaults, etc.

Selection of the danger zones is the first step in planning an emergency-lighting installation. The proper locations for lights and the amount of light required at each place then become problems for the illuminating engineer, but it is obvious that conditions are widely different. Sufficient light to enable people to move in an orderly manner to exits is wholly unrelated to the needs of a physician at the operating table where an illumination of 1000 foot-candles is wanted, with the added stipulation that it must be free from shadows.

The next step is to estimate the protection period required. That is, the time factor, which will depend in part on the reliability of the power supply and in part on the time needed to complete a course of events such as clearing an auditorium, completing a surgical operation, or repairing broken machinery. Protection periods are usually estimated at $1\frac{1}{2}$ hours, with 3 to 5 hours as the upper limit.

The 12-volt systems which were commonly used some years ago have been largely superseded. The reasons for this are probably the special high-efficiency lamps which were not always readily available and the voltage drop when lines were long. As a low-voltage replacement, 6-volt light-guards with battery and individual rectifiers can be connected directly to 115-volt a-c outlets.

In commercial installations, 32- and 110-volt systems are important. The system consists basically of the battery, the charger, and a transfer switch. In addition there are the necessary resistors, fuses, signal lights, and a switch to provide either automatic or manual control. Charging is at a trickle rate on 2.15 volts per cell or 129 volts per battery of 60 cells. After a power-line failure a high-rate charge begins automatically when the a-c service is restored, and this continues until the battery is again fully charged. At this point a TVR relay automatically terminates the high-rate charge, and the low or trickle rate is established to run until the next power failure occurs. The signal light indicates when charging is being done at the high rate. It is extinguished when this is completed.

The following emergency-lighting systems are available: (1) to operate on 115 volts, 60 cycles, single phase, 2-wire circuits; (2) to operate on 115/230 volts, 60 cycles, single phase, 3-wire circuits; (3) to operate on 115/199 volts, 60 cycles, three phase, 4-wire circuits. Other systems are doubtless available.

Lights normally "on" are the exit lights or other special lamps; those normally "off" are the emergency lamps used only at the time of a power failure.

Inspection is recommended at frequent intervals, such as once a week. The test switch should be operated and the cells having pilot balls inspected to determine whether both balls are "up," indicating that the battery is charged. Water should be added as needed, and occasionally the specific gravity of electrolyte in each cell should be measured; this should be about 1.210 when the battery is fully charged.

In the larger emergency equipments, the control unit is contained in the metal cabinet. The battery of 60 cells is mounted on a suitable acid-proofed rack. The capacity required varies with the lighting requirements and the protection period that is needed.

AUTOMOTIVE STARTING AND LIGHTING BATTERIES

Development of Starting and Lighting Systems

The application of storage batteries to starting and lighting service on passenger cars, trucks, and motorcoaches has been one of the notable developments in the battery industry. Starter systems for internal-combustion engines were first suggested about 1902, but the practical application of electric starters began about 1911. Before the electrical systems were adequately developed, various other systems employing compressed air, acetylene, and mechanical devices were in use. All these have been superseded by the electrical systems, employing batteries, motors, and generators. Aside from the fact that many of the earlier systems were not convenient to use or satisfactory in other respects, the demand for electric lights was an important factor in establishing the supremacy of electric starting. Between 1912 and 1914 development was rapid, and electric starting became the commonly accepted method. At that time systems were classified as "single-unit" or "two-unit" systems. The single-unit system employed a combined motor and generator. During the cranking period, this operated as a relatively low-speed motor, but after the engine started it functioned as a generator. Two-unit systems comprised separate motors and generators. These were smaller than the single-unit starters, and they operated at higher speeds. Occasionally reference is made to three-unit systems, the third unit being a magneto for ignition. Two-unit systems have superseded the others and are now practically universal.

Principal Parts of Starting and Lighting Systems

Any internal-combustion engine must be set in motion before it can operate on its own power. Each car must have, therefore, its individual source of energy for starting the engine. This is a battery of 6 or 12 volts on passenger cars and trucks and 12 volts on large motorcoaches. Electricity supplied by the battery operates a small electric motor which is automatically connected to the engine through a reduction gear when the starter switch is closed and automatically disconnected when the engine begins to operate under its own power. The battery also supplies current for lights and other services when the engine is not running. Current furnished by a generator, driven by the engine, maintains the battery in a charged state. The fact that a battery may need to be removed from the car at some time for extra charging is merely evidence of a maladjustment of operating conditions, in the absence of any obvious defect. At such a time 125 to 150 ampere-hours may be supplied, but this is small compared with the 2000 to 10,000 ampere-hours supplied yearly by the generator on the car. Besides charging the battery, the generator supplies most, if not all, of the current requirements of the car when the engine is running.

The load on the electrical system of automobiles has increased very greatly. In addition to the requirements of former years for starting, lighting, and ignition, modern cars have more and brighter lights, cigar lighters, multiple horns, defrosters, fans, heaters, marker lights, panel indicators, radios, and air conditioners. During the past 25 years the electrical load of passenger cars has increased from about 7 amperes to over 30 amperes and to even more for trucks and motorcoaches. Battery sizes have not been increased in proportion, but the output of generators has been materially increased.

In the early days, batteries of 6, 12, 18, and 24 volts were used on passenger cars. These are now largely 6 volts. Factors which brought about standardization of 6-volt systems undoubtedly included the

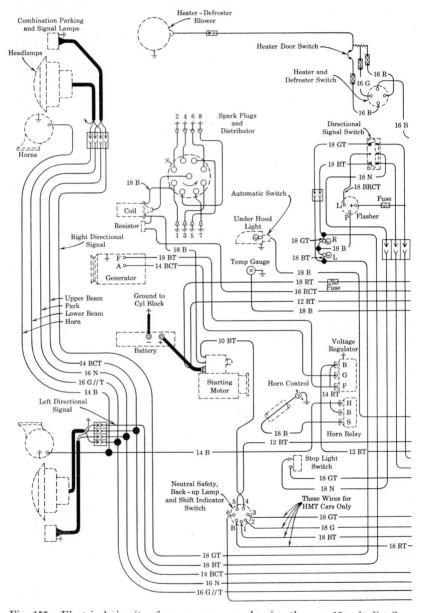

Fig. 155. Electrical circuits of a passenger car, showing the new 12-volt distribu-

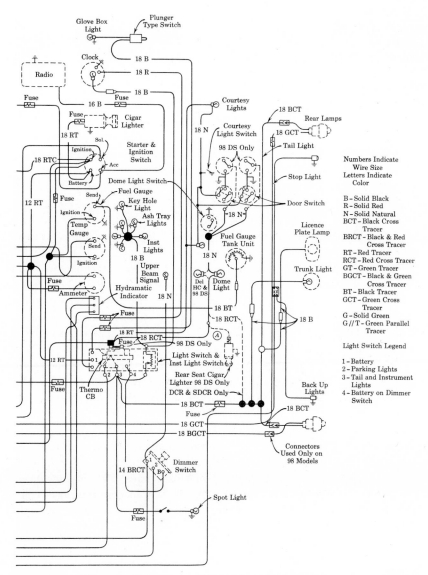

Numbers Indicate
Wire Size
Letters Indicate
Color

B – Solid Black
R – Solid Red
N – Solid Natural
BCT – Black Cross
Tracer
BRCT – Black & Red
Cross Tracer
RT – Red Tracer
RCT – Red Cross Tracer
GT – Green Tracer
BGCT – Black & Green
Cross Tracer
BT – Black Tracer
GCT – Green Cross
Tracer
G – Solid Green
G // T – Green Parallel
Tracer

Light Switch Legend

1 – Battery
2 – Parking Lights
3 – Tail and Instrument
Lights
4 – Battery on Dimmer
Switch

tor system. (Courtesy Delco-Remy Division of General Motors Corporation.)

previous development of 6-volt ignition systems, operated by 4 dry cells, and the fact that 6-volt tungsten lamps were successfully produced with short, rugged filaments which could withstand the mechanical shocks and vibration of automotive service.

Again we are in a transition stage. The 6-volt systems on some of the larger and heavier cars of 1953 are being superseded by 12-volt systems (Fig. 155).

Starting and Lighting Batteries

These batteries (sometimes called SLI batteries, meaning starting, lighting, and ignition batteries) for passenger cars and trucks consist of 3 or 6 cells of the lead-acid type in a unit case of hard rubber or bituminous composition. The size and arrangement of the cells and the location of parts are all specified in detail in the standards of the Society of Automotive Engineers (S.A.E.). Changes in this standard must be made at intervals to keep pace with advances in the automotive industry. The tendency has been to employ batteries with thinner plates and higher capacity. The shapes and general types of assembly are shown in Fig. 156.

Capacity requirements vary from about 90 ampere-hours at the 20-hour rate for the smallest passenger-car batteries to 200 ampere-hours at the same rate. Many small trucks use passenger-car batteries, but those specifically designated as truck batteries are usually found on larger and heavier vehicles. Motorcoach batteries include both 3- and 6-cell batteries.

The S.A.E. list of batteries includes the smaller sizes of the lead-acid type in general use on motorboats, tractors, and other automotive industrial applications, as well as on motor vehicles. Actually, three or four sizes are commonly used, and these are sufficient to supply a vast majority of automotive installations.

On small vehicles, having only a normal lighting and accessory load, the size of battery is determined by the cranking load. For motorcoaches, on the other hand, the lighting load is extremely heavy and the size of battery is determined by the lighting load in relation to the output of the generator. The greater the margin of generator output over the actual load, the smaller the battery can be, provided that it is not less than would be adequate for cranking the engine.

Batteries for combined starting and lighting service have a dual rating. The first is an indication of lighting ability and is the capacity in ampere-hours of the battery when it is discharged continuously at 80° F to an average final voltage of 1.75 volts per cell at the 20-hour rate, or 4-hour rate for motorcoach batteries. New batteries must equal or exceed their rated capacities on or before the third discharge. The second rating, applying only to passenger-car and motor-truck batteries, is an indication of the cranking ability under adverse conditions of low temperature. It is expressed as the number of minutes,

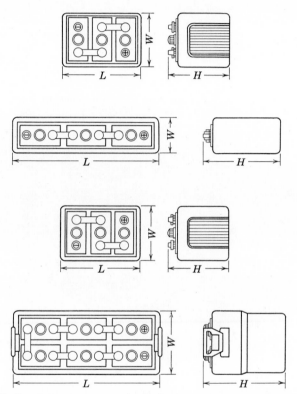

Fig. 156. Shape and location of automotive battery parts: standard assembly, long assembly, reverse assembly, and motorcoach assembly. A modification of standard assembly brings both terminals to the front side of the battery (not shown).

when the battery is discharged continuously at 300 amperes to a final terminal voltage of 1.0 volt per cell, the temperature of the battery at the beginning of such discharge being 0° F. As a part of the same low-temperature test, the terminal voltage of the battery is recorded at the end of the first 5 seconds. Minimum requirements for time and voltage of each size and type of battery are part of the standards of the Society of Automotive Engineers. These batteries are required also to meet certain minimum specifications for life tests, ranging from 234 cycles to over 500, depending on the size and type. These tests are described in Chapter 9.

Most of the automotive batteries are provided with single insulation, that is, wood or porous rubber separators between plates of opposite polarity. Some of them, however, have double insulation,

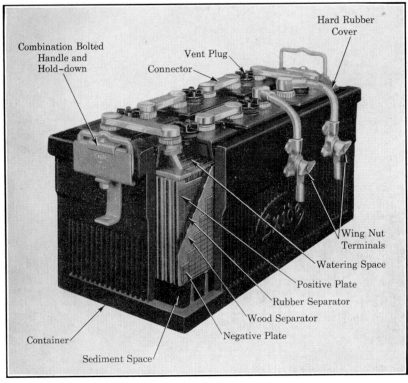

Fig. 157. Starting and lighting battery for use on motorcoaches, 12 volts. For the passenger-car type, see Fig. 21.

which is defined for purposes of the specifications as the use of a retaining sheet of porous or perforated material between the positive plates and the customary single separator (Fig. 157). The effect of double insulation on the operating characteristics of the batteries is recognized in the S.A.E. standard. The additional insulation increases the internal resistance of the battery, and a reduction of 10 per cent is allowed from the specified time and voltage requirements. The requirement for life cycles, however, is increased 15 per cent. Truck and motorcoach batteries normally are provided with double insulation and must meet the requirements specified for such batteries.

The terminals are ordinarily taper posts of different diameters to prevent wrong connections from being made. The standard for such terminals is specified by the Society of Automotive Engineers as follows:

	Inch
Small diameter, negative post	$\frac{5}{8}$
Small diameter, positive post	$\frac{11}{16}$
Taper per foot	$1\frac{1}{3}$
Minimum length of taper	$\frac{5}{8}$

The polarity of the terminals should be indicated by letters or by plus ($+$) and negative ($-$) signs.

Connections are made to these terminal posts by clamp connectors, of which there is considerable variety. These must fit well to avoid appreciable contact resistance, which would limit the amount of current delivered by the battery when cranking the engine. It is possible for the battery to supply current for the lights and other services and yet be limited by resistance at the contacts sufficiently to make starting difficult or impossible. Corrosion at the contacts requires that they be cleaned: first, all products of corrosion should be removed; second, acid on the surface and in recesses should be neutralized with dilute ammonia or soda; third, the terminals should be washed, dried, and greased; fourth, the clamp connections should be restored and tightened. Neutral greases may be used. Occasionally, some greases are found that promote corrosion. The number of patents that have been issued for various battery terminals and means of preventing corrosion is surprisingly large.

Structure. The cells for starting and lighting batteries contain plates of the pasted variety, which are burned to the connecting straps to form the plate groups. Ordinarily there are not less than 13 plates, including 6 positives and 7 negatives, in any cell. The largest sizes of cells may contain as many as 25 plates. The dimensions of the plates vary, but the ordinary sizes are about $5\frac{5}{8}$ inches in width, 5 inches in height, and 0.080 inch or less in thickness. The separators for these batteries are commonly made of wood, or wood separators are combined with retainers of perforated or slotted rubber, glass mats, etc.; but occasionally the wood separator has been superseded entirely by separators made of other materials, as for example porous-rubber or plastic separators.

The electrolyte consists of a solution of sulfuric acid of a specific gravity 1.260 to 1.280 when the battery is fully charged. The specific gravity decreases to about 1.140 when the battery is fully discharged. A high degree of purity of the electrolyte is desirable. The temperature of the electrolyte, which also represents the temperature of the cell as a whole, should not exceed 110° F, because of the increased local action within the cell and the charring of the separators.

Experiments reported by Little and Daily[5] showed the strong effect of temperature on local action within the battery as indicated by the decrease of specific gravity of the electrolyte:

Temperature	Loss in Specific Gravity per Day
120° F	0.006
100°	.003
60°	.001
0°	.0003
−40°	.0001

Cases for starting and lighting batteries are of hard rubber or bituminous compounds, molded to provide compartments for the individual cells. They are made in a variety of shapes and sizes, such as that illustrated in Fig. 16 in Chapter 2. In general, cases of hard rubber or bituminous materials are much the same in appearance, but the former are usually marked to indicate that they are hard rubber. These are preferred, because their life in service is generally longer and because they are less subject to deleterious effects of the electrolyte. Figure 17 shows a cover.

The connectors between cells are much heavier than on ordinary types of portable storage batteries. Solid connectors of lead or of lead-antimony alloy are used almost exclusively on batteries for service on passenger cars, but flexible copper connectors, heavily lead-coated, are used on some types of batteries for truck service. The voltage drop in the intercell connectors should not exceed 10 millivolts per inch of distance between post centers when the battery is discharging at the 20-minute rate. The resistance of the connectors will vary from 0.00005 to 0.0002 ohm per inch of distance between the centers of the terminal posts of adjacent cells, according to the capacity of the battery. It is obvious that if the resistance is more than a few thousandths of an ohm, the loss in voltage during the cranking period due to the IR drop in the connectors will be an appreciable part of the total voltage of the battery.

Starting and lighting batteries may be obtained in the condition

[5] J. H. Little and R. A. Daily, Storage-battery performance at low temperatures, S. A. E. Journal, 51, 149 (1943).

best suited to the needs of the purchaser.   The conditions for shipment have been specified as follows:

a. Charged and Wet.   Batteries intended for immediate use or for wet storage where suitable facilities are available shall be filled with electrolyte and fully charged.

Such batteries may be put in service after a brief inspection to determine whether the electrolyte stands at the proper height in each cell and whether the specific gravity of the electrolyte is normal.   If electrolyte has been spilled, it should be replaced by electrolyte of corresponding specific gravity.   If the battery is to be tested for capacity, or if the specific gravity of the electrolyte is low as a result of considerable time elapsing since the last charge, the battery should be given a freshening charge.   This is done at the finishing rate until the specific gravity ceases to rise, measurements being made on each cell.

b. Charged and Dry.   Batteries intended for storage in the charged condition within the time limit specified by the manufacturer shall contain dry charged plates and dry separators or double insulation.   The vents of each cell shall be closed, except as provision is made for equalizing atmospheric pressure fluctuations, and shall remain so until the battery is prepared for service.

These batteries are made ready for service by filling them with electrolyte in accordance with directions furnished by the manufacturer.   In the absence of specific instructions, the following procedure may be used: Fill the battery with electrolyte of 1.250 sp. gr. (temperature not exceeding 80° F) to about ½ inch above the plates and allow the battery to stand 1 hour.   At the end of this time the battery may be put in service after adjusting the electrolyte to the proper height in each cell.   Such batteries will give about one-half to two-thirds of their rated capacity on the initial discharge.   If the battery is not put in service within 12 hours it should be charged.

c. Uncharged.   Batteries intended for storage in the uncharged condition within the time limit specified by the manufacturer shall contain dry plates and dry or moist separators.   The vent of each cell shall be closed, except as provision is made for equalizing atmospheric pressure fluctuations, and shall remain so until the battery is prepared for service.

When the batteries are filled, electrolyte of the specific gravity recommended by the manufacturer should be employed.   The specific gravity will depend on the amount of water in the separators and sulfate in the plates and on the final specific gravity to be reached when the cells are fully charged, but it is usually within the range 1.250 to 1.280.   The temperature of the electrolyte should not exceed 80° F at

the time it is poured into the cells, but a rise in temperature will be
observed, caused by the formation of lead sulfate. Twelve hours is
usually required for the electrolyte to diffuse into the pores of the
plates and for the battery to cool. After this, and before 24 hours
has elapsed, the battery should be placed on charge. The rate of
charge suitable for this purpose is sometimes specified as the finishing
rate, but a better guide is to calculate the 20-hour rate from the rated
capacity of the battery and continue charging at this rate for about
84 hours. At the conclusion of the charge the electrolyte may be
adjusted to the proper value.

Charging System

A generator driven by the engine is the source of electricity on each
car. Small generators running at high speed can deliver as large an
output as larger and more expensive generators operating at lower
speed, but the latter are likely to last longer in service. Notwithstanding the greatly increased loads of the modern automobile, the
size of generators has not materially increased. The larger output
now required has been obtained by refinements in design and by providing ventilation. The generator capacity should equal the total
load with some reserve to keep the battery charged. The generator
must provide sufficient voltage for the lamps over a wide range of car
speeds and a variation of about 50 amperes between light- and full-load
conditions. Satisfactory operation depends in part on the selection
of the best drive ratio and on the regulation of the generator's output.

The third-brush generator without other means of regulation was
formerly in wide use. Its characteristic of maximum output at a
definite car speed was suitable for average driving conditions. It carried the load through a speed range of about 14 to 40 miles per hour,
and the charging current was not excessive.

Enlarging the battery would take care of increased charging currents
of the present day but would offer no solution for increased voltages
which came with the increased speeds and heavier loads of the modern
car. Third-brush generators of higher output can carry the maximum
load, but, if the load is light, the battery is likely to be seriously overcharged. On the other hand, when the battery becomes fully charged
and the load is light, voltages are likely to be excessive during severely
cold weather. This causes arcing at the ignition contacts, shortens the
life of lamp bulbs, and affects radio tubes adversely. The result has
been the introduction of current or voltage regulators, or both, as a
means of limiting the current and voltage supplied by the generator
which may be a shunt or compound type. Voltage regulators are

essentially electromagnets whose windings are energized so that the magnetic pull on the armature is approximately proportional to the impressed voltage. At a specified value the armature opens contacts and thereby inserts a resistance in the generator field circuit. This reduces the generator voltage and, consequently, the charging current to the battery. The armature is released, subsequently completing the cycle which is repeated often enough to maintain the voltage at the desired value. This is set for about 7.0 to 7.2 volts which is necessary for the battery.

Burning the headlights to decrease the charging rate on the battery is of little use on modern cars. The voltage relay automatically reduces the charging rate when the battery voltage rises to a predetermined value which is characteristic of a fully charged battery. At lower voltages than this the battery can safely absorb larger charging currents.

In addition there is a cut-out relay which disconnects the battery from the generator circuit when the generator voltage is too low to charge it. This relay closes at about 7 volts corresponding to a car speed of about 8 to 12 miles per hour. The cut-out relay normally remains open when the engine is stopped.

The Service That Starting and Lighting Batteries Must Perform

During the period of cranking an automobile engine, the battery is called upon to supply a large current which fluctuates rapidly because of the compression of gas in the cylinders of the engine. It has been possible to study the demands made upon starting and lighting batteries in the operation of various types of automobiles by using an oscillograph. Photographic records of the instantaneous values of current and voltage were obtained at the National Bureau of Standards.[6] In addition to the data relative to the battery requirements, the interpretation of these records brought out interesting facts with relation to the study of lubrication and engine problems.

In order to obtain records for periods of sufficiently long duration, the ordinary film drum of the oscillograph was replaced by a camera of special construction in which photographic paper in rolls of 100 feet could be used. One element was used to record the voltage at the terminals of the battery, another to record the current through the battery circuit, and the third to make the time record, which consisted of the half-second ticks of a chronometer. The general character of the

[6] G. W. Vinal and C. L. Snyder, *Natl. Bur. Standards, Technol. Paper* 186. See also Smith-Rose and Spillsbury, Some oscillograph tests on electric starters for motor cars, *J. Inst. Elec. Eng., 67,* 133 (1929).

curves is shown in Fig. 158.   This oscillogram begins at the right and
is to be read from right to left.   The time intervals are recorded at
the top of the record.   The curve next below the time record represents
the fluctuations of the voltage at the terminals of the battery when the
starter is in operation.   The last curve represents the current and
shows the fluctuations due to compression in the successive cylinders.
The zero value of this current is shown by the horizontal line in the
lower corners.

    This record was made on an old car of 4 cylinders, but it is chosen
in preference to others because it illustrates the principles without

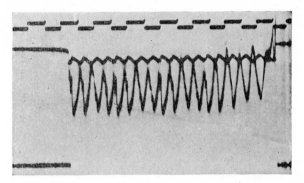

Fig. 158.   Record of the current and voltage in the battery circuit of a 4-cylinder
         car when the starter was cranking the engine, 2-unit system.

complications of overlapping cylinders, which are more difficult to
interpret.   During the first half-second after closing the starting
switch, the current fluctuated rapidly through a range of more than 100
amperes, the maximum value being about 250 amperes.   In this figure
it can be clearly seen that the minimum values of voltage correspond
to the maximum values of current.

    Temperature plays a very important role in determining the suc-
cessful operation of an electric starter system.   The low temperatures
increase the viscosity of the oils, decrease the terminal voltage and
capacity of the battery, and increase the difficulty in producing combustion of the gasoline vapor in the cylinders of the engine.   The effect
of the increased viscosity of the oil on the work that the battery has to
do is related to the current in the electric circuit.   The current that
the battery delivers to the starting motor is proportional to the torque
which the motor is able to exert in cranking the engine.   The torque
required to start any engine is obviously dependent on the friction and
therefore dependent on the viscosity of the oil the engine contains.

Both temperature and rate of discharge affect the available capacity of the battery. These matters have been discussed in preceding sections, but it is appropriate here to indicate some facts about starting and lighting batteries that may seem rather surprising. Take, for example, a battery which normally has a capacity of 100 ampere-hours at the 20-hour rate at 80° F. This same battery has a capacity of about 16 ampere-hours at a discharge rate of 300 amperes at 0° F, which is an assumed condition for starting a car, according to the S.A.E. standard. This seems very small, but to find whether it is adequate comparison must be made with the capacity needed for actually starting a car. Assume for the moment that an average current of 300 amperes is needed for cranking the engine and that the time is unusually long, 10 seconds. The battery output is 3000 ampere-seconds or 0.83 ampere-hour, which is about one-twentieth of the battery's capacity.

At extremely low temperatures the batteries do not accept charge readily. Little and Daily (*loc. cit.*) found a temperature of 40° F necessary for efficient charging. This means that at low temperatures the charging rate must be reduced. What they have called the efficiency of charging is related to temperatures as follows:

Temperature	Efficiency of Charging, %
−40° F	20
−20°	60
0°	80
+20° to +40°	90

Chubb and Harner[7] observed the sharply reduced capacities at high rates of discharge and low temperatures. The exponent n in Peukert's equation (see page 216) is materially greater under such circumstances than when the battery is operating at ordinary temperatures. Their test extended from −18° C (0° F) to 38° C (100° F).

Care of Starting and Lighting Batteries

Although starting and lighting batteries are often subjected to severe charging conditions, they will give satisfactory service provided they receive ordinary care and maintenance. This includes the proper adjustment of the charging system as well as the care of the battery itself.

The most convenient means of estimating the state of charge of the battery is by the hydrometer readings. When the battery is fully

[7] M. F. Chubb and H. R. Harner, Effect of temperature and rate of discharge on capacity of lead-acid storage batteries, *Trans. Electrochem. Soc.*, *68*, 251 (1935).

charged the specific gravity will range from 1.260 to 1.280. When the
battery is discharged the specific gravity will range from 1.140 to 1.160.
These are the specific gravities when measurements are made at ordinary temperatures. Intermediate states of charge of a battery may
be estimated by interpolating between the limits for the charged and
discharged conditions.

Batteries that are used for automobile service in the tropics require
lower specific gravities owing to the increase in chemical activity due
to the higher temperature. It is customary to adjust the electrolyte
for these batteries to a maximum value of 1.220, and this may decrease
to 1.080 when the battery is completely discharged.

The second essential in the ordinary care of the batteries is the
addition of pure water as necessary to keep the plates well covered.
The water is added to replace the so-called evaporation. By the term
"evaporation" is meant not only the evaporation that takes place in
the ordinary sense but also loss of water due to gassing when the
cells are on charge. The water should always be added after the
hydrometer readings are completed. This is because the water, being
less dense than the electrolyte, tends to remain on the top and give a
false indication of the specific gravity. The water gradually mixes
with the acid, and the electrolyte comes to a uniform density as a result
of the gassing that takes place when the cells are on charge. This
also suggests the necessity of making additions of water to cells in
extremely cold weather before running the car, rather than after, to
avoid the danger of freezing. Distilled water is much to be preferred
but is not always available.

A number of testing devices have been developed to determine
quickly the condition of starting and lighting batteries. They depend
for the most part on the interpretation of voltage data when the cells
are discharging at a high rate comparable with conditions for starting
the engine. Some of these are simple prongs applied to the terminals
of the individual cells, causing a current of several hundred amperes
to flow while the terminal voltage is read on a meter. Others which
are more elaborate provide a variety of meters for the different tests,
and these may be adjusted for the size and type of battery to be tested
and the number of plates it contains. The objective sought in making
these tests are, first, to determine whether the battery is in serviceable
condition without the necessity of opening it; and, second, to convince
the customer of the battery's condition by a demonstration.

As a preliminary, the specific gravity of each cell is read. If all
cells are uniform in this respect and if the value of specific gravity in
each cell is above 1.225, the battery is presumed to be in reasonably

good condition. Below this figure, charging is recommended. Considerable variations in specific gravity readings (50 points or more) usually indicate sources of trouble such as short circuits through separators; leakage of electrolyte through partitions between cells; worn-out plates in one or more cells; or badly contaminated electrolyte. The emphasis is placed on uniformity. Electrical tests which show voltage variations between the individual cells amounting to 0.15 volt or more when the cells are being discharged at a rate of 25 or more amperes per positive plate are usually interpreted to mean that a short circuit is present in the low cell or cells.

In the absence of evidence that short circuits exist in the battery, charging is usually recommended. This is conveniently done by using one of the various types of rectifiers designed for the purpose. As charging proceeds, the cells, if in good condition, should "come up" uniformly and finally reach a constant specific gravity. Voltage measurements made at this time should be uniform, but no definite figure can be set because the charging voltage varies with the current and temperature. Ten or twelve hours after charging has been completed and the battery disconnected from the charging circuit, equilibrium between the plates and electrolyte should be reached. There is then a definite relation between the state of charge and the terminal voltage of the individual cells.

Another essential for the satisfactory operation of a starting and lighting battery is cleanliness. Water or electrolyte which has been spilled on the top of the battery should be wiped off. A rag moistened with dilute ammonia (approximately 1 to 10) or a solution of baking soda may be used to neutralize the acid on the top of the battery.

BIBLIOGRAPHY

Brooks, Automobile storage batteries, *Trans. Am. Electrochem. Soc., 31,* 311 (1917).

Oetting, Storage batteries, *J. Cleveland Eng. Soc.,* May 7, 1918.

Oetting, Characteristics of starting and lighting batteries, *Elec. J., 16,* 134 (1919).

Vinal and Snyder, Instantaneous values of the current and voltage in the battery circuit of automobiles, *Natl. Bur. Standards, Tech. Paper* 186.

Reinhardt, The automobile storage battery, *Ind. Eng. Chem., 19,* 1124 (1927).

Critchfield, Modern automotive electrical equipment, *J. Soc. Automotive Engrs., 41,* 358, (1937).

Critchfield, Effect of application on maintenance of automotive electrical equipment, *ibid., 43,* 403 (1938).

"Storage Batteries," *Soc. Automotive Engrs., Handbook,* p. 770, 1953.

Federal Specification, Batteries: Storage, Vehicular, Ignition, Lighting, and Starting, W–B–131e, 1953.

Storage-Battery Technical Service Manual, Association of American Battery Manufacturers, Akron, Ohio, 1943.

AIRCRAFT BATTERIES

Storage batteries find important uses on aircraft for lighting, ignition, operation of various auxiliaries, and on some planes for engine starting. The weight of a storage battery is a detriment, and every effort is made to hold this to a minimum. High capacity per unit of weight is obtained by the use of thin plates and highly expanded active material. The usual installations involve the system-governed method of charging (see page 257), the battery taking care of electrical requirements until a certain engine speed is reached when control relays throw the load off the battery and onto a generator, which from then on supplies charging current to the battery.

The batteries are provided with microporous-rubber or plastic separators and often glass-mat retainers. In this condition the batteries can be shipped charged and dry. They are activated by being filled with dilute sulfuric acid of 1.275 sp. gr. at a temperature not exceeding 90° F to a height of $\frac{1}{2}$ inch above the protector covering the top of the plates. The battery should be allowed to stand for 1 hour after filling. It may then be further charged at rates specified by the manufacturer and the level of electrolyte carefully adjusted to the required height in order that the "non-spill" device may function properly if the battery is inverted.

Maximum capacity of the battery is based on (1) electrolyte of 1.285 sp. gr. when fully charged; (2) filling the cells to $\frac{1}{2}$ inch above the protector plates, and (3) a temperature of 80° F at the beginning of discharge.

Electric systems on aircraft began with 6-volt batteries and circuits somewhat similar to those on existing automobiles. With the fast-growing developments in aeronautics the electrical requirements were said to have increased from 1 kilowatt per plane in 1936 to 50 kilowatts 10 years later. The 6-volt systems were followed by 12-volt systems. About 1939 the trend toward larger aircraft brought with it the further increase to 24-volt systems. As the voltage increased the size of conductor cables could be diminished, and this made an important saving in weight. More recently 120-volt d-c systems and 208Y/120-volt a-c systems have appeared. The three-phase systems are at a frequency of 400 cycles. Detailed descriptions lag behind actual practice and will not be attempted here.

The great increase in electrical requirements has not brought about a corresponding increase in size or capacity of the batteries. The engine-driven generators have been made remarkably efficient and with forced ventilation carry the main load.   The battery on the other hand serves (1) to maintain the proper voltage level in the face of widely varying transients, (2) to supply power for short-time demands, (3) to supply power for those demands whose aggregate is small compared with those of the rotating units.

The 24-volt systems, for example, provide batteries permanently connected to a bus of about 28 volts whose exact value, usually 28.5 volts, must be determined by the needs of the battery and fixed by voltage regulators.   Accuracy in this relay adjustment is essential if the battery is to be adequately charged and not overcharged to the point of injury.

High-discharge rates are permissible.   The battery may be discharged without injury at any rate of current that it can deliver.   The maximum rate is limited by the current-carrying ability of the terminals, wiring, or apparatus to which it may be connected.   The lower operating limit of voltage is usually taken as $17\frac{1}{2}$ volts.

Exposure to low temperatures reduces the available capacity temporarily while the condition exists.   Freezing should be avoided but is not likely to occur if the battery is fully charged.   High temperatures on the other hand, above 110° F, will shorten the life of the plates. Ventilation of the battery compartment is necessary and can be adjusted to hold the temperature of the battery to safe limits provided the voltage regulators which control the charging rate are properly adjusted.

Because military airplanes fly upside down and do other "stunts" for which there is need in military combat, great emphasis has been placed on making the batteries non-spillable.   This was accomplished by the use of the so-called double chamber with stand-pipe vent plug. Above the plates is a compartment into which the electrolyte can flow from the plates if the battery is inverted.   The vent plug is elongated and is vented at the lower tip, which is so placed as to be always out of the electrolyte whatever the position of the battery may be. This requires that the electrolyte shall be adjusted to the proper height when the battery is made ready for service.   The use of the double chamber necessitates a considerable superstructure which adds to the weight of the battery and the space that it takes up in the plane.   The added fact that pressures within and without the battery are equalized for all positions of the battery is one reason for retaining the principle in a modified form.

Another type of non-spillable vent plug is illustrated in Fig. 159. A weighted cone, usually of lead, operates a small valve which is open

when the battery is upright. When the battery is inverted the cone
falls from its former position, and the valve seals off the vent from
within the cell.

To avoid radio interference, metallic shielding is used. The containers for such batteries (Fig. 160) are made of aluminum, protected
by an acid-resisting vinyl coating. The metal container extends considerably above the top of the cells and is closed by a cover of
aluminum and fastened by hold-down bolts. The cable connections

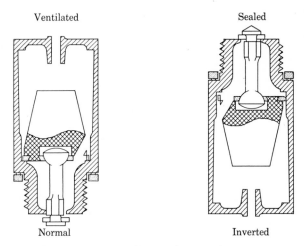

Fig. 159. Vent plug to prevent spillage of electrolyte, normal and inverted
positions.

are in shielded conduit. These are brought to a covered terminal box
on the side of the battery. All metallic shielding is electrically connected. Vents passing through the side of the box serve to equalize
the pressure inside the container with that of the outside atmosphere.

The performance of the battery is sometimes judged by the amount
of water it requires. If the amount exceeds that specified by the manufacturer, the indications are that the charging rate is too high, and
the bus voltage must be lowered accordingly. The method of charging
the batteries on the plane is essentially constant-potential charging
with some limitation of the maximum current that the batteries can
receive.

An inertia starter of 24 volts requires about 8 seconds to reach top
speed of 12,000 revolutions per minute. The current drawn from the
battery is about 320 amperes at the breakaway, but this drops rapidly
as the counterelectromotive force of the motor rises, reaching a con-

stant value of about 60 amperes.   The voltage of the battery at this time is about 23 volts, depending on the current.

For proper maintenance the batteries should be seated evenly and held firmly in place, but they should be readily accessible for inspection.

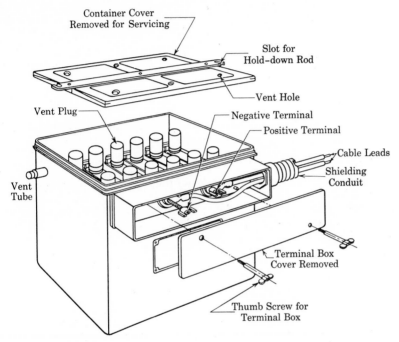

Fig. 160.   Illustrating a typical aircraft shielded battery.

Connecting cables must be flexible and sufficiently long to prevent pull on the battery case.   Considerable care in the addition of water should be taken to prevent flooding.

Airplane batteries are rated for several time intervals corresponding to the various services that they perform.   The ratings and cut-off voltages are usually those given in Table 67.

TABLE 67.   TIME RATES AND CUT-OFF VOLTAGES FOR
AIRPLANE BATTERIES

Time Rate of Discharge	Cut-off Voltages, v per cell
40 hours	1.80
5 hours	1.75
20 minutes	1.50
5 minutes	1.20

In addition a "safe rate" is specified in amperes for 1 minute. This exceeds the 5-minute rating. The purpose is to assure the safe current-carrying capacity of the intercell connectors and other parts of the cell.

BIBLIOGRAPHY

J. L. Rupp, Aircraft storage-battery design, *Trans. Am. Inst. Elec. Engrs.*, *63*, 773 (1944).

W. K. Boice and L. G. Levoy, Jr., Basic considerations in selection of electrical systems for large aircraft, *ibid.*, 279.

L. M. Cobb and H. M. Winters, Basic considerations in the selection of generators and batteries for aircraft, *ibid.*, 889.

H. C. Anderson, S. B. Crary, and N. R. Schultz, Present d-c aircraft electric supply systems, *ibid.*, 265.

H. J. Finison and R. H. Kaufman, D-C power systems for aircraft, Parts I–VII, *Gen. Elec. Rev.*, *48*, 22 (Sept.), 35 (Oct.), 56 (Dec.) (1945); *49*, 51 (March), 46 (May), 46 (June), 41 (Aug.) (1946).

J. C. Hutton, A-C power systems for large aircraft, *ibid.*, *51*, 11 (February 1948).

P. C. Bogiages, Weight analysis: basic considerations in design of aircraft electrical systems, *ibid.*, *52*, 41 (June 1949).

Cycle testing laboratory for aircraft batteries, *J. Franklin Inst.*, *234*, 63 (1947).

APPLICATIONS TO PORTABLE ELECTRIC LAMPS FOR USE IN MINES

Safe and adequate illumination for miners has been the subject of experiment and development for more than a hundred years. The early safety lamps of Clanny (1811) and Stevenson (1815), and the better known invention of Sir Humphrey Davy (1816), contributed greatly to safety in gaseous mines. The success of Davy's safety lamp was immediate; within a year its use in northern England became general. With modifications in structure and improvements in illuminants, the principle of the Davy lamp has continued to the present time. The first mention of portable electric lamps for use in mines appears to have been about 1901, but they were not used in considerable quantity until more than ten years later. In 1918, 48,000 electric lamps were reported to be in use in Pennsylvania.

Nearly all of the half million or more portable lamps now used underground in the American hemisphere are of the cap-lamp type, but a small proportion of hand lamps are used for inspection and signaling. In Europe, however, hand lamps outnumber cap lamps.

The electric cap lamp was first developed to replace the flame safety lamps employed in gaseous mines. Open-flame carbide lamps continue to be used, however, in many non-gaseous mines. The early electric cap lamps could not compete from an illumination standpoint. This condition continued for some years, but great improvements have since been made, and they are now finding use in many non-gaseous coal and metal mines.

The effective illumination provided by cap lamps is many times greater than that available from the first electric cap lamp. This improvement has resulted from (1) increased battery capacity, (2) increased lamp efficiency, (3) improved headpieces, and (4) improved distribution of the light. A notable advance in efficiency of the lamp bulb, amounting to about 20 per cent, was made by substituting krypton gas for argon. Krypton has a lower heat conductivity than either argon or nitrogen, which are commonly used in incandescent lamps. Krypton is a very rare gas whose atomic weight is 83.7. Its use is limited at present to those small lamps for which the highest efficiency is desired. If Xenon, another inert gas which is even more rare than krypton, should become available, it is possible that a still greater gain in efficiency might be made. Its atomic weight is 131.3.

Service underground is considered the most severe for portable batteries. The battery may be in almost any position when used, and it may suffer from mechanical shocks, abuse, and neglect. To prevent spillage of electrolyte, various special vent tubes, spring-actuated valves, absorbents, and jelly electrolytes have been employed. Jelly electrolytes are usually the sulfuric acid-sodium silicate type.

Tests to determine the suitability of these batteries and lamps for service in mines are specified by the U. S. Bureau of Mines. These tests cover the character of the light beam, the capacity of the battery, and safety features of lamp-mounting to avoid the possibility of igniting explosive gases if the lamp bulb should be broken.

Many types and kinds of miners' lights have been used. Some of of these employ alkaline storage batteries and others the lead-acid type. Only a few of them can be described here.

The first Edison electric cap lamp was brought out in 1914. The mechanical construction of the Edison battery together with its ability to withstand overcharge, or to remain in a discharged condition for an indefinite period of time without serious injury, makes it well adapted to this service. These batteries, although smaller than the industrial sizes, are made in the same manner and use the same sizes of tubes and pockets in the positive and negative plates, respectively. Figure 161 shows the construction of the battery, and Figure 162 shows the complete unit, battery, headpiece and lamp of the latest model, type R4 cap lamp. There are 4 cells (2 twin cells) enclosed in a nylon container. A gasketed nylon cover block is attached to the container with stainless-steel retaining bands, and through this cover only the cell filler openings protrude. During use these openings are closed

Fig. 161. Edison mine lamp, cut-away view of the battery.

Fig. 162. Edison model R4 cap lamp and battery.

with automatic safety valves, and a stainless-steel cover goes over the whole and is fastened by a magnetic lock to prevent unauthorized opening. The headpiece is molded plastic. The lamp has two filaments so that, if one burns out during the working shift, the other can be switched on.

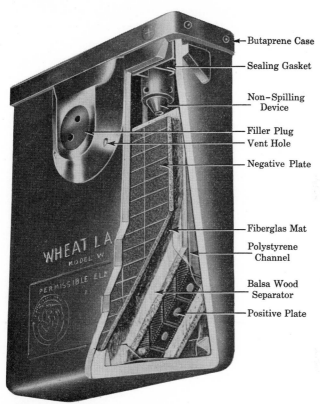

Fig. 163. Wheat mine-lamp battery of the lead-acid type, type W. A battery of 4 volts, capable of 12 hours' discharge with a 1.0-ampere bulb. In use the battery is surmounted by a locked metal cap from which the cord makes connection to the lamp. Charging is done through contacts on the lamp.

The Wheat mine lamp, "Forty-Niner," employs a battery of 2 cells of the lead-acid type. It is contained in a plastic[8] case and consists of 2 cells, giving 4 volts. The positive plates are of the Exide-Ironclad type of small size as shown in Fig. 9 of Chapter 2. The individual pencils in these plates, however, are square but otherwise like those used in motive-power batteries. The negative plates are of the flat-

[8] Butalite case (polystyrene-butadiene).

pasted type. Between the positives and negatives is a relatively thick balsa wood separator and glass wool mats. A battery of this type is shown in Fig. 163. The lamp is equipped with a krypton gas-filled bulb at the center of the reflector. A second bulb is provided as a "stand-by" bulb, which can provide sufficient illumination for working purposes. The nearly flat voltage characteristic of these cells on discharge is advantageous, as the lamps are quite sensitive to changes in voltage. Charging is accomplished on an automatic rack with selenium rectifier through safety contacts on the headpiece. The current is conducted to the battery through the cord, and it is not necessary to open or disconnect any part of the equipment in order to charge the battery. As with other electric cap lamps, the battery is worn on the belt of the miner. The complete equipment weighs 80 ounces of which the battery is 64 ounces.

BIBLIOGRAPHY

Paul, Ilsley, and Gleim, Flame safety lamps, *Bull.* 227, U. S. Bureau of Mines (1924).

Ilsley and Hooker, Permissible electric mine lamps, *Bull.* 332, Bureau of Mines (1930).

Permissible electric mine lamps, *Schedules* 6c, 10b, 11a, and subsequent issues, Bureau of Mines.

Lighting in mines, *Electrician, 121,* 418 (1938).

Maurice, Evolution of miner's hand lamp, *J. Inst. Elec. Engrs., 81,* 367 (1937).

Egeler, Illumination of mines and mining operations, *Trans. Illum. Eng. Soc., 33,* 439 (1938).

Lyon, Safety lamps, *Electrician, 116,* 93 (1936).

Index

Index 445